GENETICS AND GENETIC ENGINEERING

ISSN 1546-6426

GENETICS AND GENETIC ENGINEERING

Barbara Wexler

INFORMATION PLUS® REFERENCE SERIES
Formerly Published by Information Plus, Wylie, Texas

GALE
CENGAGE Learning·

Farmington Hills, Mich • San Francisco • New York • Waterville, Maine
Meriden, Conn • Mason, Ohio • Chicago

Genetics and Genetic Engineering

Barbara Wexler

Kepos Media, Inc.: Steven Long and
Janice Jorgensen, Series Editors

Project Editor: Laura Avery

Rights Acquisition and Management:
Ashley Maynard, Carissa Poweleit

Composition: Evi Abou-El-Seoud,
Mary Beth Trimper

Manufacturing: Rita Wimberley

Product Design: Kristine Julien

Cover photograph: © GeK/Shutterstock.com.

While every effort has been made to ensure the reliability of the information presented in this publication, Gale, a part of Cengage Learning, does not guarantee the accuracy of the data contained herein. Gale accepts no payment for listing; and inclusion in the publication of any organization, agency, institution, publication, service, or individual does not imply endorsement of the editors or publisher. Errors brought to the attention of the publisher and verified to the satisfaction of the publisher will be corrected in future editions.

Gale
27500 Drake Rd.
Farmington Hills, MI 48331-3535

ISBN-13: 978-0-7876-5103-9 (set)
ISBN-13: 978-1-4103-2550-1

ISSN 1546-6426

This title is also available as an e-book.
ISBN-13: 978-1-4103-3268-4 (set)
Contact your Gale sales representative for ordering information.

Printed in the United States of America
1 2 3 4 5   21 20 19 18 17

TABLE OF CONTENTS

PREFACE

Genetics and Genetic Engineering is part of the *Information Plus Reference Series*. The purpose of each volume of the series is to present the latest facts on a topic of pressing concern in modern American life. These topics include the most controversial and studied social issues of the 21st century: abortion, capital punishment, care of senior citizens, crime, education, the environment, health care, immigration, national security, social welfare, women, youth, and many more. Although this series is written especially for high school and undergraduate students, it is an excellent resource for anyone in need of factual information on current affairs.

By presenting the facts, it is the intention of Gale, a Cengage Company, to provide its readers with everything they need to reach an informed opinion on current issues. To that end, there is a particular emphasis in this series on the presentation of scientific studies, surveys, and statistics. These data are generally presented in the form of tables, charts, and other graphics placed within the text of each book. Every graphic is directly referred to and carefully explained in the text. The source of each graphic is presented within the graphic itself. The data used in these graphics are drawn from the most reputable and reliable sources, such as from the various branches of the U.S. government and from private organizations and associations. Every effort has been made to secure the most recent information available. Readers should bear in mind that many major studies take years to conduct and that additional years often pass before the data from these studies are made available to the public. Therefore, in many cases the most recent information available in 2017 is dated from 2014 or 2015. Older statistics are sometimes presented as well if they are landmark studies or of particular interest and no more-recent information exists.

Although statistics are a major focus of the *Information Plus Reference Series*, they are by no means its only content. Each book also presents the widely held positions and important ideas that shape how the book's subject is discussed in the United States. These positions are explained in detail and, where possible, in the words of their proponents. Some of the other material to be found in these books includes historical background, descriptions of major events related to the subject, relevant laws and court cases, and examples of how these issues play out in American life. Some books also feature primary documents or have pro and con debate sections that provide the words and opinions of prominent Americans on both sides of a controversial topic. All material is presented in an evenhanded and unbiased manner; readers will never be encouraged to accept one view of an issue over another.

HOW TO USE THIS BOOK

Genetics and genetic engineering is a hotly debated topic in the United States in the 21st century. Issues such as genetic testing to predict disease susceptibility, the implications of test results, whether to permit cloning of human organs, the moral dilemma of "playing God," and the advisability of raising genetically modified crops are discussed in coffee shops and classrooms across the nation. These topics and more are covered in this volume.

Genetics and Genetic Engineering consists of 11 chapters and three appendixes. Each chapter is devoted to a particular aspect of genetics and genetic engineering in the United States. For a summary of the information that is covered in each chapter, please see the synopses that are provided in the Table of Contents. Chapters generally begin with an overview of the basic facts and background information on the chapter's topic, then proceed to examine subtopics of particular interest. For example, Chapter 6: Genetic Testing begins by explaining that genetic testing has enabled researchers and clinicians to detect inherited traits, diagnose heritable conditions, determine and quantify the likelihood that a heritable disease will develop, and identify genetic susceptibility to familial disorders. It describes the difference between genetic testing to establish a diagnosis and testing to screen for a disease and the measures used to determine the quality of genetic tests: reliability, validity, sensitivity, specificity, positive predictive value, and negative predictive value. It describes carrier identification, preimplantation genetic diagnosis, prenatal and newborn testing, and diagnostic testing for specific genetic disorders. The chapter concludes with a discussion of direct-to-consumer marketing of genetic tests and genetic counseling. Readers can find their way through a chapter by looking for the section and subsection headings, which are clearly set off from the text. They can also refer to the book's extensive Index if they already know what they are looking for.

Statistical Information

The tables and figures featured throughout *Genetics and Genetic Engineering* will be of particular use to readers in learning about this issue. These tables and figures represent an extensive collection of the most recent and important statistics on genetics and genetic engineering and related issues—for example, graphics cover landmarks in the history of genetics, a diagram of deoxyribonucleic acid, bioengineered foods approved by the U.S. Food and Drug Administration, and the fundamentals of gene therapy. Gale, a Cengage Company, believes that making this information available to readers is the most important way to fulfill the goal of this book: to help readers understand the issues and controversies surrounding genetics and genetic engineering in the United States and to reach their own conclusions.

Each table or figure has a unique identifier appearing above it for ease of identification and reference. Titles for the tables and figures explain their purpose. At the end of each table or figure, the original source of the data is provided.

To help readers understand these often complicated statistics, all tables and figures are explained in the text. References in the text direct readers to the relevant statistics. Furthermore, the contents of all tables and figures are fully indexed. Please see the opening section of the Index at the back of this volume for a description of how to find tables and figures within it.

Appendixes

Besides the main body text and images, *Genetics and Genetic Engineering* has three appendixes. The first is the Important Names and Addresses directory. Here, readers will find contact information for a number of government and private organizations that can provide further information on genetics and genetic engineering. The second appendix is the Resources section, which can also assist readers in conducting their own research. In this section the author and editors of *Genetics and Genetic Engineering* describe some of the sources that were most useful during the compilation of this book. The final appendix is the detailed Index. It has been greatly expanded from previous editions and should make it even easier to find specific topics in this book.

COMMENTS AND SUGGESTIONS

The editors of the *Information Plus Reference Series* welcome your feedback on *Genetics and Genetic Engineering*. Please direct all correspondence to:

Editors
Information Plus Reference Series
Gale, a Cengage Company
27500 Drake Rd.
Farmington Hills, MI 48331-3535

CHAPTER 1
THE HISTORY OF GENETICS

Science seldom proceeds in the straightforward logical manner imagined by outsiders.

—James D. Watson, *The Double Helix: A Personal Account of the Discovery of the Structure of DNA* (1968)

Genetics is the biology of heredity, and geneticists are the scientists and researchers who study hereditary processes such as the inheritance of traits, distinctive characteristics, and diseases. Genetics considers the biochemical instructions that convey information from generation to generation.

Some genetic variation is related to disease, and the ability to vary genes improves the capacity of a species to survive changes in the environment. Although important advances in genetics research—such as deciphering the genetic code, isolating the genes that cause or predict susceptibility to certain diseases, and successfully cloning plants and animals—have occurred since the mid-20th century, the history of genetics study spans a period of about 150 years. As the understanding of genetics progressed, scientific research became increasingly more specific. Genetics first considered populations, then individuals, then it advanced to explore the nature of inheritance at the molecular level.

Although genetics research, especially genetic engineering of synthetic organisms, offers the tantalizing promise of preventing and combating disease and environmental threats, it also raises a host of ethical, philosophical, and political issues. For example, many people have concerns about the morality of creating new or artificial life by using stem cells to develop medical treatments, the development of synthetic components of flu vaccines, and the creation of new food products such as genetically modified salmon. There is also concern about the potential of genetic engineering to create biological weapons, the health consequences of consuming genetically engineered foods, and the environmental impact of genetic engineering.

EARLY BELIEFS ABOUT HEREDITY

From the earliest recorded history, ancient civilizations observed patterns in reproduction. Animals bore offspring of the same species, children resembled their parents, and plants gave rise to similar plants. Some of the earliest ideas about reproduction, heredity, and the transmission of information from parent to child were the particulate theories developed in ancient Greece during the fourth century BC. These theories posited that information from each part of the parent had to be communicated to create the corresponding body part in the offspring. For example, the particulate theories held that information from the parent's heart, lungs, and limbs was transmitted directly from these body parts to create the offspring's heart, lungs, and limbs.

Particulate theories were attempts to explain observed similarities between parents and their children. One reason these theories were inaccurate was that they relied on observations unaided by the microscope. Microscopy (investigation using the microscope) and the recognition of cells and microorganisms did not occur until the end of the 17th century, when the British physicist Robert Hooke (1635–1703) first observed cells through a microscope.

Until that time (and even for some time after) heredity remained poorly understood. During the Renaissance (from about the 14th to the 16th centuries) preformationist theories proposed that the parent's body carried highly specialized reproductive cells that contained whole, preformed offspring. Preformationist theories insisted that when these specialized cells containing the offspring were placed in suitable environments, they would spontaneously grow into new organisms with traits similar to the parent organism.

The Greek philosopher Aristotle (384–322 BC), who was such a keen observer of life that he is often referred to as the father of biology, noted that individuals sometimes resemble remote ancestors more closely than their immediate parents. He was a preformationist, positing that the male parent provided the miniature individual and the female provided the supportive environment in which it would grow. He also refuted the notion of a simple, direct transfer of body parts from parent to offspring by observing that animals and humans who had suffered mutilation or loss of body parts did not confer these losses to their offspring. Instead, he described a process that he called *epigenesist*, in which the offspring is gradually generated from an undifferentiated mass by the addition of parts.

Of Aristotle's many contributions to biology, one of the most important was his conclusion that inheritance involved the potential of producing certain characteristics rather than the absolute production of the characteristics themselves. This thinking was closer to the scientific reality of inheritance than any philosophy set forth by his predecessors. However, because Aristotle was developing his theories before the advent of microscopy, he mistakenly presumed that inheritance was conveyed via the blood. Nonetheless, his enduring influence is evident in the language and thinking about heredity. Although blood is not the mode of transmission of heredity, people still refer to "blood relatives," "blood lines," and offspring as products of their own "flesh and blood."

One of the most important developments in the study of hereditary processes was in 1858, when the British naturalists Charles Darwin (1809–1882) and Alfred Russel Wallace (1823–1913) announced the theory of natural selection—the idea that members of a population who are better adapted to their environment will be the ones most likely to survive and pass their traits on to the next generation. Darwin published his theories in *On the Origin of Species by Means of Natural Selection, or the Preservation of Favoured Races in the Struggle for Life* (1859). His work was not viewed favorably, especially by religious leaders who believed that it refuted the biblical interpretation of how life on Earth began. Even in the 21st century the idea that life evolves gradually through natural processes is not accepted by everyone, and the dispute over creationism and evolution continues.

CELL THEORY

In 1665, when Hooke used the microscope he had designed to examine a piece of cork, he saw a honeycomb pattern of rectangles that reminded him of cells, the chambers of monks in monasteries. His observations prompted scientists to speculate that living tissue as well as nonliving tissue was composed of cells. The French physiologist Henri Dutrochet (1776–1847) performed microscopic studies and concluded in 1824 that plant and animal tissue was composed of cells.

In 1838 the German botanist Matthias Jakob Schleiden (1804–1881) presented his theory that all plants were constructed of cells. The following year the German physiologist Theodor Schwann (1810–1882) suggested that animals were also composed of cells. Both Schleiden and Schwann theorized that cells were all created using the same process. Although Schleiden's hypotheses about the process of cell formation were not entirely accurate, both he and Schwann are credited with developing cell theory. Describing cells as the basic units of life, they asserted that all living things are composed of cells, the simplest forms of life that can exist independently. Their pioneering work enabled other scientists to understand accurately how cells live, such as the German pathologist Rudolf Virchow (1821–1902), who launched theories of biogenesis when he posited in 1858 that cells reproduce themselves.

Improvements in microscopy and the increasing study of cytology (the study of the formation, structure, and function of cells) enabled scientists to identify parts of the cell. Key cell components include the nucleus, which directs cellular activities by controlling the synthesis of proteins, and the mitochondria, which are organelles (membrane-bound cell compartments) that catalyze reactions that produce energy for the cell. Figure 1.1 is a diagram of a typical animal cell that shows its component parts, including the contents of the nucleus, where chromosomes (which contain the genes) are located.

Germplasm Theory of Heredity

Studies of cellular components, processes, and functions produced insights that revealed the connection between cytology and inheritance. The German biologist August Weissmann (1834–1914) studied medicine, biology, and zoology, and his contribution to genetics was an evolutionary theory known as the germplasm theory of heredity. Building on Darwin's idea that specific inherited characteristics are passed from one generation to the next, Weissmann asserted that the genetic code for each organism was contained in its germ cells (the cells that create sperm and eggs). The presence of genetic information in the germ cells explained how this information was conveyed and unchanged from one generation to the next.

In a series of essays about heredity published between 1889 and 1892, Weissmann observed that the amount of genetic material did not double when cells replicated, suggesting that there was some form of biological control of the chromosomes that occurred during the formation of the gametes (sperm and egg). His theory was essentially correct. Normal body growth is attributable to cell division, called mitosis, which produces cells that are genetically identical to the parent cells. The way to avoid giving offspring a double dose of heredity information is through a cell division that reduces the amount of the genetic material in the gametes by one-half. Weissmann called this process reduction division; it is now known as meiosis.

FIGURE 1.1

A typical animal cell

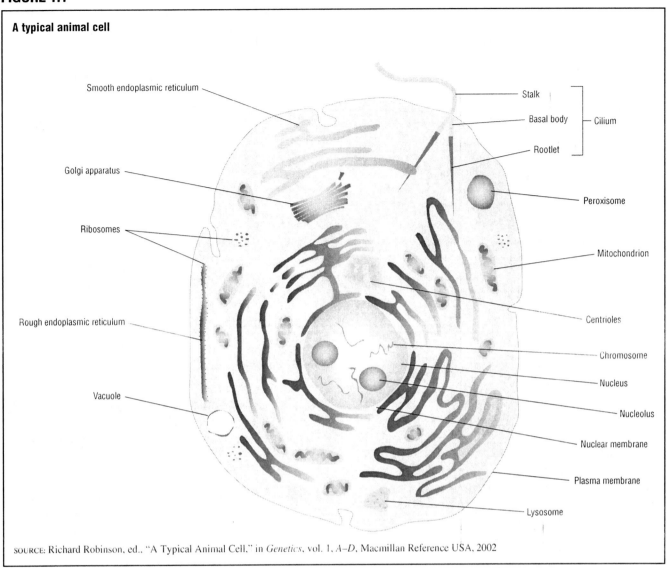

SOURCE: Richard Robinson, ed., "A Typical Animal Cell," in *Genetics*, vol. 1, *A–D*, Macmillan Reference USA, 2002

Weissmann was also the first scientist to successfully refute the members of the scientific community who believed that physical characteristics acquired through environmental exposure were passed from generation to generation. He conducted experiments in which he cut the tails off several consecutive generations of mice and observed that none of their offspring were born tailless.

A FARMER'S SON BECOMES THE FATHER OF GENETICS

Gregor Mendel (1822–1884) was born into a peasant family in what is now Hyncice, Czech Republic, and spent much of his youth working in his family's orchards and gardens. At the age of 21 he entered the Abbey of St. Thomas, a Roman Catholic monastery, where he studied theology, philosophy, and science. His interest in botany (the scientific study of plants) and an aptitude for science inspired his superiors to send him to the University of Vienna, where he studied to become a science teacher.

However, Mendel was not destined to become an academic, despite his abiding interest in science and experimentation. In fact, the man who was eventually called the father of genetics never passed the qualifying examinations that would have enabled him to teach science at the highest academic level. Instead, he instructed students at a technical school. He also continued to study botany and conduct research at the monastery, and from 1868 until his death in 1884 he served as its abbot.

Between 1856 and 1863 Mendel conducted carefully designed experiments with nearly 30,000 pea plants that he cultivated in the monastery garden. He chose to observe pea plants systematically because they had distinct, identifiable characteristics that could not be confused. Pea plants were ideal subjects for his experiments because their reproductive organs were surrounded by petals and usually matured before the flower bloomed. As a result, the plants self-fertilized, and each plant variety tended to be a pure breed. Mendel raised several generations of each type of plant to be certain that his plants were pure breeds. In this way, he confirmed that tall plants always produce tall offspring, and plants with green seeds and leaves always produce offspring with green seeds and leaves.

His experiments were designed to test the inheritance of a specific trait from one generation to the next. For example, to test the inheritance of the characteristic of plant height, Mendel self-pollinated several short pea plants, and the seeds they produced grew into short plants. Similarly, self-pollinated tall plants and their resulting seeds, called the first or F1 generation, grew to be tall plants. These results seemed logical. When Mendel bred tall and short plants together and all their offspring in the F1 generation were tall, he concluded that the shortness trait had disappeared. However, when he self-pollinated the F1 generation, the offspring, called the F2 generation (second generation), contained both tall and short plants. After repeating this experiment many times, Mendel observed that in the F2 generation there were three tall plants for every short one—a 3:1 ratio.

Mendel's attention to rigorous scientific methods of observation, large sample sizes, and statistical analyses of the data he collected bolstered the credibility of his results. These experiments prompted him to theorize that characteristics, or traits, come in pairs—one from each parent—and that one trait will assume dominance over the other. The trait that appears more frequently is considered the stronger, or dominant, trait, whereas the one that appears less often is the recessive trait. Mendel also described discrete "units of inheritance," a term that is still used today to describe genes.

Focusing on plant height and other distinctive traits, such as the color of the pea pods, seed shape (smooth or wrinkled), and leaf color (green or yellow), enabled Mendel to record and document the results of his plant breeding experiments. His observations about purebred plants and their consistent capacity to convey traits from one generation to the next represented a novel idea. The accepted belief of inheritance described a blending of traits, which, once combined, diluted or eliminated the original traits entirely. For example, it was believed that crossbreeding a tall and a short plant would produce a plant of medium height.

About the same time, Darwin was performing similar experiments using snapdragons, and his observations were comparable to those made by Mendel. Although Darwin and Mendel both explained the units of heredity and variations in species in their published works, it was Mendel who was later credited with developing the groundbreaking theories of heredity.

Mendel's Laws of Heredity

The constant characters which appear in the several varieties of a group of plants may be obtained in all the associations which are possible according to the [mathematical] laws of combination, by means of repeated artificial fertilization.

—Gregor Mendel, "Versuche über Pflanzen-Hybriden" (1865)

From the results of his experiments, Mendel formulated and published three interrelated theories in the paper "Versuche über Pflanzen-Hybriden" ("Experiments in Plant Hybridization" [1865; translated into English in 1901]). This work established the basic tenets of heredity:

- Two heredity factors exist for each characteristic or trait.
- Heredity factors are contained in equal numbers in the gametes.
- The gametes contain only one factor for each characteristic or trait.
- Gametes combine randomly, no matter which hereditary factors they carry.
- When gametes are formed, different hereditary factors sort independently.

When Mendel presented his paper, it was virtually ignored by the scientific community, which was otherwise engaged in a heated debate about Darwin's theory of evolution. Years later, well after Mendel's death in 1884, his observations and assumptions were revisited and became known as Mendel's laws of heredity.

His first principle of heredity, the law of segregation, states that hereditary units, now known as genes, are always paired and that genes in a pair separate during cell division, with the sperm and egg each receiving one gene of the pair. As a result, each gene in a pair will be present in half the sperm or egg cells. In other words, each gamete receives from a parent cell only one-half of the pair of genes it carries. Because two gametes (male and female) unite to reproduce and form a new cell, the new cell will have a unique pair of genes of its own, half from one parent and half from the other.

Diagrams of genetic traits conventionally use capital letters to represent the dominant traits and lowercase letters to represent recessive traits. Figure 1.2 uses this system to demonstrate Mendel's law of segregation. The pure red sweet pea and the pure white sweet pea each have two genes—RR for the red and rr for the white. The possible outcomes of this mating in the first generation are all hybrid (a combination of two different types) red plants (Rr)—plants that all have the same outward appearance (or phenotype) as the pure red parent but that also carry the white gene. As a result, when two of the hybrid F1 generation plants are bred, there is a 50% chance that the resulting offspring will be hybrid red, a 25% chance that the offspring will be pure red, and a 25% chance that the offspring will be pure white.

FIGURE 1.2

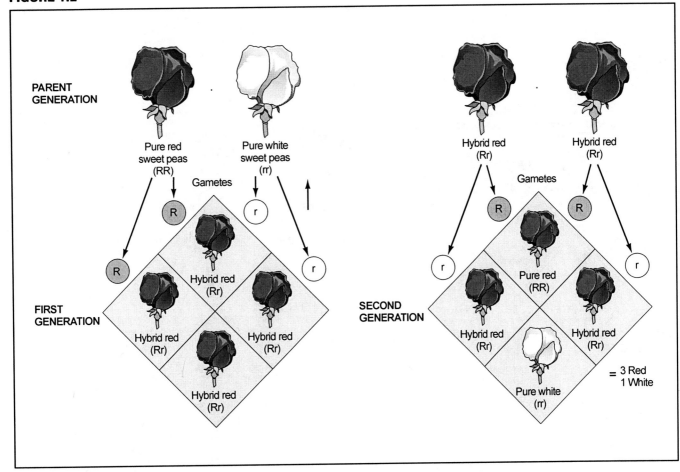

Mendel's law of segregation. *Hans & Cassidy, Cengage Gale.*

Mendel also provided compelling evidence from his experiments for the law of independent assortment. This law establishes that each pair of genes is inherited independently of all other pairs. Figure 1.3 shows the chance distribution of any possible combination of traits. The F1 generation of tall red and dwarf white sweet pea plants produced four tall hybrid red plants with the identical phenotype. Each one, however, has a combination of genetic information different from that of the original parent plants. The unique combination of genetic information is known as a genotype. The F2 generation, bred from two tall red hybrid flowers, produced four different phenotypes: tall with red flowers, tall with white flowers, dwarf with red flowers, and dwarf with white flowers. Both Figure 1.2 and Figure 1.3 demonstrate that recessive traits that disappear in the F1 generation may reappear in future generations in definite, predictable percentages.

The law of dominance, the third tenet of inheritance identified by Mendel, asserts that heredity factors (genes) act together as pairs. When a cross occurs between organisms that are pure for contrasting traits, only one trait, the dominant one, appears in the hybrid offspring. In Figure 1.2 all the F1 generation offspring are red—an

identical phenotype to the parent plant—although they also carry the recessive white gene.

Mendel's contributions to the understanding of heredity were not acknowledged during his lifetime. When he applied the knowledge gained from his pea plant studies to hawkweed plants and honeybees and failed to replicate the results, Mendel became dispirited. He set aside his botany research and returned to monastic life until his death. It was not until the early 20th century, nearly 40 years after Mendel published his findings, that the scientific community resurrected his work and affirmed its importance.

GENETICS AT THE DAWN OF THE 20TH CENTURY

During the years following Mendel's work, understanding of cell division and fertilization increased, as did insight into the component parts of cells known as subcellular structures. For example, in 1869 the Swiss physiologist Johann Friedrich Miescher (1844–1895) looked at pus he had scraped from the dressings of soldiers wounded in the Crimean War (1853–1856). In the white blood cells from the pus, and later in salmon sperm, he identified a substance he called nuclein. In 1874 Miescher

FIGURE 1.3

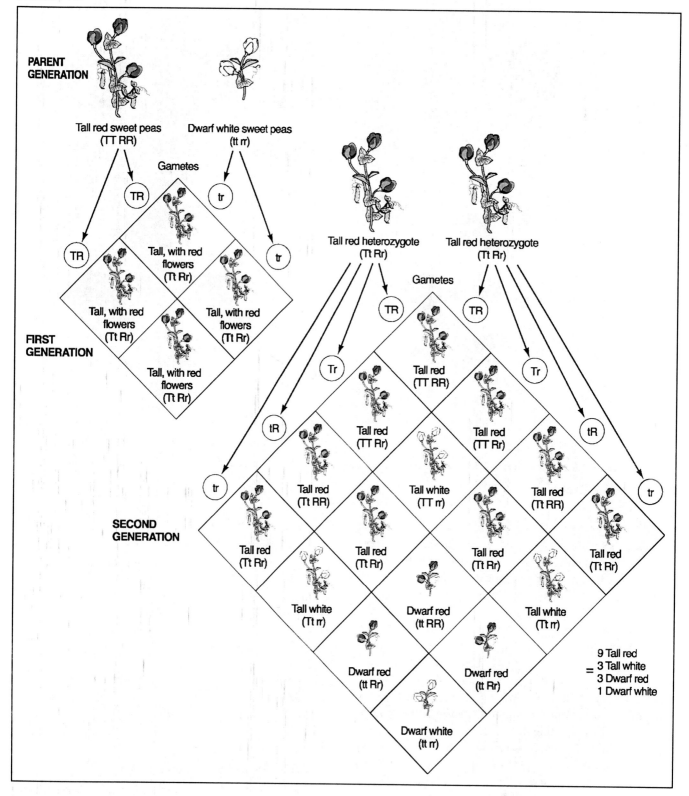

Mendel's law of independent assortment. *Hans & Cassidy, Cengage Gale.*

separated nuclein into a protein and an acid, and it was renamed nucleic acid. He proposed that it was the "chemical agent of fertilization."

In 1900 three scientists—Karl Erich Correns (1864–1933), Hugo Marie de Vries (1848–1935), and Eric von Tschermak-Seysenegg (1871–1962)—independently rediscovered and verified Mendel's principles of heredity, and Mendel's contributions to modern genetics were finally acknowledged. In 1902 Sir Archibald E. Garrod (1857–1936), a British physician and chemist, applied Mendel's principles and identified the first human disease attributable to genetic causes, which he called "inborn errors of metabolism." The disease was alkaptonuria, a condition in which an abnormal buildup of an acid (homogentisic acid or alkapton) accumulates.

Seven years later Garrod published the textbook *Inborn Errors of Metabolism* (1909), which described various disorders that he believed were caused by these inborn metabolic errors. These included albinism (a pigment disorder in which affected individuals have abnormally pale skin, hair, and eyes) and porphyria (a group of disorders resulting from abnormalities in the production of heme, a vitally important substance that carries oxygen in the blood, bone, liver, and other tissues). Garrod's was the first effort to distinguish diseases caused by bacteria from those attributable to genetically programmed enzyme deficiencies that interfered with normal metabolism.

In 1905 the British embryologist William Bateson (1861–1926) coined the term *genetics*, along with other descriptive terms used in modern genetics, including *allele* (a particular form of a gene), *zygote* (a fertilized egg), *homozygote* (an organism with genetic information that contains two identical forms of a gene), and *heterozygote* (an organism with two different forms of a particular gene). Arguably, his most important contributions to the progress of genetics were his translations of Mendel's work from German to English and his vigorous promotion of Mendel's principles.

In 1908 the British mathematician Godfrey Harold Hardy (1877–1947) and the German geneticist Wilhelm Weinberg (1862–1937) independently developed a mathematical formula that describes the actions of genes in populations. Their assumptions that algebraic formulas could be used to analyze the occurrence of, and reasons for, genetic variation became known as the Hardy-Weinberg equilibrium. It advanced the application of Mendel's laws of heredity from individuals to populations, and by applying Mendelian genetics to Darwin's theory of evolution, it improved geneticists' understanding of the origin of mutations and how natural selection gives rise to hereditary adaptations. The Hardy-Weinberg equilibrium enables geneticists to determine whether evolution is occurring in populations.

Chromosome Theory of Inheritance

> Bateson is often cited for having said, "Treasure your exceptions." I believe Sturtevant's admonition would be, "Analyze your exceptions."
>
> —Edward B. Lewis, "Remembering Sturtevant," *Genetics*, 1995

The American geneticist Walter Stanborough Sutton (1877–1916) conducted studies using grasshoppers (*Brachystola magna*) he collected at his family's farm in Kansas. Sutton was strongly influenced by reading Bateson's work and sought to clarify the role of the chromosomes in sexual reproduction. The results of his research, which were published in 1902, demonstrated that chromosomes exist in pairs that are structurally similar and proved that sperm and egg cells each have one pair of chromosomes. Sutton's work advanced genetics by identifying the relationship between Mendel's laws of heredity and the role of the chromosome in meiosis.

Along with Bateson, the American geneticist Thomas Hunt Morgan (1866–1945) is often referred to as the father of classical genetics. In 1907 Morgan performed laboratory research using the fruit fly *Drosophila melanogaster*. He chose to study fruit flies because they bred quickly, had distinctive characteristics, and had just four chromosomes. The aim of his research was to replicate the genetic variation de Vries had reported from his experiments with plants and animals.

Working in a laboratory they called the "Fly Room," Morgan and his students Calvin Blackman Bridges (1889–1938), Hermann Joseph Muller (1890–1967), and Alfred H. Sturtevant (1891–1970) conducted research that unequivocally confirmed the findings and conclusions of Mendel, Bateson, and Sutton. Breeding both red- and white-eyed fruit flies, they demonstrated that all the offspring were red eyed, indicating that the red-eye gene was dominant and the white-eye gene was recessive. The offspring carried the white-eye gene but it did not appear in the F1 generation. When, however, the F1 offspring were crossbred, the ratio of red-eyed to white-eyed flies was 3:1 in the F2 generation. (A similar pattern is shown for red and white flowers in Figure 1.2.)

The investigators also observed that all the white-eyed flies were male, prompting them to investigate sex chromosomes and hypothesize about sex-linked inheritance. The synthesis of their research with earlier work produced the chromosomal theory of inheritance, the premise that genes are the fundamental units of heredity and are found in the chromosomes. It also confirmed that specific genes are found on specific chromosomes, that traits found on the same chromosome are not always inherited together, and that genes are actual physical objects. In 1915 the four researchers published *The Mechanism of Mendelian Heredity*, which detailed the results of their research, conclusions, and directions for future research.

In *The Theory of the Gene* (1926), Morgan asserted that the ability to quantify or number genes enables researchers to predict accurately the distribution of specific traits and characteristics. He contended that the mathematical principles governing genetics qualify it as science: "That the characters of the individual are referable to paired elements [genes] in the germinal matter that are held together in a definite number of linkage groups....The members of each pair of genes separate when germ-cells mature....Each germ-cell comes to contain only one set.... These principles... enable us to handle problems of genetics in a strictly numerical basis, and allow us to predict... what will occur.... In these respects the theory [of the gene] fulfills the requirements of a scientific theory in the fullest sense."

In 1933 Morgan was awarded the Nobel Prize in Physiology or Medicine for his groundbreaking contributions to the understanding of inheritance. Muller also became a distinguished geneticist, and after pursuing research on flies to determine if he could induce genetic changes using radiation, he turned his attention to studies of twins to gain a better understanding of human genetics.

Bridges eventually discovered the first chromosomal deficiency as well as chromosomal duplication in fruit flies. He served in various academic capacities at Columbia University, the Carnegie Institution, and the California Institute of Technology and was a member of the National Academy of Sciences and a fellow of the American Association for the Advancement of Science.

Sturtevant was awarded the National Medal of Science in 1968. His most notable contribution to genetics was the detailed outline and instruction he provided about gene mapping (the process of determining the linear sequence of genes in genetic material). In 1913 he began construction of a chromosome map of the fruit fly that was completed in 1951. Because of his work in gene mapping, he is often referred to as the father of the Human Genome Project, the comprehensive map of humanity's estimated 19,000 genes. His book *A History of Genetics* (1965) recounts the ideas, events, scientists, and philosophies that shaped the development of genetics.

CLASSICAL GENETICS

Another American geneticist awarded a Nobel Prize was Barbara McClintock (1902–1992), who described key methods of exchange of genetic information. Performing chromosomal studies of maize in the botany department at Cornell University, she observed colored kernels on an ear of corn that should have been clear. McClintock hypothesized that the genetic information that normally would have been conveyed to repress color had somehow been lost. She explained this loss by seeking and ultimately producing cytological proof of jumping genes, which could be released from their original position and inserted, or transposed, into a new position. This genetic phenomenon of chromosomes exchanging pieces became known as crossing over, or recombination.

With another pioneering female researcher, Harriet Baldwin Creighton (1909–2004), McClintock published a series of research studies, including a 1931 paper that offered tangible evidence that genetic information crosses over during the early stages of meiosis. Along with the 1983 Nobel Prize in Physiology or Medicine,

McClintock received the prestigious Albert Lasker Basic Medical Research Award in 1981, making her the most celebrated female geneticist in history.

During the same period the British microbiologist Frederick Griffith (1879–1941) performed experiments with *Streptococcus pneumoniae*, demonstrating that the ability to cause deadly pneumonia in mice could be transferred from one strain of bacteria to another. Griffith observed that the hereditary ability of bacteria to cause pneumonia could be altered by a transforming principle. Although Griffith mistakenly believed the transforming factor was a protein, his observation offered the first tangible evidence linking deoxyribonucleic acid (DNA) to heredity in cells. His experiment provided a framework for researching the biochemical basis of heredity in bacteria. In 1944 the American biologist Oswald Theodore Avery (c. 1877–c. 1955), along with the American bacteriologists Colin Munro Macleod (1909–1972) and Maclyn McCarty (1911–2005), performed studies demonstrating that Griffith's transforming factor was DNA rather than simply a protein. Among the experiments Avery, Macleod, and McCarty performed was one similar to Griffith's, which confirmed that DNA from one strain of bacteria could transform a harmless strain of bacteria into a deadly strain. (See Figure 1.4.) Their findings gave credence to the premise that DNA was the molecular basis for genetic information.

FIGURE 1.4

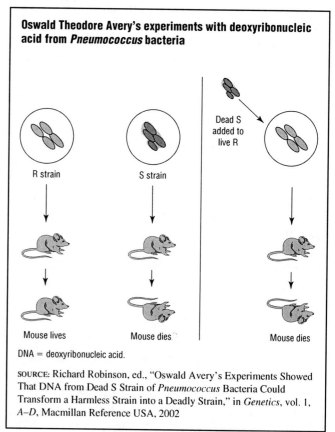

Oswald Theodore Avery's experiments with deoxyribonucleic acid from *Pneumococcus* bacteria

Dead S added to live R

R strain

S strain

Mouse lives

Mouse dies

Mouse dies

DNA = deoxyribonucleic acid.

SOURCE: Richard Robinson, ed., "Oswald Avery's Experiments Showed That DNA from Dead S Strain of *Pneumococcus* Bacteria Could Transform a Harmless Strain into a Deadly Strain," in *Genetics*, vol. 1, A–D, Macmillan Reference USA, 2002

The first half of the 20th century was devoted to classical genetics research and the development of increasingly detailed and accurate descriptions of genes and their transmission. In 1929 the American organic chemist Phoebus A. Levene (1869–1940) isolated and discovered the structure of the individual units of DNA. Called nucleotides, the molecular building blocks of DNA are composed of deoxyribose (a sugar molecule), a phosphate molecule, and four types of nucleic acid bases. (Figure 1.5 uses letters to represent the four bases: adenine [A], cytosine [C], guanine [G], and thymine [T].) Also in 1929 Theophilus Shickel Painter (1889–1969), an American cytologist, made the first estimate of the number of human chromosomes. His count of 48 was off by only two—25 years later researchers were able to stain and view human chromosomes microscopically to determine that they number 46. Analysis of chromosome number and structure would become pivotal to medical diagnosis of diseases and disorders associated with altered chromosomal numbers or structure.

Another milestone during the first half of the 20th century was the determination by the American chemist Linus Pauling (1901–1994) that sickle-cell anemia (the presence of oxygen-deficient, abnormal red blood cells that cause affected individuals to suffer from obstruction of capillaries, resulting in pain and potential organ damage) was caused by the change in a single amino acid (a building block of protein) of hemoglobin (the oxygen-bearing, iron-containing protein in red blood cells). Pauling's work paved the way for research showing that genetic information is used by cells to direct the synthesis of protein and that mutation (a change in genetic information) can directly cause a change in a protein. This explains hereditary genetic disorders such as sickle-cell anemia.

Between 1950 and 1952 the American geneticists Martha Cowles Chase (1927–2003) and Alfred Day Hershey (1908–1997) conducted experiments that provided definitive proof that DNA was genetic material. In research that would be widely recounted as the "Waring blender experiment," the investigators dislodged virus particles that infect bacteria by spinning them in a blender and found that the viral DNA, and not the viral protein, that remains inside the bacteria directed the growth and multiplication of new viruses.

FIGURE 1.5

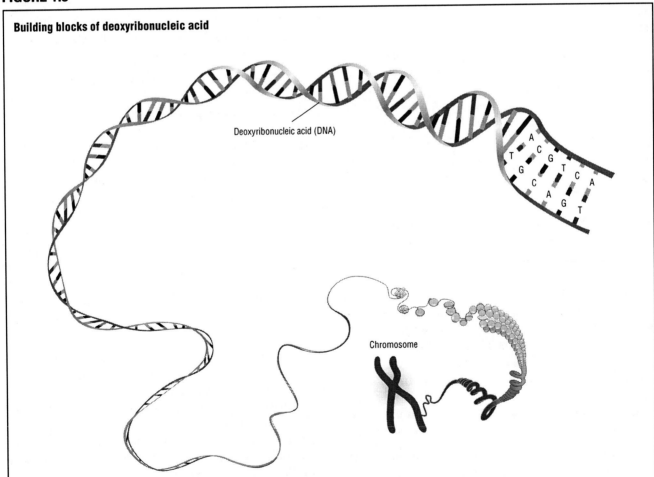

Building blocks of deoxyribonucleic acid

Deoxyribonucleic acid (DNA)

Chromosome

SOURCE: "DNA (Deoxyribonucleic Acid)," in *Talking Glossary of Genetic Terms*, U.S. Department of Health and Human Services, National Institutes of Health, National Human Genome Research Institute, Division of Intramural Research, undated, https://www.genome.gov/glossary/index.cfm?id=48 (accessed September 8, 2016)

MODERN GENETICS EMERGES

The period of classical genetics focused on refining and improving the structural understanding of DNA. By contrast, modern genetics seeks to understand the processes of heredity and how genes work.

Many historians consider 1953—the year that the American geneticist James D. Watson (1928–) and the British biophysicist Francis Crick (1916–2004) famously described the structure of DNA—as the birth of modern genetics. However, it is important to remember that Watson and Crick's historic accomplishment was not the discovery of DNA—Miescher had identified nucleic acid in cells nearly a century earlier. Similarly, although Watson and Crick earned recognition and public acclaim for their landmark research, it would not have been possible without the efforts of their predecessors and colleagues such as the British biophysicist Maurice Wilkins (1916–2004) and the British molecular biologist Rosalind Franklin (1920–1958). Wilkins and Franklin were the molecular biologists who in 1951 obtained sharp x-ray diffraction photographs of DNA crystals, revealing a regular, repeating pattern of molecular building blocks that correspond to the components of DNA. (Wilkins shared the Nobel Prize with Watson and Crick, but Franklin was ineligible to share the prize because she died in 1958, four years before it was awarded.)

Another pioneer in biochemistry, the Austrian Erwin Chargaff (1905–2002), also provided information about DNA that paved the way for Watson and Crick. Chargaff suggested that DNA contained equal amounts of the four nucleotides: the nitrogenous (containing nitrogen, a non-metallic element that constitutes almost four-fifths of the air by volume) bases adenine (A), thymine (T), guanine (G), and cytosine (C). In DNA there is always one A for each T, and one G for each C. This relationship became known as base pairing or Chargaff's rule, which also includes the observation that the ratio of AT to GC varies from species to species but remains consistent across different cell types within each species.

Watson and Crick

Watson is an American geneticist known for his willingness to grapple with big scientific challenges and his expansive view of science. In *The Double Helix: A Personal Account of the Discovery of the Structure of DNA* (1968), he chronicles his collaboration with Crick to create an accurate model of DNA. He credits his inclination to take intellectual risks and venture into uncharted territory as his motivation for this ambitious undertaking.

Watson was just 25 years old when he announced the triumph that was hailed as one of the greatest scientific achievements of the 20th century. Following this remarkable accomplishment, Watson served on the faculty of Harvard University for two decades and assumed the directorship and then the presidency of the Cold Spring Harbor Laboratory in Long Island, New York. From 1989 to 1992 he headed the National Institutes of Health's (NIH) Human Genome Project, the effort to sequence (or discover the order of) the entire human genome.

Crick was a British scientist who had studied physics before turning his attention to biochemistry and biophysics. He became interested in discovering the structure of DNA, and when in 1951 Watson came to work at the Cavendish Laboratory in Cambridge, England, the two scientists decided to work together to unravel the structure of DNA.

After his landmark accomplishment with Watson, Crick continued to study the relationship between DNA and genetic coding. He is credited with predicting the way in which proteins are created and formed, a process known as protein synthesis. During the mid-1970s Crick turned his attention to the study of brain functions, including vision and consciousness, and assumed a professorship at the Salk Institute for Biological Studies in San Diego, California. Like Watson, he received many professional awards and accolades for his work, and besides the scientific papers he and Watson coauthored, he wrote four books. Published a decade before his death in 2004, Crick's last book, *The Astonishing Hypothesis: The Scientific Search for the Soul* (1994), detailed his ideas about human consciousness.

WATSON AND CRICK MODEL OF DNA. Using the x-ray images of DNA created by Franklin and Wilkins, who also worked in the Cavendish Laboratory, Watson and Crick worked out and then began building models of DNA. Crick contributed his understanding of x-ray diffraction techniques and imaging and relied on Watson's expertise in genetics. In 1953 Watson and Crick published the paper "Molecular Structure of Nucleic Acids: A Structure for Deoxyribose Nucleic Acid" (*Nature*, vol. 171, no. 4356, April 25, 1953), which contained the famously understated first lines, "We wish to suggest a structure for the salt of deoxyribose nucleic acid (D.N.A.). This structure has novel features which are of considerable biological interest." Watson and Crick then described the shape of a double helix, an elegant structure that resembles a latticework spiral staircase. (See Figure 1.6.)

Their model enabled scientists to better understand functions such as carrying hereditary information to direct protein synthesis, replication, and mutation at the molecular level. The three-dimensional Watson and Crick model consists of two strings of nucleotides connected across like a ladder. Each rung of the ladder contains an AT pair or a GC pair, which is consistent with Chargaff's rule that there is an A for every T and a G for every C in DNA. (See Figure 1.6.) Watson and Crick posited that changes in the sequence of nucleotide pairs in the double helix would produce mutations.

FIGURE 1.6

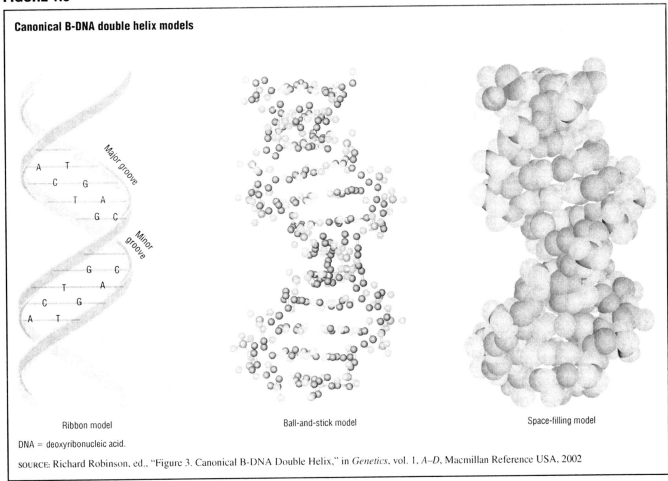

Canonical B-DNA double helix models

Major groove

Minor groove

A T
C G
T A
G C

G C
A
T
C G
A T

Ribbon model

Ball-and-stick model

Space-filling model

DNA = deoxyribonucleic acid.

SOURCE: Richard Robinson, ed., "Figure 3. Canonical B-DNA Double Helix," in *Genetics*, vol. 1, *A–D*, Macmillan Reference USA, 2002

MILESTONES IN MODERN GENETICS

Geneticists and other researchers made remarkable strides during the second half of the 20th century. In 1956 the American biochemist Vernon Ingram (1924–2006), who would be recognized as the father of molecular medicine, identified the single base difference between normal and sickle-cell hemoglobin. The implications of his finding that the mutation of a single letter in the DNA genetic code was sufficient to cause a hereditary medical disorder were far reaching. This greater insight into the mechanisms of sickle-cell disease suggested directions for research into prevention and treatment. It also prompted research that uncovered other diseases with similar causes, such as hemophilia (a bleeding disorder in which blood does not clot normally) and cystic fibrosis (an inherited disease of the mucous glands that produces problems associated with the lungs and pancreas). Just three years later the first human chromosome abnormality was identified: people with Down syndrome were found to have an extra chromosome, demonstrating that it is a genetic disorder that may be diagnosed by direct examination of the chromosomes.

Ingram's work became the foundation for current research to map genetic variations that affect human health. For example, in 1989, more than 30 years after Ingram's initial work, the gene for cystic fibrosis was identified and a genetic test for the gene mutation was developed.

Using radioactive labeling to track each strand of the DNA in bacteria, the American molecular biologists Matthew Meselson (1930–) and Franklin W. Stahl (1929–) demonstrated with an experiment in 1958 that the replication of DNA in bacteria is semiconservative. Semiconservative replication occurs as the double helix unwinds at several points and knits a new strand along each of the old strands. Meselson and Stahl's experiment revealed that one strand remained intact and combined with a newly synthesized strand when DNA replicated, precisely as Watson and Crick's model predicted. In other words, each of the two new molecules created contains one of the two parent strands and one new strand.

During the early 1960s Crick, the American biochemist Marshall Warren Nirenberg (1927–2010), the Russian American physicist George Gamow (1904–1968), and other researchers performed experiments that detected a direct relationship between DNA nucleotide sequences and the sequence of the amino acid building blocks of proteins. They determined that the four nucleotide letters (A, T, C, and G) may be combined into 64 different triplets. The triplets are code for instructions that determine the amino acid structure of proteins. Ribosomes are cellular organelles that interpret a sequence of genetic code three letters at a time and link together amino acid building blocks of proteins specified by the triplets to construct a particular protein. The 64 triplets of nucleotides that can be coded in the DNA—which are copied during cell division, infrequently mutate, and are read by the cell to direct protein synthesis—make up the universal genetic code for all cells and viruses.

Contributions of Modern Geneticists Are Recognized

In 2013 the American geneticist Mary-Claire King (1946–), who identified a genetic predisposition for breast cancer along with other disease genes, was awarded the Paul Ehrlich and Ludwig Darmstaedter Prize, one of the most highly respected international prizes awarded in the field of medicine. That same year the Nobel Prize in Physiology or Medicine was awarded to the American cell biologist James E. Rothman (1950–), the American cell biologist Randy W. Schekman (1948–), and the German American biochemist Thomas C. Südhof (1955–) for their discovery about how cells transport molecules. Disturbances in cell transport contribute to diabetes, neurological diseases, and immunological disorders.

Among the 2016 recipients of the American Society of Human Genetics prizes in human genetics was James F. Gusella (1952–), who with collaborators linked DNA markers to the mutation causing Huntington's disease (an inherited neuropsychiatric disease that affects the body and mind) and mapped and identified the genes associated with many other neurological conditions, such as amyotrophic lateral sclerosis (a progressive disease that attacks the nerve cells that control voluntary muscles such as muscles in the arms, legs, and face) and Alzheimer's disease (a brain disorder that destroys memory and thinking skills), as well as neurodevelopmental disorders, such as autism (disorders of brain development that affect social interaction and communication skills as well as behavior).

Origins of Genetic Engineering

The late 1960s and early 1970s were marked by research that would lay the groundwork for modern genetic engineering technology. In 1966 DNA was found to be present not only in chromosomes but also in the mitochondria. The first single gene was isolated in 1969, and the following year the first artificial gene was created. In 1972 the American biochemist Paul Berg (1926–) developed a technique to splice DNA fragments from different organisms and created the first recombinant DNA, or DNA molecules formed by combining segments of DNA, usually from different types of organisms. In

1980 Berg was awarded the Nobel Prize in Chemistry for this achievement, which is now referred to as recombinant DNA technology.

In 1976 an artificial gene was inserted into a bacterium, which continued to function normally. The following year DNA from a virus was fully decoded, and three researchers, working independently, developed methods to sequence DNA—in other words, they determined how the building blocks of DNA (the nucleotides A, C, G, and T) are ordered along the DNA strand. In 1978 bacteria were engineered to produce insulin, a pancreatic hormone that regulates carbohydrate metabolism by controlling blood glucose levels. Just four years later the Eli Lilly pharmaceutical company marketed the first genetically engineered drug: a type of human insulin grown in genetically modified bacteria.

In 1980 the U.S. Supreme Court decision in *Diamond v. Chakrabarty* (447 U.S. 303) permitted patents for genetically modified organisms; the first one was awarded to General Electric for bacteria to assist in clearing oil spills. The following year a gene was transferred from one animal species to another. In 1983 the first artificial chromosome was created. That same year the marker (the usually dominant gene or trait that serves to identify genes or traits linked with it) for Huntington's disease was identified; in 1993 the disease gene was identified.

In 1984 the observation that some nonfunctioning DNA is different in each individual launched research to refine tools and techniques developed by the British geneticist Alec Jeffreys (1950–) that perform genetic fingerprinting. Initially, the technique was used to determine the paternity of children, but it rapidly gained acceptance among forensic medicine specialists, who are often called on to assist in the investigation of crimes and interpret medicolegal issues.

The 1985 invention of the polymerase chain reaction (PCR), which amplifies (or produces many copies of) DNA, enabled geneticists, medical researchers, and forensic specialists to analyze and manipulate DNA from the smallest samples. PCR allowed biochemical analysis of even trace amounts of DNA. Ricki Lewis and Bernard Possidente describe in *A Short History of Genetics and Genomics* (2003) the American biochemist Kary B. Mullis's (1944–) development of PCR as the "genetic equivalent of a printing press," with the potential to revolutionize genetics in the same way that the printing press revolutionized mass communications.

In 1990 the first gene therapy (an approach that aims to prevent, treat, or cure disease by changing the pattern of gene expression) was administered. The patient was a four-year-old girl with the inherited immunodeficiency disorder adenosine deaminase deficiency. If left untreated, the deficiency is fatal. The gene therapy was given along with conventional medical therapy and was considered to be effective. The 1999 death of another gene therapy patient, as a result of an immune reaction to the treatment, tempered enthusiasm for gene therapy and prompted medical researchers to reconsider its safety and effectiveness. In 2003 the U.S. Food and Drug Administration temporarily halted gene therapy trials; however, the following year it allowed selected trials to resume.

In "Gene and Cell Therapy for Children—New Medicines, New Challenges?" (*Advanced Drug Delivery Reviews*, vol. 73, June 2014), Karen F. Buckland and H. Bobby Gaspar report that there are successful gene therapies for many inherited disorders and that patients who received the first gene therapies more than two decades earlier to correct immune deficiencies have remained healthy. Effective gene therapy has also been developed for some inherited immunodeficiency diseases. The researchers observe that gene therapy addresses a wide range of disorders, including skin and neurological diseases, and that clinical trials of gene therapy for hemophilia, muscular dystrophy (a disease that weakens muscles), and other disorders are under way. Gene therapy has proven effective for patients undergoing organ transplant, sparing them the risks that are associated with transplant and the need for lifetime enzyme replacement.

Narayanasamy Badrinath, Jeong Heo, and So Young Yoo report in "Viruses as Nanomedicine for Cancer" (*International Journal of Nanomedicine*, vol. 11, September 21, 2016) that a type of nanomedicine (engineering of tiny machines to prevent and treat disease) uses viruses equipped with antitumor molecules or anticancer drugs to infect and destroy cancer cells.

Cloning (the production of genetically identical organisms) was first performed with carrots. A cell from the root of a carrot plant was used to generate a new plant. By the early 1950s scientists had cloned tadpoles, and during the 1970s attempts were under way to clone mice, cows, and sheep. These clones were created using embryos, and many did not produce healthy offspring, offspring with normal life spans, or offspring with the ability to reproduce. In 1993 researchers at George Washington University cloned nearly 50 human embryos, but their experiment was terminated after just six days because they had not obtained the necessary permission to conduct the experiment.

In 1996 the British embryologist Ian Wilmut (1944–) and his colleagues at the Roslin Institute in Edinburgh, Scotland, successfully cloned the first adult mammal that was able to reproduce. Dolly the cloned sheep, named for the country singer Dolly Parton (1946–), focused public attention on the practical and ethical considerations of cloning.

Human Genome Project and More

The term *genetics* refers to the study of a single gene at a time, and *genomics* is the study of all genetic information contained in a cell. The Human Genome Project (HGP) set as one of its goals to map the entire nucleotide sequence of the more than 3 billion bases of DNA contained in the nucleus of a human cell. Discussions about the feasibility and value of conducting the HGP began in 1986. The following year the first automated DNA sequencer was produced commercially. Automated sequencing enabled researchers to decode millions, as opposed to thousands, of letters of genetic code per day and was a pivotal technological advance for the HGP, which began in 1987 under the auspices of the U.S. Department of Energy (DOE).

In 1988 the HGP was relocated to the NIH, and Watson was recruited to direct the project. The following year the NIH opened the National Center for Human Genome Research, and a committee composed of professionals from the NIH and the DOE was named to consider ethical, legal, and social issues that might arise from the project. In 1990 the project began working on preliminary genetic maps of the human genome and four other organisms believed to share many genes with humans.

During the early 1990s several new technologies were developed that further accelerated progress in analyzing, sequencing, and mapping sections of the genome. The advisability of granting private biotechnology firms the right to patent specific genes and DNA sequences was hotly debated. In April 1992 Watson resigned as director of the project to express his vehement disapproval of the NIH decision to patent human gene sequences. Later that year preliminary physical and genetic maps of the human genome were published.

The American geneticist Francis S. Collins (1950–) was named as the director of the HGP in 1993, and international efforts to assist were under way in England, France, Germany, Japan, and other countries. The 1995 release of DNA chip technology that simultaneously analyzes genetic information representing thousands of genes also helped speed the project to completion.

In 1995 investigators at the Institute for Genomic Research published the first complete genome sequence for any organism: the bacterium *Haemophilus influenzae*, with nearly 2 million genetic letters and 1,000 recognizable genes. In 1997 the yeast genome *Saccharomyces cerevisiae*, which consists of about 6,000 genes, was sequenced, and later that year the genome of the bacterium *Escherichia coli*, which contains approximately 4,600 genes, was sequenced.

In 1998 the genome of the first multicelled animal, the nematode worm *Caenorhabditis elegans*, was sequenced, containing approximately 18,000 genes. The following year the first complete sequence of a human chromosome (number 22) was published by the HGP. In 2000 the genome of the fruit fly *Drosophila melanogaster*, which Morgan and his colleagues had used to study genetics nearly a century earlier, was sequenced by the private firm Celera Genomics. The fruit fly sequence contains about 13,000 genes, with many sequences matching already identified human genes that are associated with inherited disorders.

In 2000 the first draft of the human genome was announced, and it was published in 2001. Also in 2000 the first plant genome, *Arabidopsis thaliana*, was sequenced. This feat spurred research in plant biology and agriculture. Although tomatoes that had been genetically engineered for longer shelf life had been marketed during the mid-1990s, agricultural researchers were beginning to see new possibilities to enhance crops and food products. For example, in 2000 plant geneticists developed genetically engineered rice that manufactured its own vitamin A. Many researchers believe the genetically enhanced strain of rice can help prevent vitamin A deficiency in developing countries.

The 2001 publication of the human genome estimated that humans have between 30,000 and 35,000 genes. The HGP was completed in 2003, the same year that the Cold Spring Harbor Laboratory held educational events to commemorate and celebrate the 50th anniversary of the discovery of the double helical structure of DNA. In 2004 human gene count estimates were revised downward to between 20,000 and 25,000. In 2005 and 2006 sequencing of more than 10 human chromosomes was completed, including the human X chromosome, which is one of the two sex chromosomes; the other is the Y chromosome.

In 2006 the American structural biologist Roger D. Kornberg (1947–) was awarded the Nobel Prize in Chemistry for determining the intricate way in which information in the DNA of a gene is copied to provide the instructions for building and running a living cell. In 2007 the Nobel Prize in Physiology or Medicine was awarded jointly to Mario R. Capecchi (1937–), Martin J. Evans (1941–), and Oliver Smithies (1925–2017), acknowledging their landmark discovery about the use of embryonic stem cells (undifferentiated cells from the embryo that have the potential to become a wide variety of specialized cell types) to introduce gene modifications in mice, and their development of a technique known as gene targeting that is used to inactivate single genes.

In 2008 Jeffrey M. Kidd et al. published in "Mapping and Sequencing of Structural Variation from Eight Human Genomes" (*Nature*, vol. 453, no. 7191, May 1, 2008) the first high-resolution genomic maps that detailed genetic structural variations among eight individuals of varied ancestry. These variations are important

in terms of health and understanding the risk of developing genetic diseases. Furthermore, these eight genomes may be used as benchmarks for comparison with other human genomes that are sequenced.

The HGP is one of the principal achievements in recent biomedical history, but because it involved the time, energy, and efforts of thousands of researchers, some observers question whether any individual researchers should be awarded the Nobel Prize for the completion of the sequencing effort. In "The Nobel Prize as a Reward Mechanism in the Genomics Era: Anonymous Researchers, Visible Managers and the Ethics of Excellence" (*Journal of Bioethical Inquiry*, vol. 7, no. 3, September 2010), Hub Zwart of Radboud University Nijmegen observes that it is difficult to ignore the magnitude of the accomplishment of the HGP, but opines, "If a Nobel Prize is to be awarded for deciphering the human genome, therefore, it is difficult to see how this can be done in a manner that is both meaningful and fair." Zwart suggests that the Nobel Prize should be awarded to the researchers Francis S. Collins; J. Craig Venter (1946–), a biologist and entrepreneur involved in the HGP as well as the lead researcher responsible for the 2010 creation of a cell with a synthetic genome; and Eric Steven Lander (1957–), a professor of biology at the Massachusetts Institute of Technology, a cochair of the President's Council of Advisors on Science and Technology, and the lead author of the first article presenting the genome sequence.

A 2014 analysis of data from seven studies suggests that the human genome may be even smaller than previously thought—containing fewer than 20,000 genes. Iakes Ezkurdia et al. posit in "Multiple Evidence Strands Suggest That There May Be as Few as 19,000 Human Protein-Coding Genes" (*Human Molecular Genetics*, vol. 23, no. 22, June 15, 2014) that of the estimated 19,000 to 20,000 genes in the human genome, fewer than 10 protein-coding genes may separate mice from humans.

The Postgenomic Era and Epigenetics

Epigenetic changes are heritable changes in gene function that occur without a change in the sequence of DNA. Many of these changes occur in response to environmental influences such as diet, exposure to chemicals and toxins, or psychological stress. This means that an individual's actions today—to drink alcohol, smoke, or eat a diet of fast food—might not only affect that individual's health but also the health of future generations—children, grandchildren, and great-grandchildren.

One example of epigenetic mechanisms that are used to control gene expression is DNA methylation (the process in which methyl groups are added to nucleotides in genomic DNA). DNA methylation governs many important functions including decreasing gene expression, especially viral genes and other microbial threats that may have entered the host's genome. DNA methylation is vital for normal growth and development. Alteration of DNA methylation also plays a key role in the development of diseases, including cancer.

The epigenome is a code of chemicals that modify DNA by tightly wrapping certain genes to inactivate them and relaxing others to enable their expression. Epigenetics is the study of these markers, which switch parts of the genome on and off at specific times and locations, and the factors that influence them, which in some cases may be passed down to successive generations. Epigenetic modifications to DNA exert powerful influences on gene activity; they do not alter genes but instead affect how they are expressed. Epigenetics is thought to explain much of the diversity seen in nature and promises to deepen the understanding of health and disease. Epigenetics considers single genes or sets of genes, whereas epigenomics is the global analysis of epigenetic changes across the entire genome.

The postgenomic era began with a firestorm of controversies about genetic and stem cell research and genetically modified food and crops. It also ushered in the epigenomic era, which promises new ways to prevent, diagnose, and treat diseases that threaten human health. In "Epigenomics Starts to Make Its Mark" (Nature.com, April 2, 2014), Ewen Callaway describes the promise and challenges of this new era. Genome-wide association studies compare genetic variations in sick and healthy people to identify variations associated with disease. Epigenome-wide association studies look at differences in the distribution of methyl groups (the addition of methyl groups tends to suppress gene activity) to identify arrangements that are common in a disease or associated with variation in a trait. Callaway asserts, "Because epigenetic alterations can occur over a person's lifetime, they could offer clues about the biology of diseases that genetics cannot. Many of the common genetic variants associated with disease that have been identified by [genome-wide association studies] have failed to explain fully why one person develops a condition and another does not."

CHAPTER 2
UNDERSTANDING GENETICS

Genetics (the inheritance of traits in living organisms) is a basic concept in biology. The same processes that provide the mechanism for organisms to pass genetic information to their offspring lead to the gradual change of species over time, which in turn produces biodiversity (the variety of life and the genetic differences among living organisms) and the evolution of new species. An understanding of genetics is increasingly important as genetics research and technology—and the controversies surrounding them—gain greater influence on social trends and individual lives.

Rapid and revolutionary advances in genetics, medicine, and biotechnology have created new opportunities as well as a complex and far-reaching range of ethical, legal, and social issues. Genetic testing to screen for disease susceptibility, moral questions about the cloning of human organs, and the debate about genetically modified foods affect many lives. To fully appreciate the benefits, risks, and ramifications of genetics research and to critically evaluate the related issues raised by ethicists, scientists, and other stakeholders, it is vital to understand the basics of genetics, a discipline that integrates biology, mathematics, sociology, medicine, and public health.

To understand genetics, the development of organisms, and the diversity of species, it is important to learn how deoxyribonucleic acid (DNA) functions as the information-bearing molecule of living organisms. This chapter defines basic genetics terms such as *DNA*, *chromosomes*, and *proteins*, as well genetic concepts and processes, including reproduction and inheritance.

DNA CODES FOR PROTEINS

From the perspective of genetics, the DNA molecule has two major attributes. The first is that it is able to replicate—that is, to make an exact copy of itself that can be passed to another cell, thereby conveying its precise genetic characteristics. Figure 2.1 shows how DNA replication produces two completely new and identical daughter strands of DNA. The second critical attribute is that it stores detailed instructions to manufacture specific proteins (molecules that are essential to every aspect of life). DNA is a blueprint or template for making proteins, and much of the behavior and physiology (life processes and functions) of a living organism depend on the repertoire of proteins its DNA molecules know how to manufacture.

The function of DNA depends on its structure. The double strand of DNA is composed of individual building blocks called nucleotides that are paired and connected by chemical bonds. A nucleotide contains one of four nitrogenous (containing nitrogen, a nonmetallic element that constitutes almost four-fifths of the air by volume) bases: the purines (nitrogenous bases with two rings) adenine (A) and guanine (G) and the pyrimidines (nitrogenous bases with one ring; pyrimidines are smaller than purines) cytosine (C) and thymine (T). The two strands of DNA lie side by side to create a predictable sequence of nitrogenous base pairs. (See Figure 2.2.) A stable DNA structure is formed when the two strands are a constant distance apart, which can occur only when a purine (A or G) on one strand is paired with a pyrimidine (T or C) on the other strand.

Proteins are molecules that perform all the chemical reactions necessary for life and provide structure and shape to cells. The properties of each protein depend primarily on its shape, which is determined by the sequence of its building blocks, known as amino acids. Proteins may be tough like collagen, the most abundant protein in the human body, or they may be stretchy like elastin, a protein that mixes with collagen to make softer, more flexible tissues such as skin. Figure 2.3 shows the shapes of four types of protein structures.

FIGURE 2.1

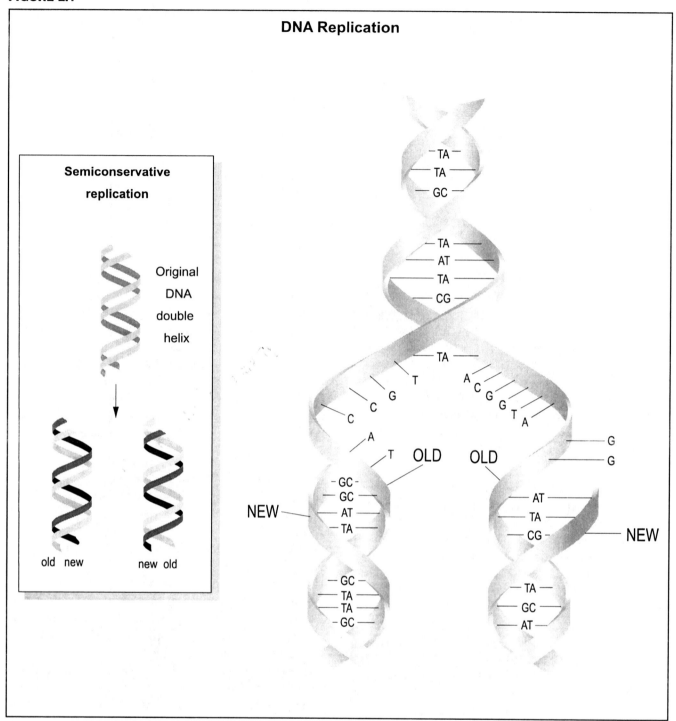

DNA Replication

Deoxyribonucleic acid replication. *Argosy Publishing, Cengage Gale.*

Proteins can act as structural components by building the tissues of the body. For example, some of the proteins in an egg include a bond that acts like an axle, allowing other parts of the molecule to spin around like wheels. However, when the egg is heated, these bonds break or denature, locking the "wheels" of the molecule in place. This is why an egg gets hard when it is cooked.

Enzymes such as lactase, which helps in the digestion of lactose (milk sugar), and hormones such as insulin (a pancreatic hormone that regulates carbohydrate metabolism by controlling blood glucose levels) are proteins that facilitate and direct chemical reactions. Defense proteins, which are able to combat invasion by bacteria or viruses, are embedded in the walls of cells and act as channels, determining which substances to let into the cell and which to block. Some bacteria know how to make proteins that protect them from antibiotics (drugs such as penicillin and streptomycin that inhibit the growth of or destroy microorganisms), while the

FIGURE 2.2

Deoxyribonucleic acid

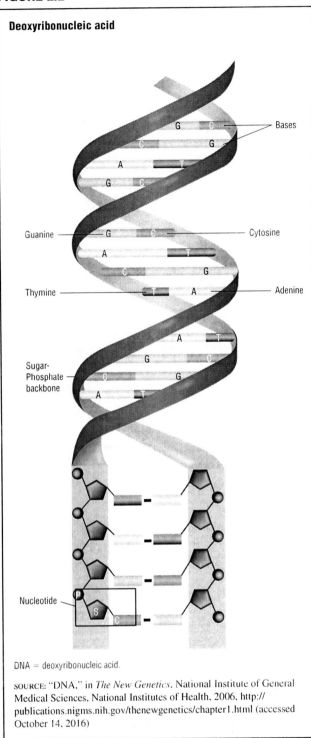

Bases

Guanine — Cytosine

Thymine — Adenine

Sugar-Phosphate backbone

Nucleotide

DNA = deoxyribonucleic acid.

SOURCE: "DNA," in *The New Genetics*, National Institute of General Medical Sciences, National Institutes of Health, 2006, http://publications.nigms.nih.gov/thenewgenetics/chapter1.html (accessed October 14, 2016)

FIGURE 2.3

Four protein structures

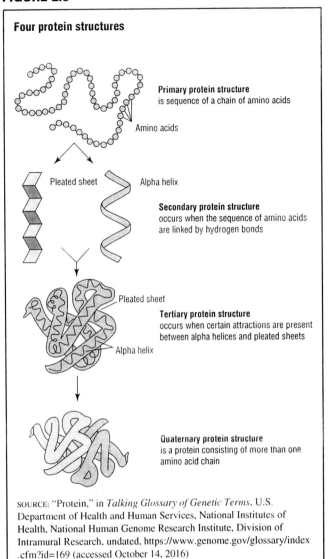

Primary protein structure is sequence of a chain of amino acids

Amino acids

Pleated sheet   Alpha helix

Secondary protein structure occurs when the sequence of amino acids are linked by hydrogen bonds

Pleated sheet

Tertiary protein structure occurs when certain attractions are present between alpha helices and pleated sheets

Alpha helix

Quaternary protein structure is a protein consisting of more than one amino acid chain

SOURCE: "Protein," in *Talking Glossary of Genetic Terms*, U.S. Department of Health and Human Services, National Institutes of Health, National Human Genome Research Institute, Division of Intramural Research, undated, https://www.genome.gov/glossary/index.cfm?id=169 (accessed October 14, 2016)

human immune system can make proteins that target bacteria for destruction. Many essential biological processes depend on the highly specific functions of proteins.

Amino Acids and Proteins

Amino acids are organic compounds that contain the amine and carboxylic acid functional groups, along with a side group (sometimes called R groups) that makes each type of amino acid different from one another. Amino acids are the building blocks of proteins. (See Figure 2.4.) There are 20 different kinds of amino acids, and each has a slightly different chemical composition. The structure and function of each protein depends on its amino acid sequence—in a protein containing a hundred amino acids, a change in a single one may dramatically affect the function of the protein. Amino acids act like jigsaw puzzle pieces, linking together in a chain to make up the protein. Each amino acid links to its neighbor with a special kind of covalent bond (covalent bonds hold atoms together) called a peptide bond. Many amino acids link together, side by side, to make a protein. Figure 2.5 is a diagram of a protein structure. Proteins range in length from 50 to 500 amino acids that are linked head to tail.

Besides their ability to form peptide bonds to their neighbors, amino acids also contain molecular appendages called side groups. Depending on the particular

FIGURE 2.4

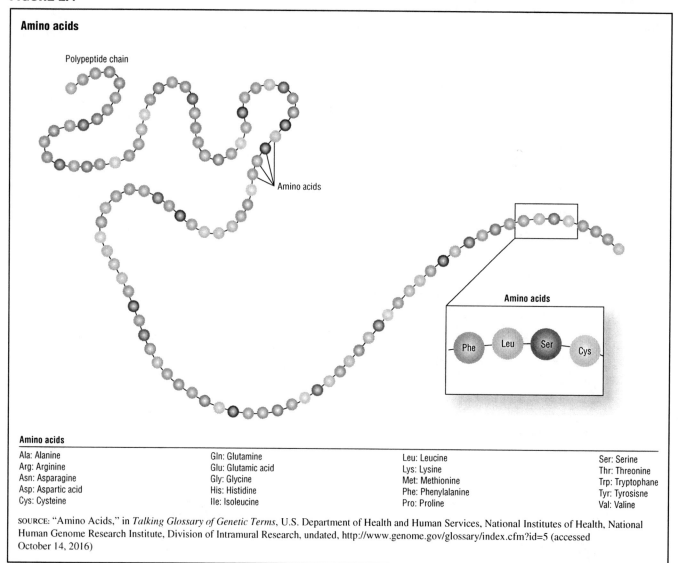

Amino acids

Polypeptide chain

Amino acids

Amino acids

Phe   Leu   Ser   Cys

Amino acids

Ala: Alanine	Gln: Glutamine	Leu: Leucine	Ser: Serine
Arg: Arginine	Glu: Glutamic acid	Lys: Lysine	Thr: Threonine
Asn: Asparagine	Gly: Glycine	Met: Methionine	Trp: Tryptophane
Asp: Aspartic acid	His: Histidine	Phe: Phenylalanine	Tyr: Tyrosisne
Cys: Cysteine	Ile: Isoleucine	Pro: Proline	Val: Valine

SOURCE: "Amino Acids," in *Talking Glossary of Genetic Terms*, U.S. Department of Health and Human Services, National Institutes of Health, National Human Genome Research Institute, Division of Intramural Research, undated, http://www.genome.gov/glossary/index.cfm?id=5 (accessed October 14, 2016)

FIGURE 2.5

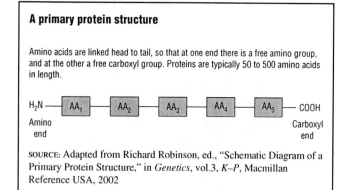

A primary protein structure

Amino acids are linked head to tail, so that at one end there is a free amino group, and at the other a free carboxyl group. Proteins are typically 50 to 500 amino acids in length.

H_2N — AA$_1$ — AA$_2$ — AA$_3$ — AA$_4$ — AA$_5$ — COOH

Amino end

Carboxyl end

SOURCE: Adapted from Richard Robinson, ed., "Schematic Diagram of a Primary Protein Structure," in *Genetics*, vol.3, *K–P*, Macmillan Reference USA, 2002

atomic arrangement of the side group, neighboring amino acids experience different pushes and pulls as they attract or repel one another. The combination of the side-by-side peptide bond linking the amino acids into a chain, along with the extra influences of the side groups, twists a protein into a specific shape. This shape is called the protein's conformation, which determines how the protein interacts with other molecules.

Genes and Proteins

Along the length of a DNA molecule there are regions that hold the instructions to manufacture specific proteins. These regions are called protein-encoding genes and are an essential element of the modern understanding of genetics. Like an MP3 player that holds thousands of songs but plays only the one that is selected at any given moment, a DNA molecule can contain anywhere from a dozen to several thousand of these protein-encoding genes. However, as with the MP3 player, at any given time only some of these genes will be expressed—that is, switched on to actively produce the protein they know how to make.

The protein-synthesizing instructions present in DNA are interpreted and acted on by ribonucleic acid (RNA). RNA, as its name suggests, is similar to DNA, except that the sugar in RNA is ribose (instead of deoxyribose), the base uracil (U) replaces thymine (T), and RNA molecules are usually single stranded and shorter than DNA molecules. (See Figure 2.6.) RNA is used to transcribe and translate the genetic code contained in DNA. Transcription is the process by which a molecule of messenger RNA (mRNA) is made, and translation is the synthesis of a protein using mRNA code. Figure 2.7 shows the transcription of mRNA and how mRNA is involved in protein synthesis (translation).

Genetic Synthesis of Proteins

How does a gene make a protein? The process of protein synthesis is quite complex. The basic sequence of protein synthesis includes the following steps:

1. A gene is triggered for expression—to synthesize a protein.

2. Half of the gene is copied into a single strand of mRNA in a process called transcription.

FIGURE 2.6

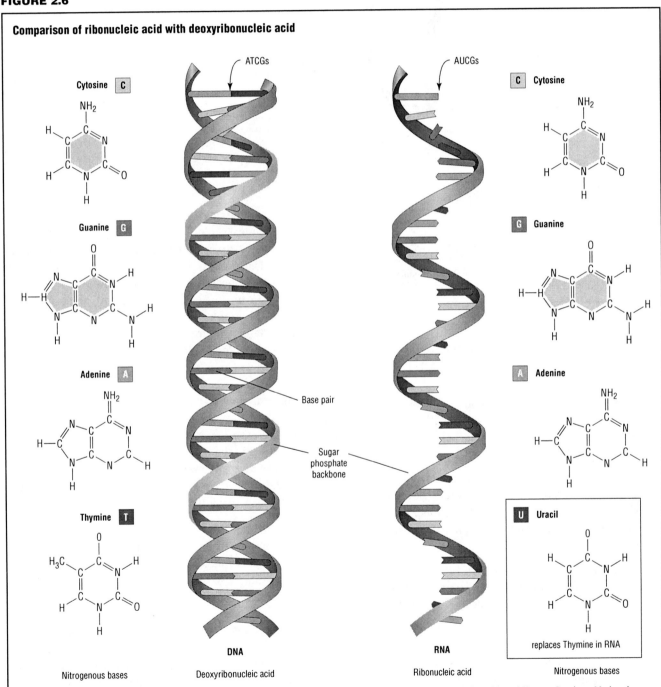

Comparison of ribonucleic acid with deoxyribonucleic acid

SOURCE: Adapted from "Ribonucleic Acid (RNA)," in *Talking Glossary of Genetic Terms*, U.S. Department of Health and Human Services, National Institutes of Health, National Human Genome Research Institute, Division of Intramural Research, undated, http://www.genome.gov/Glossary/index.cfm?id=180 (accessed October 14, 2016)

FIGURE 2.7

Messenger RNA

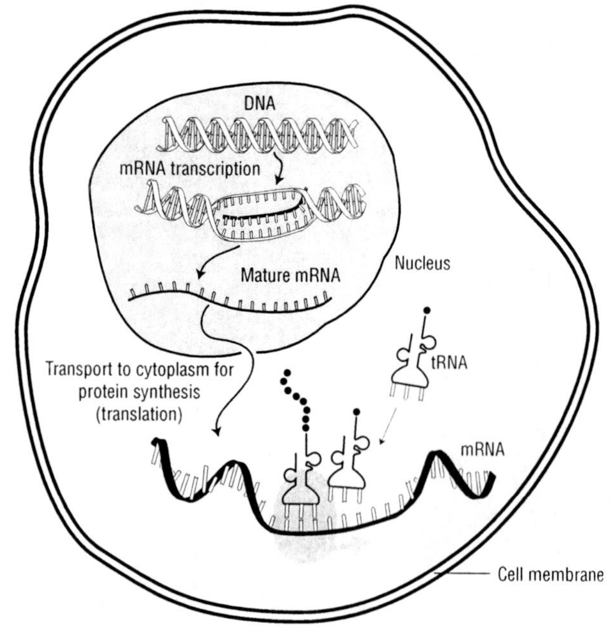

DNA

mRNA transcription

Mature mRNA

Nucleus

Transport to cytoplasm for protein synthesis (translation)

tRNA

mRNA

Cell membrane

DNA = deoxyribonucleic acid.
mRNA = messenger RNA.
tRNA = transfer RNA.

SOURCE: R. C. Geer and D. J. Messersmith, "Transcription," in "Molecular Biology Review, *Introduction to Molecular Biology Resources*, National Center for Biotechnology Information, revised 2007, http://www.ncbi.nlm.nih.gov/Class/MLACourse/Modules/MolBioReview/transcription.html (accessed October 14, 2016)

FIGURE 2.8

Codon

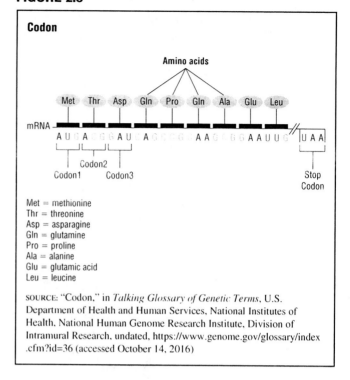

Amino acids

Met Thr Asp Gln Pro Gln Ala Glu Leu

mRNA

A U G A C G G A U C A G C C G   A A G   G G A A U U G   U A A

Codon1   Codon2   Codon3   Stop Codon

Met = methionine
Thr = threonine
Asp = asparagine
Gln = glutamine
Pro = proline
Ala = alanine
Glu = glutamic acid
Leu = leucine

SOURCE: "Codon," in *Talking Glossary of Genetic Terms*, U.S. Department of Health and Human Services, National Institutes of Health, National Human Genome Research Institute, Division of Intramural Research, undated, https://www.genome.gov/glossary/index.cfm?id=36 (accessed October 14, 2016)

3. The mRNA anchors to a ribosome, an organelle (membrane-bound cell compartment) where protein synthesis occurs.

4. Each sequence of three bases on the mRNA, called a codon, uses the right type of transfer RNA (tRNA) to pick up a corresponding amino acid from the cell. (See Figure 2.8.)

5. A string of amino acids is assembled on the ribosome, side by side, in the same order as the codons of the mRNA, which, in turn, correspond to the sequence of bases from the original DNA molecule in a process called translation.

6. When the codons have all been read and the entire sequence of amino acids has been assembled, the protein is released to twist into its final form.

First, the DNA molecule receives a trigger telling it to express a particular gene. Many influences may trigger gene expression, including chemical signals from hormones and energetic signals from light or other electromagnetic energy. For example, the spiral backbone of the DNA molecule can actually carry pulsed electrical signals that participate in activating gene expression.

Once a gene has been triggered for expression, a special enzyme system causes the DNA's double spiral to spring apart between the beginning and end of the gene sequence. The process is similar to a zipper with teeth that remain connected above and below a certain area, but pop open to create a gap along part of the zipper's length. At this point, the DNA base pairs that make up the gene sequence are separated. DNA replication is based on the understanding that any exposed nucleotide thymine (T) will pick up an adenine (A) and vice versa, whereas an exposed cytosine (C) will connect to an available guanine (G) and vice versa. Here, the same process takes place except that instead of unzipping and copying the entire length of the DNA molecule, only the region between the beginning and end of the gene is copied. Instead of making a new double spiral, only one side of the gene is replicated, creating a special, single strand of bases called mRNA.

Once the mRNA has made a copy of one side of the gene, it separates from the DNA molecule and eventually connects to an organelle in the cell called a ribosome, and the DNA returns to its original state. (Figure 2.9 is a drawing of a ribosome, which is a tiny particle of RNA and protein found in the cell's cytoplasm.) Here, it anchors to another type of RNA called ribosomal RNA (rRNA). This is where the actual protein synthesis takes place. Using a special genetic coding system, each sequence of three bases on the mRNA copied from the gene is used to catch a corresponding amino acid floating inside the cell. These sequences of three bases are the codons, and they link to tRNA. A different form of tRNA

FIGURE 2.9

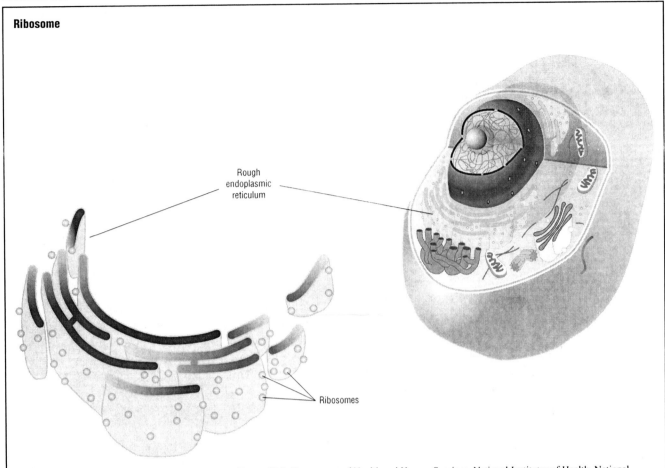

Ribosome

Rough
endoplasmic
reticulum

Ribosomes

SOURCE: "Ribosome," in *Talking Glossary of Genetic Terms*, U.S. Department of Health and Human Services, National Institutes of Health, National Human Genome Research Institute, Division of Intramural Research, undated, https://www.genome.gov/glossary/index.cfm?id=178 (accessed October 14, 2016)

is used to catch each different type of amino acid. The tRNA then catches the appropriate amino acid.

One after another, the mRNA's codons cause the corresponding tRNAs to be captured and linked, side by side, using the peptide bonds to connect them. When the last codon has been read and the entire peptide sequence is complete, the newly formed protein molecule is released from the ribosome. When this happens, all the side groups are able to interact, twisting the protein into its final shape. In this way, a protein-encoding gene is able to manufacture a protein from a series of DNA bases.

THE CELL IS THE BASIC UNIT OF LIFE

Ever since Matthias Jakob Schleiden (1804–1881) and Theodor Schwann (1810–1882) put forth their theories—Schleiden in 1838 and Schwann in 1839—that all plants and animals are composed of cells, there has been continuous refinement of cell theory. Schleiden studied plants and by examining them under a microscope determined that they were composed of cells. Schwann applied Schleiden's theory and came to the same conclusion about the evolution of animal life. The idea that the cell is the basic unit of life became known as "cell theory."

The early view that cells were made up of protoplasm (a jellylike substance) has given way to the more sophisticated understanding that cells are highly complex organizations of even smaller molecules and substructures. Cytology (the study of the formation, structure, and function of cells) has benefited from ever-improving technology, including powerful microscopes that enable researchers to identify the organelles and determine their roles in inheritance.

Cells are the basic units and building blocks of nearly every organism. (One exception is viruses, which are simple organisms that are not composed of cells.) Each cell of an organism contains the same genetic information, which is passed on faithfully when cells divide. Different types of cells arise because they use different parts of the information, as determined by the cell's history and the immediate environment. Different cell types may be organized into tissues and organs.

Cell Structure and Function

Plants and animals, as well as other organisms such as fungi, are composed of eukaryotic cells, or eukaryotes, because they have nuclei and membrane-bound structures known as organelles. In eukaryotic cells the organelles within the cell sustain, support, and protect it by creating a barrier between the cell and its environment, acting to build and repair cell parts, storing and releasing energy, transporting material, disposing of waste, and increasing in number.

Each organelle functions like an organ system for the cell. For example, the nucleus is the command center, masterminding protein synthesis within the cell. (See Figure 1.1 in Chapter 1.) The ribosomes work as protein factories, the Golgi apparatus is a protein sorter, and the endoplasmic reticulum operates as a protein processor. Lysosomes and peroxisomes serve as the cell's digestive system, and mitochondria convert energy in the cell. The surface membrane of the cell acts like skin by selectively permitting molecules in and out of the cell.

The nuclei of eukaryotes contain the chromosomes, which are chains of genetic material coded in DNA. The threadlike chromosomes are contained in the nucleus of a typical animal cell. Genes are segments of DNA that carry a basic unit of hereditary information in coded form. (See Figure 2.10.) They contain instructions for making proteins.

The eukaryotic chromosome is composed of chromatin (a combination of nuclear DNA and protein) and contains a linear array of genes. It is visible just before and during cell division. (See Figure 2.11.) Human cells normally contain 23 pairs of chromosomes, or a total of 46 chromosomes, that may be examined using a process known as karyotyping (the organization of a standard picture of the chromosomes). Figure 2.12 is a karyotype—that is, a photo of an individual's chromosomes.

Cells without nuclei, such as bacteria and blue-green algae, are called prokaryotic cells or prokaryotes. Prokaryotic cells are smaller than eukaryotic cells, contain less genetic information, and are able to grow and divide more quickly. They perform these functions without organelles. A prokaryotic cell's DNA is not contained in one location; instead, it floats in different regions of the cell. The DNA of prokaryotes also contains jumping genes that are able to bind to other genes and transfer gene sequences from one site of a chromosome to another.

FIGURE 2.10

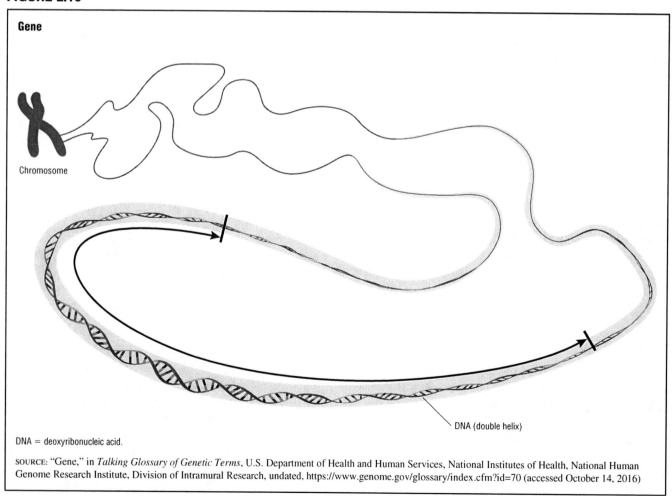

Gene

Chromosome

DNA (double helix)

DNA = deoxyribonucleic acid.

SOURCE: "Gene," in *Talking Glossary of Genetic Terms*, U.S. Department of Health and Human Services, National Institutes of Health, National Human Genome Research Institute, Division of Intramural Research, undated, https://www.genome.gov/glossary/index.cfm?id=70 (accessed October 14, 2016)

FIGURE 2.11

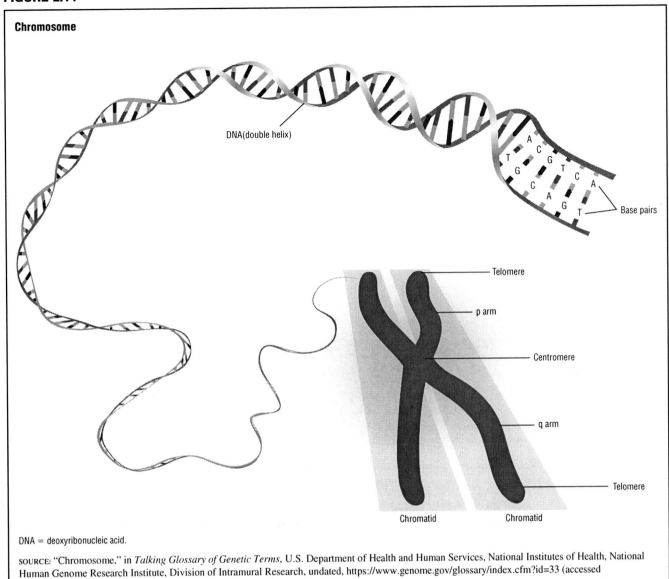

Chromosome

DNA(double helix)

Base pairs

Telomere

p arm

Centromere

q arm

Telomere

Chromatid   Chromatid

DNA = deoxyribonucleic acid.

SOURCE: "Chromosome," in *Talking Glossary of Genetic Terms*, U.S. Department of Health and Human Services, National Institutes of Health, National Human Genome Research Institute, Division of Intramural Research, undated, https://www.genome.gov/glossary/index.cfm?id=33 (accessed October 14, 2016)

GROWTH AND REPRODUCTION

The growth of an organism occurs as a result of cell division in a process known as mitosis. Many cells are relatively short lived, and mitosis allows for regular renewal of these cells. It is also the process that generates the millions of cells needed to grow an organism, or in the case of a human being, the trillions of cells needed to grow from birth to adulthood.

Mitosis is a continuous process that occurs in several stages. Between cell divisions the cells are in interphase, during which there is cell growth, and the genetic material (DNA) contained in the chromosomes is duplicated so that when the cell divides, each new cell has a full-scale version of the same genetic material. The process of mitosis involves exact duplication—gene by gene—of the cell's chromosomal material and a systematic method for evenly distributing this material. It concludes with the

physical division known as cytokinesis, when the identical chromosomes pull apart and each heads for the nucleus of one of the new daughter cells. Mitosis occurs in all eukaryotic cells, except the gametes (sperm and egg), and always produces genetically identical daughter cells with a complete set of chromosomes.

If, however, mitosis occurred in the gametes, then when fertilization (the joining of sperm and egg) took place, the offspring would receive a double dose of hereditary information. To prevent this from occurring, the gametes undergo a process of reduction division known as meiosis. Meiosis reduces the number of chromosomes in the gametes by half, so that when fertilization occurs the normal number of chromosomes is restored. For example, in humans the gametes produced by meiosis are haploid—they have just one copy of each of the 23 chromosomes. Besides preventing the number

FIGURE 2.12

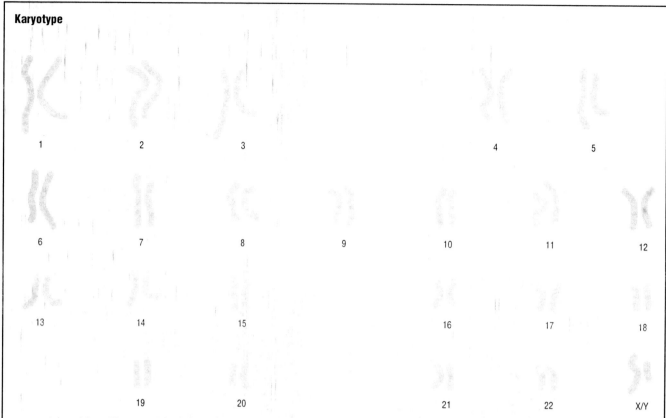

Karyotype

SOURCE: Adapted from "Karyotype," in *Talking Glossary of Genetic Terms*, U.S. Department of Health and Human Services, National Institutes of Health, National Human Genome Research Institute, Division of Intramural Research, undated, https://www.genome.gov/glossary/index.cfm?id=114 (accessed October 14, 2016)

of chromosomes from doubling with each successive generation, meiosis also provides genetic diversity in offspring.

During meiosis the chromosomal material replicates and concentrates itself into homologous chromosomes (doubled chromosomes), each of which is joined at a central spot called the centromere. Figure 2.11 shows chromosomes joined at the centromere, and Figure 2.13 shows the location of the centromere in the chromosome. Pairing up along their entire lengths, they are able to exchange genetic material in a process known as crossing over. Figure 2.14 shows homologous chromosomes crossing over during meiosis to create new gene combinations. Crossing over results in much of the genetic variation that is observed among parents and their offspring. The pairing of homologous chromosomes and crossing over occur only during meiosis.

The process of meiosis also creates another opportunity to generate genetic diversity. During one phase of meiosis, called metaphase, the arrangement of each pair of homologous chromosomes is random, and different combinations of maternal and paternal chromosomes line up with varying orientations to create new gene combinations on different chromosomes. This action is called independent assortment. Figure 2.15 shows the process of meiosis in an organism with six chromosomes. Because recombination and independent assortment of parental chromosomes takes place during meiosis, the daughter cells are not genetically identical to one another.

Gamete Formation

In male animals gamete formation, known as spermatogenesis, begins at puberty and takes place in the testes. Spermatogenesis involves a sequence of events that begins with the mitosis of primary germ cells to produce primary spermatocytes. (See Figure 2.16.) Four primary spermatocytes undergo two meiotic divisions, and as they undergo spermatogenesis they lose much of their cytoplasm and develop the spermatozoa's characteristic tail, the motor apparatus that provides the propulsion necessary to reach the egg cell. Like oocytes (egg cells) produced by the female, the mature sperm cells will be haploid (possessing just one copy of each chromosome).

In female animals gamete formation, known as oogenesis, takes place in the ovaries. Primary oocytes are produced by mitosis in the fetus before birth. Unlike the male, which continues to produce sperm cells throughout life, in the female the total number of eggs ever to be produced is present at birth. At birth, or shortly

FIGURE 2.13

Centromere

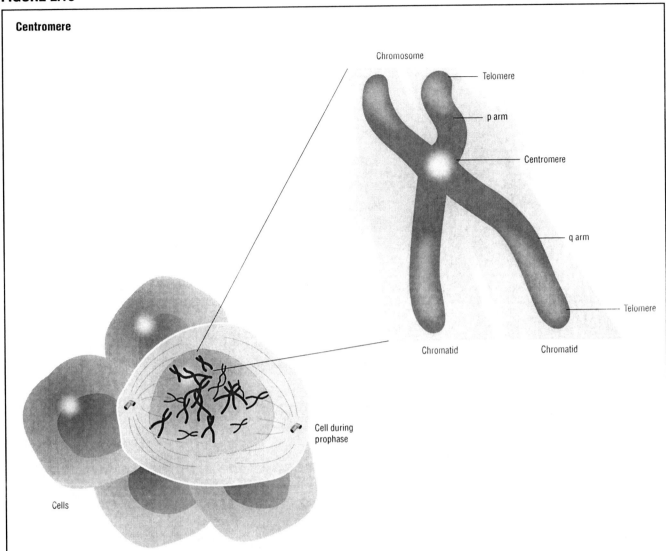

SOURCE: "Centromere," in *Talking Glossary of Genetic Terms*, U.S. Department of Health and Human Services, National Institutes of Health, National Human Genome Research Institute, Division of Intramural Research, undated, https://www.genome.gov/glossary/index.cfm?id=29 (accessed October 14, 2016)

before, meiosis begins and primary oocytes remain in the prophase of meiosis until puberty. At puberty the first meiotic division is completed, and a diploid cell becomes two haploid daughter cells; one large cell becomes the secondary oocyte and the other the first polar body. The secondary oocyte undergoes meiosis a second time but the meiosis does not continue to completion without fertilization. Figure 2.17 shows the sequence of events leading to the production of a mature ovum.

Fertilization

Like gamete formation, fertilization is a process, as opposed to a single event. It begins when the sperm and egg first come into contact and fuse together and culminates with the intermingling of two sets of haploid genes to reconstitute a diploid cell with the potential to become a new organism.

Of the millions of sperm released during an ejaculation, less than 1% survive to reach the egg. Of the few hundred sperm that reach the egg, it is only possible for one to successfully fertilize it. While the sperm are in the female reproductive tract, swimming toward the egg, they undergo a process known as capacitation, during which they acquire the capacity to fertilize the egg. As the sperm approach the egg they become hyperactivated, and in a frenzy of mechanical energy the sperm attempt to burrow their way through the outer shell of the egg called the zona pellucida.

The cap of the sperm, known as the acrosome, contains enzymes that are crucial for fertilization. These acrosomal enzymes dissolve the zona pellucida by making a tiny hole in it, so that one sperm can swim through and reach the surface of the egg. When the sperm comes into contact with the egg, the egg transforms the zona

FIGURE 2.14

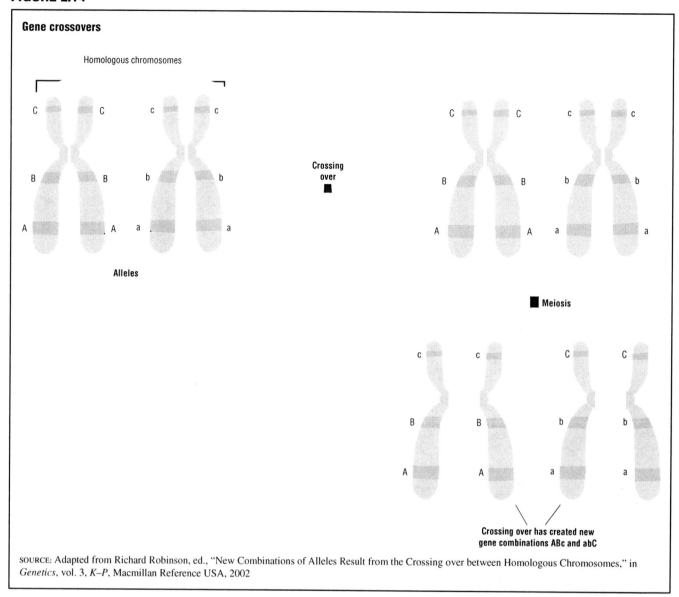

Gene crossovers

SOURCE: Adapted from Richard Robinson, ed., "New Combinations of Alleles Result from the Crossing over between Homologous Chromosomes," in *Genetics*, vol. 3, *K–P*, Macmillan Reference USA, 2002

pellucida by creating an impenetrable barrier, so that no other sperm may enter.

Sperm penetration triggers the second meiotic division of the egg. With this division the chromosomes of the sperm and egg form a single nucleus. The resulting cell—the first cell of an entirely new organism—is called a zygote. The zygote then divides into two cells, which, in turn, continue to divide rapidly, producing a ball of cells now called the blastocyst. The blastocyst is an early stage of embryogenesis, the process that describes the development of the fertilized egg as it becomes an embryo. In humans the developing baby is considered an embryo until the end of the eighth week of pregnancy.

GENETIC INHERITANCE

For inheritance of simple genetic traits, the two inherited copies of a gene determine the phenotype (the observable characteristic) for that trait. When genes for a particular trait exist in two or more different forms that may differ among individuals and populations, they are called alleles. For example, brown and blue eye colors are different alleles for eye color. For every gene, the offspring receives two alleles, one from each parent. The combination of inherited alleles is the genotype of the organism, and its expression (the observable characteristic) is its phenotype. Figure 2.18 is a graphic example of phenotype.

For many traits the phenotype is a result of an interaction between the genotype and the environment. Some of the most readily apparent traits in humans, such as height, weight, and skin color, result from interactions between genetic and environmental factors. In addition, there are complex phenotypes that involve multiple gene-encoded proteins; the alleles of these particular genes are influenced by other factors, either genetic or environmental. So even

FIGURE 2.15

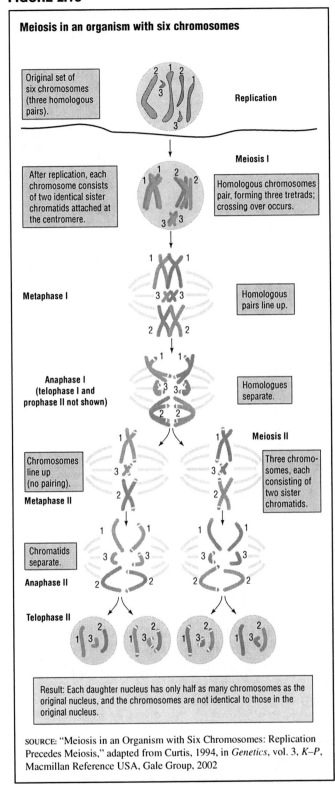

Meiosis in an organism with six chromosomes

Original set of six chromosomes (three homologous pairs).

Replication

After replication, each chromosome consists of two identical sister chromatids attached at the centromere.

Meiosis I

Homologous chromosomes pair, forming three tretrads; crossing over occurs.

Metaphase I

Homologous pairs line up.

Anaphase I (telophase I and prophase II not shown)

Homologues separate.

Meiosis II

Chromosomes line up (no pairing).
Metaphase II

Three chromosomes, each consisting of two sister chromatids.

Chromatids separate.
Anaphase II

Telophase II

Result: Each daughter nucleus has only half as many chromosomes as the original nucleus, and the chromosomes are not identical to those in the original nucleus.

SOURCE: "Meiosis in an Organism with Six Chromosomes: Replication Precedes Meiosis," adapted from Curtis, 1994, in *Genetics*, vol. 3, *K–P*, Macmillan Reference USA, Gale Group, 2002

a recessive allele is expressed only when both alleles are recessive. Recessive genes continue to pass from generation to generation, but they are only expressed in individuals who do not inherit a copy of the dominant gene for the specific trait. Figure 2.19 shows the inheritance of a recessive trait, in this example a recessive mutation that produces offspring with sickle-cell anemia (the presence of oxygen-deficient, abnormal red blood cells that cause affected individuals to suffer from obstruction of capillaries, resulting in pain and potential organ damage) or cystic fibrosis (an inherited disease of the mucous glands that produces problems with the lungs and pancreas).

There are also some instances, known as incomplete dominance, when one allele is not completely dominant over the other, and the resulting phenotype is a blend of both traits. Skin color in humans is an example of a trait often governed by incomplete dominance, with offspring appearing to be a blend of the skin tones of each parent. Furthermore, some traits are determined by a combination of several genes (multigenic or polygenic), and the resulting phenotype is determined by the final combination of alleles of all the genes that govern the particular trait.

Some multigenic traits are governed by many genes, each contributing equally to the expression of the trait. In such instances a defect in a single gene pair may not have a significant impact on expression of the trait. Other multigenic traits are predominantly directed by one major gene pair and only mildly influenced by the effects of other gene pairs. For these traits the impact of a defective gene pair depends on whether it is the major pair governing expression of the trait or one of the minor pairs influencing its expression.

A range of other factors enters into whether a trait will be evidenced and the extent to which it is expressed. For example, different individuals may express a trait with different levels of severity. This phenomenon is called variable expressivity.

Determining Genetic Probabilities of Inheritance

Conventionally, geneticists use uppercase letters to represent dominant alleles and lowercase letters to stand for recessive alleles. An organism with a pair of identical alleles for a trait is described as a homozygote, or homozygous for that particular trait. When organisms are homozygous for a dominant trait, all uppercase letters symbolize the trait, whereas those that are homozygous for a recessive trait are represented by all lowercase letters. A heterozygote is an organism with different alleles for a trait, one donated from each parent, and when one is dominant and the other is recessive the trait is shown using a combination of uppercase and lowercase letters.

Although the combination of alleles is a random event, it is possible to predict the probability that an

though the presence of certain genes indicates susceptibility or likelihood to develop a certain trait, it does not guarantee expression of the trait.

For a specific trait some alleles may be dominant whereas others are recessive. The phenotype of a dominant allele is always expressed, whereas the phenotype of

FIGURE 2.16

Process of spermatogenesis

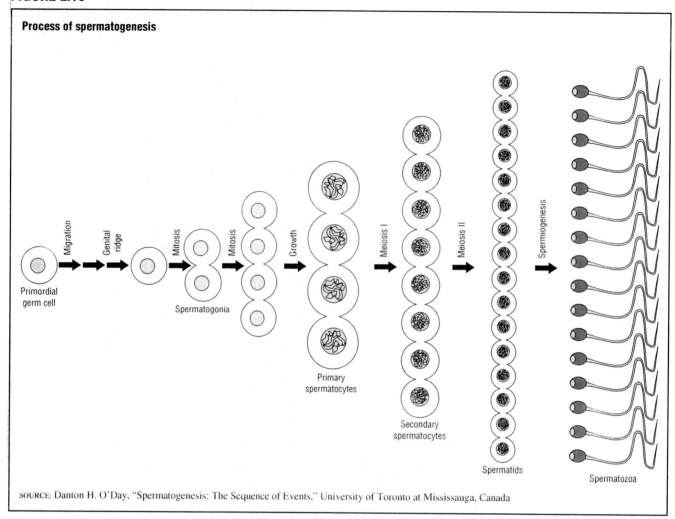

SOURCE: Danton H. O'Day, "Spermatogenesis: The Sequence of Events," University of Toronto at Mississauga, Canada

FIGURE 2.17

Process of oogenesis

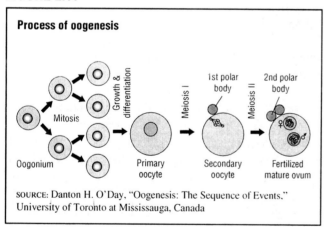

SOURCE: Danton H. O'Day, "Oogenesis: The Sequence of Events," University of Toronto at Mississauga, Canada

offspring will have the same or a different phenotype from its parents when the genotypes with respect to the specific trait of both parents and the phenotype associated with each possible combination of alleles are known. The formula used to determine genetic possibilities was developed by the British geneticist Reginald Crundall Punnett (1875–1967). The Punnett square is a grid configuration that depicts genotype and phenotype. In Punnett squares the genotypes of parents are represented with four letters. There are two alleles for each trait. Genotypes of haploid gametes are represented with two letters. Gametes will contain one allele for each trait in every possible combination, and all possible fertilizations are calculated. Punnett squares are used to compute the cross of a single gene or two genes and their alleles, but they become extremely complicated when used to predict the offspring of more than two alleles.

To construct a Punnett square, the alleles in the gametes of one parent are placed along the top, and the alleles in the gametes of the other parent are placed along the left side of the grid. The number of characteristics considered determines the size of the Punnett square. A monohybrid cross looks at one trait and has a two-by-two structure, with four possible genotypes resulting from the cross. A dihybrid cross looks at two traits and has a four-by-four structure, providing 16 possible genotypes. Figure 2.20 is a simple Punnett square that shows one parent as being homozygous for the AA allele and another that is heterozygous for the Aa allele, with the four possible offspring: 50% AA and 50% Aa. This is an example of

FIGURE 2.18

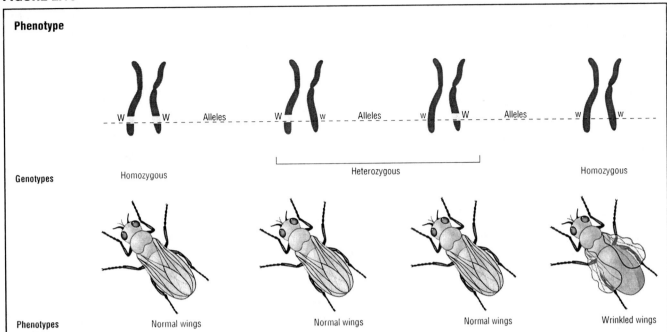

Phenotype

Genotypes

Homozygous   Heterozygous   Homozygous

Phenotypes   Normal wings   Normal wings   Normal wings   Wrinkled wings

SOURCE: "Phenotype," in *Talking Glossary of Genetic Terms*, U.S. Department of Health and Human Services, National Institutes of Health, National Human Genome Research Institute, Division of Intramural Research, undated, http://www.genome.gov/Glossary/index.cfm?id=152 (accessed October 14, 2016)

FIGURE 2.19

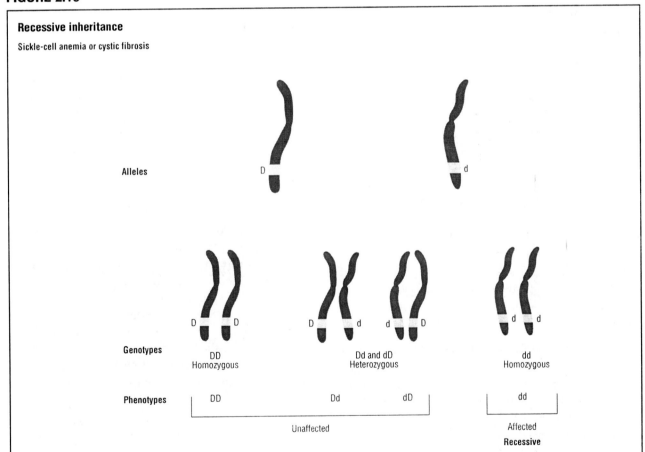

Recessive inheritance

Sickle-cell anemia or cystic fibrosis

Alleles

Genotypes

DD
Homozygous

Dd and dD
Heterozygous

dd
Homozygous

Phenotypes

DD   Dd   dD   dd

Unaffected   Affected
                                        Recessive

SOURCE: "Recessive," in *Talking Glossary of Genetic Terms*, U.S. Department of Health and Human Services, National Institutes of Health, National Human Genome Research Institute, Division of Intramural Research, undated, http://www.genome.gov/Glossary/index.cfm?id=172 (accessed October 14, 2016)

FIGURE 2.20

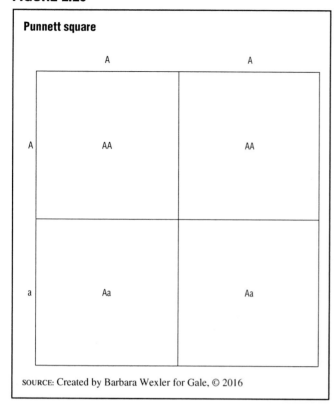

Punnett square

	A	A
A	AA	AA
a	Aa	Aa

SOURCE: Created by Barbara Wexler for Gale, © 2016

offspring that shows just one phenotype even though two genotypes are present. Figure 2.21 is a Punnett square that shows the predicted patterns of recessive inheritance. When both parents are carriers of the trait and dominant inheritance, the offspring is likely to arise from an affected parent and a normal parent.

Mitochondrial Inheritance

Mitochondria are the organelles involved in cellular metabolism and energy production and conversion. Mitochondria are inherited exclusively from the mother and contain their own DNA, known as mtDNA. Because some of the mtDNA multiplies during the organism's growth and development, it is more susceptible to mutation (changes in DNA sequence). The fact that mitochondria contain their own DNA has prompted scientists to speculate that they originally existed as independent one-celled organisms that over time developed interdependent relationships with more complex, eukaryotic cells.

Mitochondrial inheritance looks much like Mendelian inheritance (or genetic inheritance as described by Gregor Mendel's [1822–1884] laws), with two important exceptions. First, all maternal offspring are usually affected, whereas even in autosomal dominant disorders (those not related to the sex genes) only 50% of offspring are expected to be affected. (See Figure 2.22.) Second, mitochondrially inherited traits are never passed through the male parent. Males are as likely to be affected as females, but their offspring are not at risk. In other words, when

there is a mutation in a mitochondrial gene, it is passed from a mother to all of her children; sons will not pass it on, but daughters will pass it on to all their children.

Mutations in mtDNA have been linked to the development of disease in humans. Leber's hereditary optic neuropathy, a painless loss of vision that afflicts people between the ages of 12 and 30 years, was the first human disease to be associated with a mutation in mtDNA. Many diseases that are linked to mtDNA affect the nervous system, heart or skeletal muscles, liver, or kidneys—sites of energy usage.

DETERMINING SEX

From the moment of fertilization, the new organism is assigned a sex, and its growth will proceed to develop either as a male or a female organism. The first clues that prompted scientists to consider that the determination of sex was influenced by genetics came from two key observations. The first was the fact that there is a general tendency toward a one-to-one ratio of males to females in all species. The second was the realization that the determination of sex followed the principles of Mendelian genetics—that gender was predictable as expected when individuals pure for a recessive trait were crossed with individuals that were hybrid.

The determination of sex occurs in all complex organisms, but the processes vary, even among animals. In humans 22 of the 23 pairs of chromosomes are as likely to be found in males as in females. These 22 chromosomes are known as the autosomes and the 23rd chromosome is called the sex chromosome. The sex chromosomes of females are identical and are called X chromosomes. (See Figure 2.23.) In males the sex chromosomes consist of an X chromosome and a smaller Y chromosome.

Chromosome Theory of Sex Determination

Because there is a genetic influence on sex and gender, it is important to understand the difference between these terms. Generally, the term *sex* refers to the biological differences between males and females, including genetic differences. The term *gender* is used to describe the social and cultural roles of males and females in society or people's personal sense of themselves, which is called "gender identity." When their sex assigned at birth is different from their gender identity, people may consider themselves "transgender," "non-binary," or "gender-nonconforming."

The chromosome theory of sex determination states that:

- Sex is determined by the sex chromosome.

- In females the sex chromosomes are identical—both are X chromosomes.

FIGURE 2.21

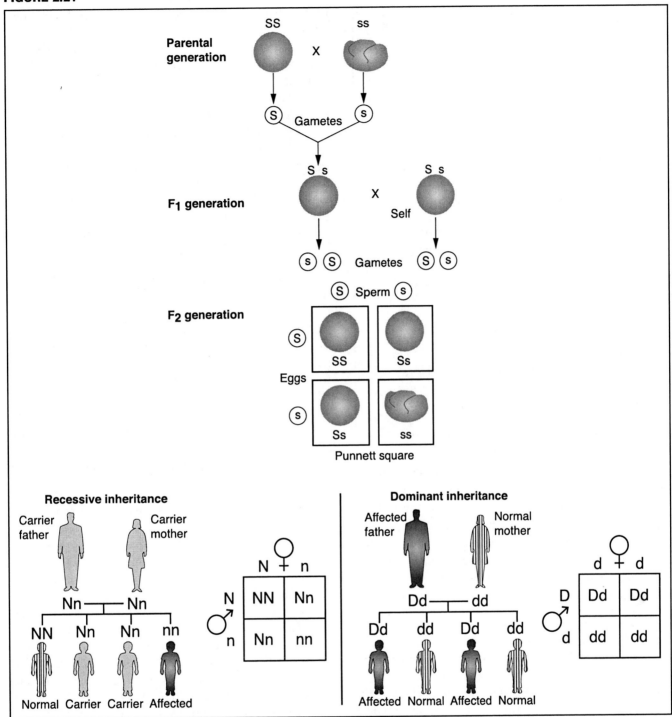

Punnett square with predicted patterns of recessive inheritance. *Argosy Publishing, Cengage Gale.*

- Because females have an XX genotype, all egg cells contain an X chromosome.

- In males the sex chromosomes are not identical; one is X and one is Y.

- Because males have an XY genotype, half of all sperm cells contain an X chromosome and half contain a Y chromosome.

- After fertilization, the egg may receive either an X or a Y chromosome from the sperm. Because all egg cells

contain an X chromosome, the determination of gender is wholly dependent on the chromosomal composition of the sperm. Sperm carrying the Y chromosome are known as androsperm; those containing the X chromosome are called gynosperm. If the sperm carries the Y chromosome, the offspring will be male (XY); if it carries the X chromosome, the offspring will be female (XX).

The determination of sex occurs at conception with the designation of chromosomal composition that is

FIGURE 2.22

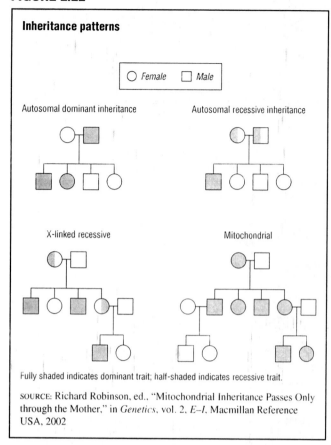

Inheritance patterns

○ Female   □ Male

Autosomal dominant inheritance   Autosomal recessive inheritance

X-linked recessive   Mitochondrial

Fully shaded indicates dominant trait; half-shaded indicates recessive trait.

SOURCE: Richard Robinson, ed., "Mitochondrial Inheritance Passes Only through the Mother," in *Genetics*, vol. 2, *E–I*, Macmillan Reference USA, 2002

either XX or XY. However, a number of other genetic and environmental influences determine sex differentiation (the way in which the genetically predetermined sex becomes a reality). Differentiation translates the genetically coded message for sex into the physical traits, such as the hormones that influence the development of male and female genitalia, body functions, and behaviors that are associated with sex-linked and gender identity.

Interestingly, humans have an inherent tendency toward female development. Research conducted during the 1940s and 1950s confirmed that many animals with just a single X chromosome developed as females, although in many instances they did not develop completely and were sterile (unable to reproduce). The absence of the Y chromosome results in female development, whereas the presence of the Y chromosome sets in motion the series of events that result in male development. These findings led to the premise that female development is the default option in the process of sex determination.

Distribution of Males and Females in the Population

Because human males produce equal numbers of sperm bearing either the X or the Y chromosome, and fertilization is a random event, then it stands to reason that in each generation equal numbers of males and females should be born. However, an examination of birth statistics in the United States and in other countries where reliable statistics have been compiled over time show that every year there are more births of males than females. For example, according to Brady E. Hamilton et al. of the Centers for Disease Control and Prevention, in "Births: Final Data for 2014" (*National Vital Statistics Reports*, vol. 64, no. 12, December 23, 2015), the ratio of males per 1,000 females for births to mothers of all races in the United States was 1,048 in 2014. Table 2.1 shows that births by sex varied by race, from a low of 1,029 males per 1,000 females among African Americans to a high of 1,063 males per 1,000 females among Asians or Pacific Islanders. The annual distribution of births by sex has remained essentially unchanged over the past seven decades, with only small annual variations.

For years this difference was attributed to the idea that males were inherently stronger than females and better able to survive pregnancy and birth. This theory was dispelled when researchers found that nearly three times as many male fetuses spontaneously abort (dying before birth). In fact, male life expectancy is less than female life expectancy at every age, from conception to adulthood. (See Table 2.2.) The explanation for the higher proportion of male births appears to be that more male offspring are conceived—possibly as many as 125 to every 100—because the prenatal death rate for males is so high; at birth the gap closes to about 105 to every 100.

One explanation for the higher number of males conceived is that the smaller and stronger Y sperm are better able to swim quickly and successfully reach the egg cell. Along with androsperm size and mobility, environmental conditions influence gender determination and the chances of fetal survival. For example, the mother's age and general health are strongly linked to favorable outcomes of conception and pregnancy and are less strongly linked to, but are associated with, sex of her offspring. Younger mothers are more likely to conceive male offspring, by a ratio as high as 120 to 100, and unfavorable prenatal conditions such as poor health or maternal illness are more likely to compromise the survival of the male fetus than the female fetus.

Sex-Linked Characteristics

The two sex chromosomes also differ in terms of the genes they contain, which relate to many traits other than sex. The Y chromosome is quite small and carries few genes other than the one that determines male gender. One of the few confirmed traits linked to the Y chromosome is the hairy ear trait, a characteristic that is distinctive but largely unrelated to health. Because this trait is located exclusively on the Y chromosome, it only appears in males.

FIGURE 2.23

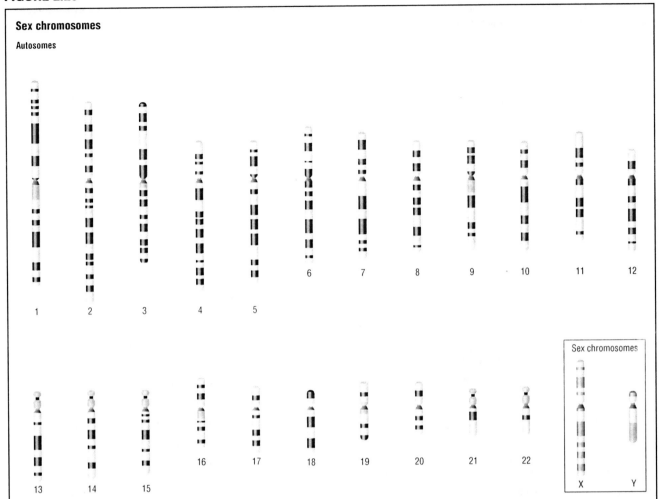

Sex chromosomes

Autosomes

Sex chromosomes

SOURCE: "Sex Chromosome," in *Talking Glossary of Genetic Terms*, U.S. Department of Health and Human Services, National Institutes of Health, National Human Genome Research Institute, Division of Intramural Research, undated, http://www.genome.gov/glossary/index.cfm?id=181 (accessed October 14, 2016)

The X chromosome is larger and holds many genes that are as necessary for males as they are for females. The genes on the X chromosome are called X-linked, and traits or conditions arising from these genes are called X-linked traits or conditions. Most people, male and female, likely have several so-called defective genes with the potential to produce harmful characteristics or conditions, but these genes are usually recessive and are not expressed in the phenotype unless they are combined with a similar recessive gene on the corresponding chromosome. For this to occur, both parents would have to contribute the same defective gene. Species are also protected from the harmful effects of single defective genes by virtue of the fact that most traits are multigenic (controlled by more than one gene).

Although X-linked dominant alleles affect males and females, males are more strongly affected because they inherit just one X chromosome and do not have a counterbalancing normal allele. Huntington's disease (an inherited neuropsychiatric disease that affects the body and mind) is an example of a disorder caused by an X-linked dominant allele. Males are affected more frequently and more severely than females by X-linked recessive alleles. The male receives his X chromosome from his mother. Because males have just one X chromosome, all the alleles it contains are expressed, including those that cause serious and sometimes lethal medical disorders. Examples of disorders caused by X-linked recessive alleles range from relatively harmless conditions such as red-green color blindness to the always-fatal Duchenne muscular dystrophy (DMD), one of a group of muscular dystrophies, which is characterized by the enlargement of selected muscles, particularly in the calves. This enlargement is called pseudohypertrophy and occurs when fat and connective tissue replace muscle, giving the impression of enlarged muscles. DMD is one of the most prevalent types of muscular dystrophy and involves rapid progression of muscle degeneration early in life. It is X-linked, affects mainly

TABLE 2.1

Selected demographic characteristics of births, by race of mother, 2014

Characteristic	All races	White	Black	American Indian or Alaskan Native	Asian or Pacific Islander
			Number		
Births	3,988,076	3,019,863	640,562	44,928	282,723
			Rate		
Birth rate	12.5	12.0	14.5	9.9	14.6
Fertility rate	62.9	63.2	64.6	44.8	60.7
Total fertility rate	**1,862.5**	**1,875.5**	**1,872.0**	**1,288.5**	**1,715.5**
Sex ratio[a]	1,048	1,051	1,029	1,031	1,063
All births			Percent		
Births to mothers under 20	6.3	6.0	9.5	11.3	1.7
4th- and higher-order births[b]	12.2	11.7	16.1	21.2	6.6
Births to unmarried mothers	40.2	35.7	70.4	65.7	16.4
Mothers born in the 50 states and District of Columbia	78.1	81.8	83.4	94.2	23.3
			Mean		
Age of mother at first birth	26.3	26.4	24.2	23.1	29.5

[a]Male births per 1,000 female births.
[b]Based on live-birth order.
Notes: Race and Hispanic origin are reported separately on birth certificates. Race categories are consistent with 1977 Office of Management and Budget standards. Forty-nine states and the District of Columbia reported multiple-race data for 2014 that were bridged to single-race categories for comparability with other states. In this table, all women, including Hispanic women, are classified only according to their race. Birth rates are births per 1,000 population. Fertility rates are computed by relating total births, regardless of age of mother, to women aged 15–44. Total fertility rates are sums of birth rates for 5-year age groups multiplied by 5. Populations estimated as of July 1. Mean age at first birth is the arithmetic average of the age of mothers at the time of birth, computed directly from the frequency of first births by age of mother.

SOURCE: Brady E. Hamilton et al., "Table 13. Selected Demographic Characteristics of Births, by Race of Mother: United States, 2014," in "Births: Final Data for 2014," *National Vital Statistics Reports*, vol. 64, no. 12, December 2015, http://www.cdc.gov/nchs/data/nvsr/nvsr64/nvsr64_12.pdf (accessed October 14, 2016)

males, and is the most common neuromuscular disease of childhood, with an estimated overall prevalence (the number of cases of a disease that are present in a particular population at a given time) of 63 cases per 1 million births. The National Institute of Neurological Disorders and Stroke indicates in "Muscular Dystrophy: Hope through Research" (August 2013, https://www.ninds.nih.gov/Disorders/Patient-Caregiver-Education/Hope-Through-Research/Muscular-Dystrophy-Hope-Through-Research) that one out of every 3,500 to 6,000 boys worldwide is diagnosed with it. Another X-linked recessive allele causes Tay-Sachs disease, a disease that is most common among people of Jewish descent and results in neurological disorders and death in childhood.

All the male offspring of females who carry X-linked recessive alleles will be affected by the recessive allele. Female children are not as likely to express harmful recessive X-linked traits because they have two X chromosomes. Fifty percent of the female offspring will receive the recessive allele from a mother who carries the allele and an unaffected father. (See Figure 2.22.)

COMMON MISCONCEPTIONS ABOUT INHERITANCE

There are many myths and misunderstandings about genetics and inheritance. For example, some people mistakenly believe that in any population dominant traits are inevitably more common than recessive traits. This is simply not true, as evidenced by the observation that, among humans, the allele that produces six fingers and six toes on each hand and foot, respectively, is dominant over the allele for five fingers and five toes, but the incidence of polydactyly (extra digits) is actually quite low.

Another lingering misconception is that sex-linked diseases occur only in males. This is untrue but it is easy to understand the source of the misunderstanding. For years it was thought that hemophilia (a disease that is characterized by uncontrolled bleeding) did not occur in females. The observation seemed reasonable because there were no reported cases of the disease among females. Although it was true that there were no females with the disease, the reasoning was incorrect. For a female to suffer from hemophilia, she would require a defective recessive gene on both of her X chromosomes, meaning her mother was carrying the gene and her father was affected with the disease. Because most people with hemophilia died young, few lived to produce offspring. In other words, female hemophiliacs were rare because the pairings that might produce them were infrequent. During the 1950s the first cases of hemophilia in females were documented, and the theory was discarded.

Finally, the idea that humans are entirely unique in their genetic makeup is false. In fact, human beings share much of their genetic composition with other organisms

TABLE 2.2

Life expectancy at selected ages, by race and sex, selected years 1900–2014

Specified age and year	All races			White			Black or African American[a]		
	Both sexes	Male	Female	Both sexes	Male	Female	Both sexes	Male	Female
At birth				Life expectancy, in years					
1900[b, c]	47.3	46.3	48.3	47.6	46.6	48.7	33.0	32.5	33.5
1950[c]	68.2	65.6	71.1	69.1	66.5	72.2	60.8	59.1	62.9
1960[c]	69.7	66.6	73.1	70.6	67.4	74.1	63.6	61.1	66.3
1970	70.8	67.1	74.7	71.7	68.0	75.6	64.1	60.0	68.3
1980	73.7	70.0	77.4	74.4	70.7	78.1	68.1	63.8	72.5
1990	75.4	71.8	78.8	76.1	72.7	79.4	69.1	64.5	73.6
1995	75.8	72.5	78.9	76.5	73.4	79.6	69.6	65.2	73.9
2000	76.8	74.1	79.3	77.3	74.7	79.9	71.8	68.2	75.1
2001	77.0	74.3	79.5	77.5	74.9	80.0	72.0	68.5	75.3
2002	77.0	74.4	79.6	77.5	74.9	80.1	72.2	68.7	75.4
2003	77.2	74.5	79.7	77.7	75.1	80.2	72.4	68.9	75.7
2004	77.6	75.0	80.1	78.1	75.5	80.5	72.9	69.4	76.1
2005	77.6	75.0	80.1	78.0	75.5	80.5	73.0	69.5	76.2
2006	77.8	75.2	80.3	78.3	75.8	80.7	73.4	69.9	76.7
2007	78.1	75.5	80.6	78.5	76.0	80.9	73.8	70.3	77.0
2008	78.2	75.6	80.6	78.5	76.1	80.9	74.3	70.9	77.3
2009	78.5	76.0	80.9	78.8	76.4	81.2	74.7	71.4	77.7
2010	78.7	76.2	81.0	78.9	76.5	81.3	75.1	71.8	78.0
2011	78.7	76.3	81.1	79.0	76.6	81.3	75.3	72.2	78.2
2012	78.8	76.4	81.2	79.1	76.7	81.4	75.5	72.3	78.4
2013	78.8	76.4	81.2	79.1	76.7	81.4	75.5	72.3	78.4
2014	78.8	76.4	81.2	79.0	76.7	81.4	75.6	72.5	78.4
At 65 years									
1950[c]	13.9	12.8	15.0	14.1	12.8	15.1	13.9	12.9	14.9
1960[c]	14.3	12.8	15.8	14.4	12.9	15.9	13.9	12.7	15.1
1970	15.2	13.1	17.0	15.2	13.1	17.1	14.2	12.5	15.7
1980	16.4	14.1	18.3	16.5	14.2	18.4	15.1	13.0	16.8
1990	17.2	15.1	18.9	17.3	15.2	19.1	15.4	13.2	17.2
1995	17.4	15.6	18.9	17.6	15.7	19.1	15.6	13.6	17.1
2000	17.6	16.0	19.0	17.7	16.1	19.1	16.1	14.1	17.5
2001	17.9	16.2	19.2	18.0	16.3	19.3	16.2	14.2	17.7
2002	17.9	16.3	19.2	18.0	16.4	19.3	16.3	14.4	17.8
2003	18.1	16.5	19.3	18.2	16.6	19.4	16.5	14.5	18.0
2004	18.4	16.9	19.6	18.5	17.0	19.7	16.8	14.9	18.3
2005	18.4	16.9	19.6	18.5	17.0	19.7	16.9	15.0	18.3
2006	18.7	17.2	19.9	18.7	17.3	19.9	17.2	15.2	18.6
2007	18.8	17.4	20.0	18.9	17.4	20.1	17.3	15.4	18.8
2008	18.8	17.4	20.0	18.9	17.5	20.0	17.5	15.5	18.9
2009	19.1	17.7	20.3	19.2	17.7	20.3	17.8	15.9	19.2
2010	19.1	17.7	20.3	19.2	17.8	20.3	17.8	15.9	19.3
2011	19.2	17.8	20.3	19.2	17.8	20.4	18.0	16.2	19.4
2012	19.3	17.9	20.5	19.3	18.0	20.4	18.1	16.2	19.5
2013	19.3	17.9	20.5	19.3	18.0	20.5	18.1	16.3	19.5
2014	19.3	18.0	20.5	19.3	18.0	20.5	18.2	16.3	19.6
At 75 years									
1980	10.4	8.8	11.5	10.4	8.8	11.5	9.7	8.3	10.7
1990	10.9	9.4	12.0	11.0	9.4	12.0	10.2	8.6	11.2
1995	11.0	9.7	11.9	11.1	9.7	12.0	10.2	8.8	11.1
2000	11.0	9.8	11.8	11.0	9.8	11.9	10.4	9.0	11.3
2001	11.2	9.9	12.0	11.2	10.0	12.1	10.5	9.0	11.5
2002	11.2	10.0	12.0	11.2	10.0	12.1	10.5	9.1	11.5
2003	11.3	10.1	12.1	11.3	10.2	12.1	10.7	8.7	11.6
2004	11.5	10.4	12.4	11.6	10.4	12.4	10.9	9.4	11.2
2005	11.5	10.4	12.3	11.5	10.4	12.3	10.9	9.4	11.2
2006	11.7	10.6	12.5	11.1	10.6	12.5	11.1	9.1	12.0
2007	11.9	10.7	12.6	11.9	10.8	12.6	11.2	9.8	12.1
2008	11.8	10.7	12.6	11.8	10.7	12.6	11.3	9.8	12.2
2009	12.1	11.0	12.9	12.1	10.4	12.9	11.6	10.2	12.5
2010	12.1	11.0	12.9	12.1	11.0	12.8	11.6	10.2	12.5
2011	12.1	11.1	12.9	12.1	11.0	12.8	11.7	10.4	12.5
2012	12.2	11.2	12.9	12.1	11.1	12.9	11.8	10.4	12.7
2013	12.2	11.2	12.9	12.1	11.1	12.9	11.8	10.4	12.7
2014	12.2	11.2	13.0	12.2	11.2	12.9	11.8	10.5	12.7

in the natural world. For example, much medical research is performed using mice because mice and humans share about 98% of their genomes. Furthermore, most human genetic variation is relatively insignificant. Even variations that alter the sequence of amino acids in a protein often produce no discernible influence on the action of the protein. Differences in some portions of DNA with as yet unknown functions appear to have no impact at all.

TABLE 2.2

Life expectancy at selected ages, by race and sex, selected years 1900–2014 [CONTINUED]

Specified age and year	White, not Hispanic			Black, not Hispanic			Hispanic[d]		
	Both sexes	Male	Female	Both sexes	Male	Female	Both sexes	Male	Female
At birth				Life expectancy, in years					
2006	78.2	75.7	80.6	73.1	69.5	76.4	80.3	77.5	82.9
2007	78.4	75.9	80.8	73.5	69.9	76.7	80.7	77.8	83.2
2008	78.4	76.0	80.7	73.9	70.5	77.0	80.8	78.0	83.3
2009	78.7	76.3	81.1	74.3	70.9	77.4	81.1	78.4	83.5
2010	78.8	76.4	81.1	74.7	71.4	77.7	81.2	78.5	83.8
2011	78.8	76.4	81.1	74.9	71.7	77.8	81.4	78.8	83.7
2012	78.9	76.6	81.2	75.1	71.8	78.1	81.6	79.1	83.9
2013	78.9	76.5	81.2	75.1	71.8	78.1	81.6	79.1	83.8
2014	78.8	76.5	81.1	75.2	72.0	78.1	81.8	79.2	84.0
At 65 years									
2006	18.7	17.2	19.9	17.1	15.1	18.5	20.2	18.5	21.5
2007	18.8	17.4	20.0	17.2	15.3	18.7	20.5	18.7	21.7
2008	18.8	17.4	20.0	17.4	15.4	18.8	20.4	18.7	21.6
2009	19.1	17.7	19.5	17.7	15.8	19.1	20.7	19.0	21.9
2010	19.1	17.7	20.3	17.7	15.8	19.1	20.6	18.8	22.0
2011	19.1	17.8	20.3	17.9	16.1	19.2	20.7	19.1	21.8
2012	19.3	17.9	20.4	18.0	16.1	19.4	21.0	19.5	22.1
2013	19.3	17.9	20.4	18.0	16.1	19.4	20.9	19.3	22.0
2014	19.3	18.0	20.5	18.1	16.2	19.5	21.1	19.6	22.2
At 75 years									
2006	11.7	10.6	12.5	11.1	9.6	12.0	13.0	11.7	13.7
2007	11.8	10.7	12.6	11.2	9.7	12.1	13.1	11.8	13.8
2008	11.8	10.7	12.6	11.3	9.8	12.2	13.0	11.7	13.8
2009	12.0	11.0	12.9	11.6	10.1	12.4	13.3	12.0	13.8
2010	12.0	11.0	12.8	11.6	10.1	12.5	13.2	11.7	14.1
2011	12.0	11.0	12.8	11.7	10.4	12.5	13.2	12.0	13.9
2012	12.1	11.1	12.9	11.7	10.4	12.6	13.5	12.3	14.2
2013	12.1	11.1	12.9	11.7	10.4	12.6	13.4	12.3	14.1
2014	12.1	11.1	12.9	11.8	10.4	12.6	13.6	12.4	14.3

[a]Data shown for 1900–1960 are for the nonwhite population.

[b]Death registration area only. The death registration area increased from 10 states and the District of Columbia (D.C.) in 1900 to the coterminous United States in 1933.

[c]Includes deaths of persons who were not residents of the 50 states and D.C.

[d]Hispanic origin was added to the U.S. standard death certificate in 1989 and was adopted by every state in 1997. Life expectancies for the Hispanic population are adjusted for underreporting on the death certificate of Hispanic ethnicity, but are not adjusted to account for the potential effects of return migration. To address the effects of age misstatement at the oldest ages, the probability of death for Hispanic persons older than 80 years is estimated as a function of non-Hispanic white mortality with the use of the Brass relational logit model.
Notes: Populations for computing life expectancy for 1991–1999 are 1990-based postcensal estimates of the U.S. resident population. Starting with *Health, United States, 2012*, populations for computing life expectancy for 2001–2009 were based on intercensal population estimates of the U.S. resident population. Populations for computing life expectancy for 2010 were based on 2010 census counts. Life expectancy for 2011 and beyond was computed using 2010-based postcensal estimates. In 1997, life table methodology was revised to construct complete life tables by single years of age that extend to age 100. (Anderson RN. Method for constructing complete annual U.S. life tables. NCHS. Vital Health Stat 2(129). 1999.) Previously, abridged life tables were constructed for 5-year age groups ending with 85 years and over. In 2000, the life table methodology was revised. The revised methodology is similar to that developed for the 1999–2001 decennial life tables. In 2008, the life table methodology was further refined. Starting with 2003 data, some states allowed the reporting of more than one race on the death certificate. The multiple-race data for these states were bridged to the single-race categories of the 1977 Office of Management and Budget standards, for comparability with other states. The race groups, white and black include persons of Hispanic and non-Hispanic origin. Persons of Hispanic origin may be of any race.

SOURCE: "Table 15. Life Expectancy at Birth, at Age 65, and at Age 75, by Sex, Race, and Hispanic Origin: United States, Selected Years 1900–2014," in *Health, United States, 2015: With Special Feature on Racial and Ethnic Health Disparities*, U.S. Department of Health and Human Services, Centers for Disease Control and Prevention, National Center for Health Statistics, May 2016, http://www.cdc.gov/nchs/data/hus/hus15.pdf (accessed October 15, 2016)

CHAPTER 3
GENETICS AND EVOLUTION

The essence of Darwinism lies in a single phrase: natural selection is the major creative force of evolutionary change. No one denies that natural selection will play a negative role in eliminating the unfit. Darwinian theories require that it create the fit as well.

—Stephen Jay Gould, "The Return of Hopeful Monsters" (*Natural History*, vol. 86, no. 6, June–July 1977)

The term *evolution* is generally used to describe the theory that all organisms are linked via descent to a common ancestor. Evolution also refers to the gradual process during which change occurs. In biology it is the theory that groups of organisms, such as species, change or develop over long periods so that their descendants differ from their ancestors in form, structure, and physiology as well as in terms of their life processes, activities, and functions. (Species are the smallest groups into which most living things that share common characteristics are divided. A key characteristic of a species is that its members can breed within the group but not outside it.)

It is important to understand that not all change is considered to be evolution; evolution encompasses only those changes that may be passed to the next generation. For example, evolution does not explain why humans are taller and bigger today than they were a century ago. This phenotypic (observable) change is attributable to changes in the environment—that is, improvements in nutrition and medicine—and is not inherited. Similarly, although evolution leads to increasing complexity, it does not necessarily signify progress, because an adaptation, trait, or strategy that is successful at one time may be unsuccessful at another time.

In genetic terms evolution may be defined as any change in the gene pool of a population over time or changes in the frequency of alleles in populations of organisms from generation to generation. Evolution requires genetic variation, and the incremental and often uneven changes described by the process of evolution arise in response to an organism's or species' genetic reaction to environmental influences.

Evidence of evolution has been derived from fossil records, genetics studies, and changes observed among organisms over time. The process produces the transformations that generate new species only able to survive if they can respond quickly and favorably enough to environmental changes. Population genetics is the discipline that considers variation and changing ratios of genetic types within populations to explain how populations evolve. Changes within a population are called microevolution. In contrast, macroevolution describes larger-scale changes that produce entirely new species. Although some researchers speculate that the two processes are different, many scientists believe macroevolutionary change is simply the final outcome of the collected effects of microevolution.

Molecular evolution is the term used to describe the period before cellular life developed on Earth. Scientists hypothesize that specific chemical reactions occurred that created information-containing molecules that contributed to the origin of life on the planet. Theories about molecular evolution presume that these early information-containing molecules were precursors to genetic structures capable of replication (duplication of deoxyribonucleic acid [DNA] by copying specific nucleic acid sequences) and mutation (change in DNA sequence).

Advances in genetics have strengthened support for evolution. For example, genetic research has confirmed that all species on Earth are related and that genetic fingerprinting (also known as DNA profiling, because it relies on the individuality of DNA to distinguish between organisms) can determine how closely species are related. Furthermore, DNA evidence confirms the British naturalist Charles Darwin's (1809–1882) observations that humans and apes are closely related and that human evolution began in Africa.

NATURAL SELECTION

Natural selection is a mechanism of evolution. Darwin described the principles of organic evolution by means of natural selection. Much of his early research focused on geology, and he developed theories about the origin of different land formations when he went on a five-year expedition around the world. During his travels he developed an interest in population diversity.

When Darwin identified 12 different species of finches on the Galápagos Islands off the coast of Ecuador, he speculated that the birds must have descended from a common ancestor even though they differed in beak shape and overall size. The birds became known as "Darwin's finches" and are examples of a process called adaptive radiation, in which species from a common ancestor successfully adapt to their environment via natural selection. Darwin suspected that the finches had become geographically isolated from one another and after years of adapting to their distinctive environments had gradually evolved into separate species that were incapable of interbreeding.

To explain this occurrence, Darwin relied on his own observations of the existence of variation in and between species, his knowledge of animal breeding, and the results of zoological research conducted by the French naturalist Jean Baptiste Lamarck (1744–1829). Lamarck suggested four laws to explain how animal life might change:

- The life force tends to increase the volume of the body and to enlarge its parts.

- New organs can be produced in a body to satisfy a new need.

- Organs develop in proportion to their use.

- Changes that occur in the organs of an animal are transmitted to that animal's progeny.

Darwin famously took issue with the fourth law, Lamarck's theory of acquired traits, particularly his suggestion that giraffes that make their necks longer by stretching to reach the uppermost leaves on tall trees would then pass on longer necks to their offspring. However, although Darwin discredited the specifics of Lamarck's theories concerning evolution, he agreed with Lamarck's idea that species change over time, and he acknowledged Lamarck as an important forerunner and influence on his own work.

Darwin's ideas were also influenced by *An Essay on the Principle of Population, as It Affects the Future Improvement of Society* (1798), written by the British economist Thomas Robert Malthus (1766–1834). Malthus hypothesized that unchecked population growth always exceeds the growth of the means of subsistence (the food supply needed to sustain it). In other words, if there were no outside factors stopping population growth, there would inevitably be more people than food. According to Malthus, actual population growth is kept in line with food supply growth by "positive checks," such as starvation and disease, which increase the death rate, and "preventive checks," such as postponement of marriage, which reduces the birthrate. Malthus's hypothesis suggested that the human population always tended to rise above the food supply, but that historically overpopulation had been prevented by wars, famine, and epidemics of disease.

Darwin knew that farmers had been able to modify species of domestic animals for hundreds of years. Cattle breeders produced breeds that yielded exceptional milk production by selectively breeding their best milk producers. Superior egg-laying hens had been bred using the same technique. Because it was possible for farmers to modify a species by artificially selecting those members permitted to reproduce, Darwin hypothesized that nature might have a comparable mechanism for determining which characteristics might be passed on to future generations.

He also realized that although individual organisms in every species have the potential to produce many offspring, the natural population of any species remains relatively constant over time. Darwin concluded that the natural environment acts as a natural selector by determining over long periods which variations are best suited to survive and, by virtue of their survival, reproduce and pass on traits and adaptations that improve health and longevity.

Applying the principles of natural selection to the question of giraffes' neck lengths provides an explanation that is different from the one proposed by Lamarck. Short-necked giraffes were less able to obtain food, so they faced starvation. As such, the genes linked to the potential to develop long necks were more likely to be passed to the next generation than the genes for short necks. Over time, the process of natural selection resulted in a population of giraffes with long necks.

In "Giraffe Genome Sequence Reveals Clues to Its Unique Morphology and Physiology" (*Nature Communications*, vol. 7, May 17, 2016), Morris Agaba et al. report that by sequencing the giraffe genome and comparing it to the genome of an okapi, a close relative of the giraffe, they identified some of the specific genes involved in regulating skeletal development. Mutations in a small number of these genes may account for giraffes' adaptations—long necks and the strong pumping mechanism in their heart that prevents them from fainting when they lower their heads to drink water.

Laboratory research and observation also refuted the theory of inheritance of acquired characteristics. When white mice had their tails cut off and were permitted to reproduce, each new generation was born with tails. Children of parents who had suffered amputations or disfiguring accidents did not share their parents' disabilities. Darwin's belief in evolution by natural selection was based on four premises:

- Individuals within a species are variable.

- Some of these variations are passed on to offspring.

- In every generation more offspring are produced than can survive.

- The survival and reproduction of individuals are not random. The individuals who survive and reproduce or reproduce the most are those with the most favorable variations. They are naturally selected.

As support for the theory of acquired inheritance diminished, appreciation of the underlying assumptions for the role of natural selection in evolution grew.

Attacks on Darwin's Theories

Darwin's theories were met with criticism from scientists and members of the clergy. Even some scientists who subscribed to evolutionary theory took issue with the concept of natural selection. Lamarck's followers were among the most outspoken opponents of Darwin's theories. This was especially ironic because it was Lamarck's work that had inspired Darwin.

Other objections raised by scientists were related to how poorly inheritance was understood at that time. The notion of blending inheritance—that an organism blends together the traits it inherits from its parents—was popular. Those who endorsed it observed that, according to Darwin's assumptions, any new variation would mix with existing traits and would no longer exist after several generations. Although Gregor Mendel (1822–1884) had published the paper "Versuche über Pflanzen-Hybriden" ("Experiments in Plant Hybridization") in 1865 that proposed particulate as opposed to blended inheritance, his theory was not widely accepted until 1900, when it was revisited and confirmed.

The other objection to Darwin's theories was the argument that variation within species was limited and that, once the existing variation was exhausted, natural selection would cease abruptly. In 1907 the American geneticist Thomas Hunt Morgan (1866–1945) and his colleagues effectively dispelled this objection. Their experiments with fruit flies demonstrated that new hereditary variation occurs in every generation and in every trait of an organism.

The clergy were even more vociferous adversaries. Darwin's major works, *On the Origin of Species by Means of Natural Selection, or the Preservation of Favoured Races in the Struggle for Life* (1859) and *The Descent of Man, and Selection in Relation to Sex* (1871), were published during a period of heightened religious fervor in England. Many religious leaders were aghast at Darwin's assertion that all life had not been created by God in one fell swoop. Moral outrage and opposition to Darwinian theory persisted into the next century, as many fundamentalist Christian denominations became more vocal about their creationist beliefs.

In some instances opponents protested the teaching of evolution in schools and continued to defend creationism. In 1925 the science teacher John Thomas Scopes (1900–1970) was tried in a Tennessee court for teaching his high school class Darwin's theory of evolution. Scopes had violated the Butler Act, which prohibited teaching evolutionary theory in public schools in Tennessee. Dubbed the "Monkey Trial" because of the simplified interpretation of Darwin's idea that humans evolved from apes, the courtroom drama pitted the defense attorney Clarence Darrow (1857–1938) against the prosecutor William Jennings Bryan (1860–1925) in a debate that began over the teaching of evolution but became a conflict of deeply held intellectual, religious, and social values. In his acerbic account of the trial proceedings, the American writer H. L. Mencken (1880–1956) wrote in "'The Monkey Trial': A Reporter's Account" (2016, http://www.ucl.ac.uk/USHistory/Making/Mencken.htm):

> The Scopes trial, from the start, has been carried on in a manner exactly fitted to the anti-evolution law and the simian imbecility under it. There hasn't been the slightest pretense to decorum. The rustic judge, a candidate for re-election, has postured the yokels like a clown in a ten-cent side show, and almost every word he has uttered has been an undisguised appeal to their prejudices and superstitions.... Darrow has lost this case. It was lost long before he came to Dayton. But it seems to me that he has nevertheless performed a great public service by fighting it to a finish and in a perfectly serious way. Let no one mistake it for comedy, farcical though it may be in all its details. It serves notice on the country that Neanderthal man is organizing in these forlorn backwaters of the land, led by a fanatic, rid of sense and devoid of conscience. Tennessee, challenging him too timorously and too late, now sees its courts converted into camp meetings and its Bill of Rights made a mock of by its sworn officers of the law. There are other States that had better look to their arsenals before the Hun is at their gates.

At the end of deliberations, Darrow requested a guilty verdict so that the case could be heard before the Tennessee Supreme Court on appeal. The jury complied, and the presiding judge fined Scopes $100. A year later the Tennessee Supreme Court overturned the verdict on a technicality. Wanting to close the case once and for all, the lower court dismissed it altogether.

Misuse of Darwin's Theories

After Darwin's theories became well known, some people made use of his terminology and concepts to argue that certain groups of people were naturally superior to others. The term *social Darwinism* is used to refer to these ideas, but it is important to note that Darwin himself did not believe in social Darwinism.

One example of social Darwinist thinking would be arguing that the rich and successful members of society are fitter or superior or are in some way more highly evolved than the poor. Social Darwinism has also been used to justify racism and colonialism, as in the 19th and early 20th centuries, when many white Europeans and Americans asserted that they were naturally superior to Africans and Asians and used this claim to justify taking control of their land and resources. Some argued further that it was not just a right but an obligation—the "white man's burden"—for Europeans and Americans to rule over and "civilize" people in less industrialized parts of the world.

None of these social Darwinist theories is scientific in nature, and all are false. Modern genetics shows that there is no group of people that is more evolved or otherwise better than the rest of humanity. Despite having been discredited by scientific research, social Darwinism continues to be used in attempts to justify various prejudices and inequalities.

THE MODERN EVOLUTION-CREATIONISM DEBATE

Arguments over the accuracy and importance of Darwin's theories continue in the 21st century. Creationists believe the biblical account of Earth's creation as it appears in the book of Genesis. Some acknowledge microevolution (changes in a species over time in response to natural selection), but they generally do not believe in speciation (that one species can beget or become another over time). There are various gradations of creationist beliefs, but all reject evolution and its argument that the interaction of natural selection and environmental factors explains the diversity of life on Earth.

At the core of the conflict is the observation that evolution threatens the view that human beings have a special place in the universe. Many creationists find it disturbing to contemplate the idea that human existence is a random occurrence or that universal order is a chance occurrence rather than a response to a divine decree or plan.

One-Third of Americans Believe Humans Did Not Evolve

More than a century and a half after Darwin's publication of *On the Origin of Species by Means of Natural Selection*, his theory remains highly controversial. Scientists assert that evolution is well established by scientific evidence, but surveys repeatedly reveal that a substantial portion of Americans do not believe that the theory of evolution best explains the origins of human life.

For example, in *For Darwin Day, 6 Facts about the Evolution Debate* (February 10, 2017, http://www.pewresearch.org/fact-tank/2016/02/12/darwin-day/), David Masci of the Pew Research Center indicates that in 2014 just 33% of Americans said "humans and other living things evolved solely due to natural processes," whereas 34% completely rejected evolution, asserting that "humans and other living things have existed in their present form since the beginning of time." Twenty-five percent of Americans said a supreme being "guided" evolution.

According to Masci, of the U.S. religious groups examined in 2014, 74% of Jehovah's Witnesses rejected evolution, as did 57% of evangelical Protestants and 52% of Mormons. By contrast, 67% of Buddhists, 63% of people who identified as being religiously "unaffiliated," 62% of Hindus, and 58% of Jews accepted evolution.

Masci reports that compared with the United States, larger percentages of people in other countries reject evolution. For example, about 40% of residents of Ecuador, Nicaragua, and the Dominican Republic believe humans have always existed in their present form as do the majority of Muslims living in Afghanistan, Indonesia, and Iraq.

A 2012 poll reported that more people in Great Britain and Canada endorsed the theory of evolution than did Americans. In *Britons and Canadians More Likely to Endorse Evolution than Americans* (September 5, 2012, http://angusreidglobal.com/wp-content/uploads/2012/09/2012.09.05_CreEvo.pdf), Angus Reid notes that just 30% of Americans said they believe that human beings evolved from less advanced life forms over the course of millions of years, compared with 69% of Britons and 61% of Canadians. Angus Reid also indicates that U.S. survey respondents in the Northeast (37%) were more likely to think human beings evolved than those living in the South (24%).

Some States Move to Teach Creationism in Public Schools

Because so many Americans reject evolution, it is not surprising that the decision about which theory—evolution or creationism—should be taught in public schools has been hotly contested. The American Institute of Biological Sciences (AIBS), a nonprofit organization that is dedicated to advancing biological research and education, reports in "AIBS State News on Teaching Evolution" (2017, https://www.aibs.org/public-policy/evolution_state_news.html) that during the 2015–16 academic year there were active controversies over this issue in Alabama, Arizona, Indiana, Louisiana, Mississippi, Missouri, Montana, Oklahoma, Pennsylvania, and South Dakota. The issue remained unresolved in other states, as debates favoring and opposing antievolution and anti-creationism legislation continued.

Some opponents of evolutionary theory propose an alternate theory known as intelligent design (ID), an explanation that credits intelligence, rather than an undirected process such as natural selection, as the source of life on Earth. ID proponents disagree with a basic tenet of evolutionary theory: Darwin's claim that the complex design of biological systems resulted by chance. They contend that direction from an intelligent designer (a supernatural being) is necessary to explain adequately the origins and complexity of life on Earth, particularly human life. U.S. organizations that promote ID include the Discovery Institute, a nonprofit public policy think tank; the Foundation for Thought and Ethics, a Christian nonprofit organization; and the Thomas More Law Center, a Christian nonprofit public interest organization.

Throughout the United States school boards have considered whether they want to teach students evolution, creationism, or ID. In 2005, 80 years after the Scopes trial, the school board in Dover, Pennsylvania, became the first in the nation to require that students be taught an alternative explanation to evolution. In 2006 this mandate was overturned, but its brief success emboldened other states to follow suit. As of February 2017, no state school board had mandated the teaching of ID exclusively; however, many school boards had moved to include it in the science curricula. The AIBS also reports that despite a ban on teaching creationism in public school science classes, a 2013 survey found that 310 private schools in nine states and the District of Columbia that received public funding through school voucher programs taught creationism in science classes.

Even some staunch advocates of evolutionary theory do not necessarily wish to exclude the teaching of creationism or ID—they simply prefer that these alternative explanations be taught in the context of religion classes rather than in science classes. There are even those who advocate both theories. The Vatican has stated that it does not consider evolution to be in conflict with the Christian faith. Francis S. Collins (1950–), who served as the director of the Human Genome Project at the National Institutes of Health until August 2008, has repeatedly expressed his view that God created the universe and chose the remarkable mechanism of evolution to create plants, animals, and humans.

In "Science, Evolution, and Intelligent Design" (2017, http://www.ucsusa.org/scientific_integrity/what_you_can _do/why-intelligent-design-is-not.html#.WKxQ_vkrKUl), the Union of Concerned Scientists avers, "There is no debate about evolution among the vast majority of scientists, and no credible alternative scientific theory exists. Debates within the community are about specific mechanisms within evolution, not whether evolution occurred."

Other organizations that oppose ID include the American Association for the Advancement of Science, an international nonprofit organization; the American Association of University Professors, an organization of professors and other academics; and the National Center for Science Education.

Glenn Branch of the National Center for Science Education notes in "Darwin Day Approaches" (January 12, 2016, https://ncse.com/news/2016/01/darwin-day-approaches -0016860) that "Darwin Day," which is celebrated on February 12, acknowledges the importance of Darwin's contributions to science. Besides recognizing the birthday of the father of evolutionary biology, 310 congregations in 48 states and 11 foreign countries also celebrated "Evolution Weekend" in February 2016 by presenting sermons and hosting discussion groups about "the compatibility of faith and science."

VARIATION AND ADAPTATION

Effective variations and adaptations are perpetuated in a species and tend to be incorporated into the normal or predominant phenotype for most individuals in the species. Variation persists, but it ranges around an evolutionarily determined norm. This is called adaptive radiation. For example, over time Darwin's finches developed beaks best suited to their functions. The finches that eat grubs have long, thin beaks to enter holes in the ground and pull out the grubs. The finches that eat buds and fruit have clawlike beaks to grind their food, giving them a survival advantage in environments where buds are the only available food source. In another example of adaptive radiation, present-day giraffes have necks of varying lengths, but most tend to be long.

Ancient humans underwent many evolutionary changes. An example of adaptive radiation in humans is the development and refinement of the upper limbs to perform fine motor skills necessary to make and use complex tools. Another adaptation is the quantity of the pigment melanin present in the skin. People from areas near the equator have more melanin, an adaptation that darkens their skin and protects them from the sun. Anatomical structure also appears to have responded to the environment. Those who thrive in colder climates produce offspring that are shorter and broader than those who live in warmer climates. This adaptive stature, with its relatively low surface area, enables those people to conserve rather than lose body heat. For example, Alaskan Natives, who are generally of short stature, are well suited to their cold climate.

Natural selection does not produce uniformity or perfection. Instead, it generates variability that persists when it helps a species to adapt to, and thrive in, its environment. It acts on outward appearance (phenotypes), not on internal coding (genotypes), and it is not a process that always discards individual genes in favor of others that might produce traits better suited for survival. For traits attributable to multiple genes, many different combinations of gene pairs may produce the same or comparable phenotypes. Multiple phenotypes may be neutral or even beneficial in terms of survival in a given environment, and there is no reason for such variation to be eliminated by natural selection.

Furthermore, natural selection does not completely or rapidly eliminate genes that produce traits unsuited for adaptation or survival. Although some individuals with harmful traits die young or do not reproduce, some do reproduce and pass their genes and traits to the next generation. Culling out these genes may require several generations. In other instances, seemingly harmful genes may be retained in the gene pool because there may be circumstances or environments in which their presence would improve survival. For example, the recessive allele that causes sickle-cell diseases (sickle-cell anemia and sickle B-thalassemia, in which the red blood cells contain abnormal hemoglobin) may have had a role in survival in some parts of the world. People who are pure recessive for this trait become ill and die prematurely, but those who are hybrid for the trait may have retained a survival advantage in areas where malaria is present because they are immune to the effects of malaria. This phenomenon is called balanced polymorphism and is an example of the seemingly counterintuitive actions of natural selection. Although the sickle-cell trait is not beneficial on its own, historically it was advantageous in areas where malaria was a greater threat to survival than sickle-cell diseases.

By definition, natural selection is an unending, continuous process. The popular understanding of natural selection as "survival of the fittest" is somewhat misleading because organisms and species with phenotypes most suited to survive in their environments are not necessarily the "fittest." Some examples of natural selection include the evolution of bacteria that are antibiotic resistant and insects that resist extermination with pesticides.

Increasing and Decreasing Genetic Variation

A gene pool encompasses the alleles for all the genes in a population. Gene pools in natural populations contain considerable variation, and for evolution to proceed there must be mechanisms to create and increase genetic variation. Mutation (a change in a gene) serves to create or increase genetic variation. Recombination, a process that creates new alleles and new combinations of alleles, also increases genetic variation. Another way for new alleles to enter a gene pool is by migrating from another population. In closely related species new organisms can enter a population, mate within it, and produce fertile hybrids. This action is known as gene flow.

The tendency toward increased genetic variation within a population is balanced by other mechanisms that act to decrease it. For example, some variations have the effect of limiting an organism's ability to reproduce. Any such variations, as well as those incidentally paired with them, will be nonadaptive because they will have a lower probability of being passed to the next generation. Under such circumstances, the process of natural selection serves to reduce genetic variation. In fact, differences in reproductive capability are often called natural selection. Whenever natural selection weeds out nonadaptive alleles by limiting an organism's reproductive capability, it is depleting genetic variation within the population.

Genetic Drift

Other evolutionary mechanisms contribute to genetic variation. Genetic drift is random change in the genetic composition of a population. It may occur when two groups of a species are separated and as a result cannot reproduce with each other. The gene pool of these groups will naturally differ over time. When the two groups are reunited and reproduce, gene migration occurs as their genetic differences combine, serving to increase genetic diversity. If they remain separate, their genetic differences may become so great that they develop into two separate species, such as what happened to Darwin's finches.

There are two broad types of genetic drift: population bottleneck and the founder effect. An example of population bottleneck is the population of bison in the United States. Hunted nearly to extinction during the late 1800s, the sudden decrease in population size resulted in a population with little genetic variation. The founder effect occurs when a few members of an original population start a new colony. For example, the Amish in the United States exhibit the founder effect. This is because the Amish population arose from a small group of founders, one of whom had Ellis-van Creveld syndrome (a rare form of dwarfism that causes short stature, extra fingers, abnormal teeth and nails, and heart defects). Because the Amish tend to marry within their community, this syndrome is more common among the Amish than in the U.S. population.

In small populations genetic drift can cause relatively rapid change because each individual's alleles constitute a large proportion of the gene pool, and when an individual does not reproduce, the results are felt more acutely in smaller, rather than larger, populations. When small populations are affected by genetic drift, they may suffer a loss of valuable diversity. This is why scientists and others involved in conservation, such as zoo curators of endangered species, make every effort to ensure that populations are large enough to withstand the effects of genetic drift.

MUTATION

Variation is the essence of life, and mutations are the source of all genetic variation. A staggering number and variety of alterations, rearrangements, and duplications of genetic material have occurred since the first living cells, in which there is an incredible range of life forms, from amoebas and fruit flies to whales and humans. These dramatically different life forms were all produced using genetic material that was present and reproduced from the first living cells. They resulted from the process of mutation, an alteration in genetic material.

Perhaps because of their negative depiction in science fiction and horror films, mutations are widely considered to be sudden, harmful, and dramatic. Most mutations are not harmful, and the same limited number of mutations has probably been recurring in each species for millions of years. Many helpful mutations have already been incorporated into the normal genotype through natural selection. Harmful mutations do occur, and while they tend to be eliminated through natural selection, they do recur randomly. It is a mistake, however, to consider mutation as sudden or exclusively harmful. Mutation cannot generate sudden, drastic changes—it requires many generations to generalize throughout a species population. Furthermore, if the changes caused by mutation did not favor survival, natural selection would work against their generalization throughout the species population. Without mutation, there would have been no development of life and no evolution would have occurred.

The effects of mutation vary greatly. Although mutations are changes in genetic material, they do not necessarily affect an individual organism's phenotype. When phenotype is affected, it is because the code for protein synthesis has been changed. Whether mutation will affect phenotype and the extent to which it will be influenced depends on how protein manufacture is affected, when and where the mutation occurs, and the complexity of the genetic controls governing the selected trait.

When a trait is governed by the interaction of many genes, with each exerting about the same influence, the effects of a mutation might be negligible. In traits controlled by a single pair of genes, mutation is likely to exert a much greater influence. For traits governed by the interaction between a major controlling gene pair and several other less influential gene pairs, the location of the mutation determines its impact. Other considerations such as whether the mutated gene is dominant or recessive also determine whether it will directly act on an individual's phenotype—a mutated recessive gene might not appear in the phenotype for several generations.

Mutation can occur in any cell in an organism's body, but only germinal mutations (those that affect the cells that give rise to sperm or eggs) are passed to the next generation. When mutation occurs in somatic cells in the body such as muscle, liver, or brain cells, only cells that derive from mitotic division of the affected cell will contain the mutation. Although mutations in somatic cells can cause disease, most do not have a significant impact, because if the mutated gene is not involved in the specialized function of the affected cell, these "silent mutations" will not be detected. Furthermore, mutations that appear only in somatic cells disappear when the organism dies; they are not passed on to subsequent generations and do not enter the gene pool that is the source of genetic variation for the species.

Another reason that many mutations are not expressed in the phenotype is that they affect only one copy of the gene, leaving diploid organisms with an intact copy of the gene. These types of mutations have a recessive inheritance pattern and do not affect phenotype unless an individual inherits two copies of the mutation. There is also the question of the probability of a mutation in a gamete affecting offspring. The mutation may occur in a single gamete and so may only be passed on if that particular gamete is involved in conception. For example, human semen contains more than 50 million sperm per ejaculate, so it is unlikely that a mutation carried by a single sperm will be passed on. When mutation occurs during embryonic (before birth) development and all gametes are affected, there is a greater chance that it might influence the phenotype of future generations. Alternatively, if, like many mutations, the gene defect occurs with advancing age, and the affected individuals are beyond their reproductive years, then it will have no impact on future generations.

How Different Types of Genetic Mutations Occur

Mutation is a normal and fairly frequent occurrence, and the opportunity for a mutation to take place exists every time a cell replicates. In general, the cells that divide many times throughout the course of an organism's life (e.g., skin cells, bone marrow cells, and the cells that line the intestines) are at greater risk for mutation than those that divide less frequently (e.g., adult brain and muscle cells).

DNA nearly always reproduces itself accurately, but even a minor alteration produces a mutation that may alter a protein, prevent its production, or have no effect at all. There are five broad classes, and within them many varieties, of mutations, and each is named for the error or action that causes it:

- Point mutations are substitutions, deletions, or insertions in the sequence of DNA bases in a gene. The most common point mutation in mammals is called a base substitution and occurs when an A-T pair replaces a G-C pair. Base substitutions are further classified as either transitions or transversions. Transitions occur when one pyrimidine (C or T) is substituted for the other or one purine (A or G) is substituted on the other strand of DNA. Transversions occur when a purine replaces a pyrimidine. Sickle-cell anemia results from a transversion in which T replaces A in the gene for a component of hemoglobin.

- Structural chromosomal aberrations occur when the DNA in chromosomes is broken. The broken ends may remain loose or join those occurring at another break to form new combinations of genes. When movement of a chromosome section from one chromosome to another takes place, it is called translocation. Translocation between human chromosomes 8 and 21 has been implicated in the development of a specific type of leukemia (cancer of the white blood cells). It has also been shown to cause infertility (inability to sexually reproduce) by hindering the distribution of chromosomes during meiosis.

- Numerical chromosomal aberrations are changes in the number of chromosomes. In a duplication mutation, genes are copied, so the new chromosome contains all of its original genes plus the duplicated one. Polyploidy is a numerical chromosomal aberration in which the entire genome has been duplicated and an individual who is normally diploid (having two of each chromosome) becomes tetraploid (containing four of each chromosome). Polyploidy is responsible for the creation of thousands of new species because it increases genetic diversity and produces species that are bigger, stronger, and more able to resist disease.

- Aneuploidy refers to occasions when just one or a few chromosomes are involved and generally describes the loss of a chromosome. Examples of aneuploidy are Down syndrome (multiple, characteristic physical and cognitive disabilities), in which there is an extra chromosome 21 (usually caused by an error in cell division called nondisjunction), and Turner's syndrome, in which there is only one X chromosome. Common characteristics of Turner's syndrome include short stature and lack of ovarian development as well as increased risk of cardiovascular problems, kidney and thyroid problems, skeletal disorders such as scoliosis (curvature of the spine) or dislocated hips, and hearing problems.

- Transposon-induced mutations involve sections of DNA that copy and insert themselves into new locations on the genome. Transposons usually disrupt and inactivate gene function. In humans selected types of hemophilia have been linked to transposon-induced mutations.

Frequency and Causes of Mutation

In the absence of environmental influences, mutations occur rarely and are seldom expressed because many forms of mutation are expressed by a recessive allele. The most common naturally occurring mutations arise simply as accidents. Susceptibility to mutation varies during the life cycle of an organism. For example, among humans mutation of egg and sperm cells increase with advancing age—the older the parents the more likely they are to carry gametes with mutations. Susceptibility also varies among members of a species such as humans based on their geographic location and ethnic origin.

The mutation rate is the frequency of new mutations per generation in an organism or a species. Mutation rates vary widely from one gene to another within an organism and between organisms. The mutation rate for bacteria is 1 per 100 million genes per generation. Despite this relatively low rate, the enormous number of bacteria (there are more than 20 billion produced in the human intestines each day) translates into millions of new mutations to the bacteria population every day. The human mutation rate is estimated at 1 per 10,000 genes per generation, and human mutation rates are comparable throughout the world. The only exceptions are populations that have been exposed to factors known as mutagens that cause a change in DNA structure and as a result increase the mutation rate. With approximately 19,000 to 20,000 genes, a typical human contains two mutations. Considering the fact that human genes mutate approximately once every 30,000 to 50,000 times they are duplicated, and in view of the complexity of the gene replication process, it is surprising that so few "mistakes" are made.

Gene size, gene-base composition, and the organism's capacity to repair DNA damage are closely linked to how many mutations occur and remain in the genome. Larger genes are more susceptible than smaller ones because there are more opportunities and potential sites for mutation. Organisms better able to repair and restore DNA sequences, such as many kinds of yeast and bacteria, will be less likely to have high mutation rates.

The overwhelming majority of human mutations arise in the father, as opposed to the mother. When compared with egg cells, there are more opportunities for mutation during the many cell divisions needed to produce sperm. Sperm are produced later in the life of males, whereas females produce eggs earlier (during embryonic development) and are born with their full complement of eggs. Thus, the frequency of mutations that are passed to the next generation increases with parental age.

The mutation rate is partly under genetic control and is strongly influenced by exposure to environmental mutagens. Some mutagens act directly to alter DNA, and others act indirectly, by triggering chemical reactions in the cell that result in the breakage of a gene or a group of genes. Radiation (radioactive material in the earth or from artificial sources such as x-rays) or ultraviolet light (from natural sources such as the sun) can significantly accelerate the rate of mutations.

The link between radiation and mutation was first identified during the 1920s by Hermann Joseph Muller (1890–1967), a student of Thomas Hunt Morgan who worked in the famous "Fly Room." Muller discovered that he could increase the mutation rate of fruit flies more than a hundredfold by exposing reproductive cells to high doses of radiation. His pioneering work prompted medical and dental professionals to minimize patients' exposure to radiation. Because mutagens in the reproductive cells are likely to affect heredity, special precautions such as donning a lead apron to block exposure are used when dental x-rays or other diagnostic imaging studies are performed. Similarly, special care is taken to prevent radiation exposure to the embryo or fetus because a mutation during this period of development when cells are rapidly proliferating might be incorporated in many cells and could result in birth defects.

MODERN SYNTHESIS OF EVOLUTIONARY GENETICS

Twenty-first-century theories of evolutionary genetics are indebted to Darwin for his groundbreaking descriptions of organisms, individuals, and speciation. Modern theory differs considerably in that it addresses evolutionary mechanisms at the level of populations, genes, and phenotypes and incorporates understanding of actions, such as genetic drift, that Darwin had not considered.

The modern synthesis of evolutionary theory differs from Darwinism by identifying mechanisms of evolution that act in concert with natural selection, such as random genetic drift. It asserts that characteristics are inherited as discrete entities called genes and that variation within a population results from the presence of multiple alleles of a gene. The most controversial tenet of modern evolutionary theory is its contention that speciation is usually the result of small, gradual, and incremental genetic changes. In other words, macroevolution is simply the cumulative effect of microevolution. There are, however, several other theories that have been suggested, including punctuated equilibrium, macromutation, and phyletic gradualism.

Punctuated equilibrium suggests that long periods of stasis (stability) are followed by rapid speciation. It posits that new species arose rapidly during a period of a few thousand years and then remained essentially unchanged for millions of years before the next period of adaptation. This theory also proposes that change occurred in a small portion of the population, rather than uniformly throughout the population. Although it is different from Darwin's theory of speciation, it is not inconsistent with natural selection; it simply presumes different mechanisms and timetables for the development of new species.

Macromutation proposes that rather than an accumulation of small changes, evolutionary change occurs rapidly and with few transitional forms. In this fast-paced theory, mutation produced new radically different species, sometimes called "hopeful monsters" in "bursts" of evolutionary change.

Phyletic gradualism posits that evolution occurs at a fairly constant rate and that new species arise by slow, measured transformation of ancestral species. This theory does not agree that evolution occurs in bursts. Instead, it explains that observed bursts of evolution are simply illusions—that species diverged slowly and that transitional forms were simply not preserved. Among the as yet unanswered questions evolutionary biologists seek to answer is: Is evolution a slow, steady, gradual process or does it occur in quick jumps?

CHAPTER 4
GENETICS AND THE ENVIRONMENT

Nature prevails enormously over nurture when the differences in nurture do not exceed what is commonly to be found among persons of the same rank in society and in the same country.

—Francis Galton, "The History of Twins, as a Criterion of the Relative Powers of Nature and Nurture" (*Journal of the Anthropological Institute*, vol. 5, 1876)

Although genetics controls many of an organism's traits, it is simplistic and incorrect to assume that organisms, including humans, are completely defined by their genes. There are some phenotypes (observable traits) that are exclusively controlled by either genetics or environment, but most are influenced by a complex interaction of the two. Modern genetics defines environment as every influence other than genetic—such as air, water, diet, radiation, and exposure to infection—and it subscribes to the overarching assumption that every trait of every organism is the product of some set of interactions between genes and the environment. Epigenetic changes, which are inherited factors that alter the physical properties of deoxyribonucleic acid (DNA) but do not change the DNA sequence of nucleotides, also influence how genes are expressed.

NATURE VERSUS NURTURE

The debate over the relative contributions of genetics and environment—nature versus nurture—remains unresolved in many fields of study, from education to animal behavior to human disease. Historically, scientists assumed opposing viewpoints, choosing to favor either nature or nurture, rather than exploring the ways in which both play critical and complementary roles. Nature proponents argued that human characteristics are uniquely and primarily conditioned by genetics, whereas nurture proponents asserted that environmental influences and experiences determined differences between individuals and populations.

Genes versus Environment

When researchers analyze the origins of disease, the terms used to describe causation are *genetic* versus *environmental*, but the issues are the same as those in the nature-versus-nurture debate. Conditions considered to be primarily genetic are ones in which the presence or absence of genetic mutations determines whether an individual or population will develop a disease, independent of environmental exposures or circumstances. A disease considered to be primarily environmental is one in which people of virtually any genetic background can develop the disease when they are exposed to the specific environmental factors that cause it. For example, chemicals in cigarettes and other tobacco products can cause cancer.

Even the conditions and diseases once believed to be at either end of the continuum (caused by either purely genetic or purely environmental factors) may not be exclusively attributable to one or the other. For example, an automobile accident that results in an injury might be deemed entirely environmentally caused, but many geneticists would contend that risk-taking behaviors such as the propensity to exceed the speed limit are probably genetically mediated. Furthermore, the course and duration of rehabilitation and recovery from an injury or illness is also likely genetically influenced.

At the other end of the continuum are diseases believed to be predominantly genetic in origin, such as sickle-cell anemia. Although this disease does not have an environmental cause, there are environmental triggers that may determine when and how seriously the disease will strike. For example, sickle-cell attacks are more likely when the body has an insufficient supply of oxygen, so people with the sickle-cell trait who live at high altitudes or those who engage in intense aerobic exercise may be at increased risk of attacks. There are many more conditions for which the risk of developing the disease is strongly influenced by both genetic and environmental factors. Multiple genes and environmental factors may be involved in causing a given condition and its expression.

Asthma: A Disease with Genetic and Environmental Causes

Asthma (a disorder of the lungs and airways that causes wheezing and other breathing problems) is a good example of how genes and the environment interact to cause diseases. The scientific evidence that asthma has a genetic basis came from studies that were conducted during the late 20th century, which found patterns of inheritance in families and genetic factors that were involved in the severity and triggers for asthma attacks. For example, in "Genetic Factors in the Presence, Severity, and Triggers of Asthma" (*Archives of Disease in Childhood*, vol. 73, no. 2, August 1995), a study of pairs of twins in which at least one twin had asthma, Edward P. Sarafino and Jarrett Goldfedder discovered that genetics and environment both made strong contributions to the development of the illness. If asthma was directed solely by genes, then 100% of the monozygotic (identical) twins, who are exactly the same genetically, would be expected to have asthma (the concordance rate is the rate of agreement, when both members of a pair of twins have the same trait). Instead, the researchers found that 59% of twins both had asthma. If asthma was entirely environmentally caused, then genes should make no difference at all—the concordance rate would be the same for monozygotic and dizygotic (fraternal, meaning nonidentical) twins. Sarafino and Goldfedder noted that the concordance rate of 59% was more than twice as high in monozygotic twins as in dizygotic twins (24%). This research confirms that asthma has a significant genetic component.

Nobuyuki Hizawa of the University of Tsukuba reports in "Genetics of Asthma" (*Journal of General and Family Medicine*, vol. 16, no. 4, 2015) that twin and family studies indicate that between 60% and 70% of asthma is heritable. The varied symptoms, asthma severity, and the ways in which people with asthma respond to treatment reflect the interaction of individual genetic profiles and environmental exposures.

Asthma is an example of a condition that is polygenic (controlled by more than one gene). Familial studies demonstrate that asthma does not conform to simple Mendelian patterns of inheritance and that multiple independent segregating genes are required for phenotypic expression (polygenic inheritance). Nevertheless, it has long been known that the complex phenotypes of asthma have a significant genetic contribution. Researchers speculate that several genes combine to increase susceptibility to asthma. There are also genes that lower susceptibility to developing the condition. Hizawa suggests that the genes involved in the susceptibility to and inception of asthma are probably different from those involved in asthma severity.

Michael E. Wechsler and Elliot Israel conclude in "The Genetics of Asthma" (*Seminars in Respiratory and Critical Care Medicine*, vol. 23, no. 4, August 2002) that although a small percentage of cases of asthma may result from a single gene defect or a single environmental factor, asthma is a complex genetic disease that cannot be explained by single-gene models. In most instances it appears to result from the interaction of multiple genetic and environmental factors. Like other diseases such as diabetes and hypertension (high blood pressure), the complexity of asthma genetics may be characterized by the contribution of different genes and different environments in different populations. Population studies of asthma face challenges comparable to genetics studies of other common complex traits. They are complicated by genetic factors such as incomplete penetrance (transmission of disease genes without the appearance of the disease), genetic heterogeneity (mutations in any one of several genes that may result in similar phenotypes), epistasis (when the effects of multiple genes have a greater effect on phenotype than individual effects of single genes), polygenic inheritance (mutations in multiple genes simultaneously producing the affected phenotype), and gene-environment interactions.

Environmental triggers for asthma include allergens such as air pollution, tobacco smoke, dust, and animal dander. Other environmental factors linked to its development are a diet high in salt, a history of lung infections, and the lack of siblings living at home. Researchers theorize that younger children in families with older siblings are less likely to develop asthma because their early exposures to foreign substances (such as dirt and germs) brought into the home by older siblings heightened their immune system. The enhanced immune response is believed to have a protective effect against asthma.

In "Gene-Environment Interactions: The Road Less Traveled by in Asthma Genetics" (*Journal of Allergy and Clinical Immunology*, vol. 123, no. 1, January 2009), Donata Vercelli of the University of Arizona, Tucson, reports that the genetics of asthma made great strides in 2007 with the publication of the results of the first genome-wide association study (GWAS) of childhood asthma. GWASs look at sets of single nucleotide polymorphisms (SNPs) throughout the genome to identify the most frequently occurring genetic variants that are associated with a disease and the heritable traits that are associated with a risk of developing the disease. Figure 4.1 shows how a GWAS compares DNA from people with a disease and those without by looking for SNP markers of genetic variation. The GWAS for childhood asthma found that variants of chromosome 17q21 were associated with an increased risk of asthma. Vercelli also observes that it is now well understood that the environment strongly influences the effect of genetic variants and suggests that there are gene-environment interactions that have not been identified.

FIGURE 4.1

How genome-wide association studies are conducted

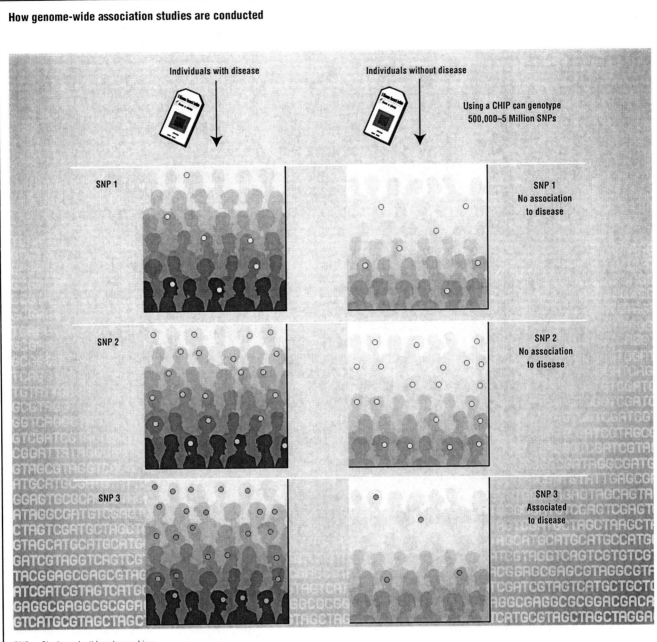

SNP = Single nucleotide polymorphism.

SOURCE: "GWAS—Genome-Wide Association Studies," in *Genome-Wide Association Studies Fact Sheet*, National Human Genome Research Institute, August 27, 2015, https://www.genome.gov/20019523/ (accessed October 17, 2016)

Shuk-Mei Ho of the University of Cincinnati College of Medicine observes in "Environmental Epigenetics of Asthma: An Update" (*Journal of Allergy and Clinical Immunology*, vol. 126, no. 3, September 2010) that asthma is influenced by a complex interplay of genetic and environmental factors now known to be mediated by epigenetics. Asthma risk is influenced by genetics: having one parent with asthma doubles a child's risk of asthma and having two affected parents increases the risk fourfold. Furthermore, multiple studies of the genetic association of asthma have identified 43 asthma genes. Nonetheless, Ho opines that genetics alone is insufficient to fully explain the dramatic increase in the prevalence of asthma during the past three decades and the marked increase in the frequency of occupational asthma. Ho posits that these increases are attributable to gene-environment interactions that are mediated by epigenetic mechanisms and that epigenetics regulate the immune responses that are associated with asthma.

For example, living in a developed country is a strong risk factor for asthma. Ho explains that this increased risk may be in part due to outdoor air pollutants, such as environmental tobacco smoke, diesel exhaust particles, pollens, molds, and viral and bacterial pathogens (disease-causing agents), that have been shown to trigger asthma; to indoor pollutants; or to other environmental exposures such as diet. According to Ho, researchers are expanding their understanding of the epigenetic actions of some of these environmental risk factors. For example, DNA methylation (the process in which methyl groups are added to nucleotides in genomic DNA) may be the epigenetic mechanism that explains how a mother's lifelong smoking may influence her unborn child's asthma risk. Figure 4.2 shows how DNA methylation silences gene expression in one region, while the region without methylation remains active.

In "Stress and Asthma: Novel Insights on Genetic, Epigenetic, and Immunologic Mechanisms" (*Journal of Allergy and Clinical Immunology*, vol. 134, no. 5, November 2014), Stacy L. Rosenberg et al. observe that some ethnic minorities and economically disadvantaged people are often exposed to chronic psychosocial stress and are disproportionately affected by asthma. The researchers posit that the association between chronic stress and asthma may be due to changes in the methylation and expression of genes that regulate responses to stress.

Genetic Susceptibility

Genetic susceptibility is the concept that the genes an individual inherits affect how likely he or she is to develop a particular condition or disease. When an individual is genetically susceptible to a particular disease, his or her risk of developing the disease is higher. Genetic susceptibility interacts with environmental factors to produce disease, but genes and environment do not necessarily make equal contributions to causation. Figure 4.3 shows how genes and environment interact to increase the susceptibility, development, and progression of asthma. Genes can cause a slight or a strong susceptibility. When the genetic contribution is weak, the environmental influence must be strong to produce disease, and vice versa.

In most instances a susceptibility gene strongly influences the risk of developing a disease only in response to a specific environmental exposure. If the environmental exposure occurs infrequently, the gene will be of low penetrance, and it may seem that the environmental exposure is the primary cause of the disease, even though the gene is required for developing the disease. For this reason, even when environmental agents are suspected to

FIGURE 4.2

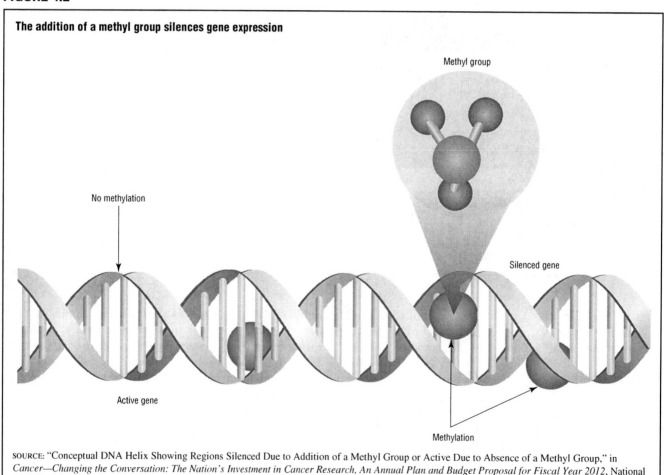

The addition of a methyl group silences gene expression

Methyl group

No methylation

Silenced gene

Active gene

Methylation

SOURCE: "Conceptual DNA Helix Showing Regions Silenced Due to Addition of a Methyl Group or Active Due to Absence of a Methyl Group," in *Cancer—Changing the Conversation: The Nation's Investment in Cancer Research, An Annual Plan and Budget Proposal for Fiscal Year 2012*, National Cancer Institute, 2012, https://www.cancer.gov/about-nci/budget/annual-plan/nci-plan-2012.pdf (accessed October 17, 2016)

Genetics and Genetic Engineering

FIGURE 4.3

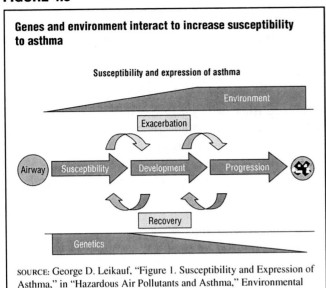

Genes and environment interact to increase susceptibility to asthma

Susceptibility and expression of asthma

Environment

Exacerbation

Airway | Susceptibility | Development | Progression

Recovery

Genetics

SOURCE: George D. Leikauf, "Figure 1. Susceptibility and Expression of Asthma," in "Hazardous Air Pollutants and Asthma," Environmental Health Perspectives Supplements, vol. 110, no. S4, August 2002, http://www.ncbi.nlm.nih.gov/pmc/articles/PMC1241200/pdf/ehp110s-000505.pdf (accessed October 17, 2016)

be a major cause of a particular disease, there is still the possibility that genetic factors also play a major part, particularly genetic mutations with low penetrance. Similarly, a critical mix of nature and nurture is likely to determine individual traits and characteristics. Genetic factors may be considered as the foundation on which environmental agents exert their influence. For example, individuals who inherit a certain array of multiple alleles of susceptibility genes have an increased risk from birth of developing asthma. This susceptibility may become evident when initial sensitization and exacerbation occur in early childhood and when immunity develops to allergens. Whether asthma diminishes in frequency and severity or progresses depends largely on each individual's environmental exposures. Based on this premise, it is now widely accepted that although certain environmental factors alone and certain genetic factors alone may explain the origins of some traits and diseases, most of the time the interaction of both genetic and environmental factors will be required for their expression.

Since the 1990s researchers have identified more and more genes that influence an individual's susceptibility to disease. Scientists have already linked specific DNA variations with increased risk of common diseases and conditions, including cancer, diabetes, hypertension, and Alzheimer's disease (a progressive neurological disease that causes impaired thinking, memory, and behavior). The questions that persist in the genes-versus-environment debate no longer focus on whether a particular trait or disease is caused exclusively by a specific gene. Instead, researchers continue to explore the extent to which genes and environment influence the development of specific

traits, especially conditions that are linked to health and susceptibility to disease.

TWIN AND FAMILIAL STUDIES

The British anthropologist Francis Galton (1822–1911), a cousin of Charles Darwin (1809–1882), conducted some of the first reported twin studies. Advancing his cousin's theory of evolution, he performed twin studies in 1876 to investigate the extent to which the similarity of twins changes over the course of development. Galton is considered the originator of the field of medical genetics because of his considerable contributions to the nature-versus-nurture debate and the research methods he developed for evaluating heritability. Genetic determination is the combination of genes that creates a trait or characteristic, and heritability is what causes differences in those characteristics.

Twin studies, meaning comparisons of monozygotic twins to dizygotic twins, are performed to estimate the relative contributions of genes and environment—that is, the extent to which genetic-versus-environmental influences operate on specific traits. Twin studies also help determine the proportion of the variability in a trait that might be due to genetic factors. The studies aim to identify the causes of familial resemblance by comparing the concordance rates of monozygotic and dizygotic twins. Monozygotic twins share the same genetic material (they have 100% of their genes in common) and dizygotic twins share only half their genetic material (they have 50% of their genes in common). Most twin studies report their results in terms of pairwise concordance rates, which measure the number of pairs, or probandwise concordance rates, which count the number of individuals.

Monozygotic twins serve as excellent subjects for controlled experiments because they share prenatal environments, and those reared together also share common family, social, and cultural environments. Furthermore, twin studies can point to hereditary effects and estimate heritability, a term that describes the magnitude of the genetic effect. The limitations of such studies include the potential to confuse or mistakenly attribute to genetics the divergent impact of environmental influences. In addition, in some studies it has been difficult to control for other potential causes or sources of variation.

Some of the most conclusive twin study research analyzed monozygotic and dizygotic twins who were raised apart. Researchers sought to establish whether characteristics such as personality traits, aptitudes, and occupational preferences are the products of nature or nurture. Similar characteristics among monozygotic twins reared apart might indicate that their genes played a major role in developing that trait. Different characteristics might indicate the opposite—that environmental influences assume a much stronger role. By comparing monozygotic and dizygotic twins, investigators test their hypotheses and confirm the findings of earlier research. For example, if monozygotic twins raised in different homes have many similarities, but dizygotic twins raised apart have little in common, researchers may conclude that genes are more important than environment in determining specific characteristics, traits, susceptibilities, and diseases.

Paul Lichtenstein et al. indicate in "Environmental and Heritable Factors in the Causation of Cancer—Analyses of Cohorts of Twins from Sweden, Denmark, and Finland" (New England Journal of Medicine, vol. 343, no. 2, July 13, 2000) that they examined nearly 45,000 pairs of twins to determine if the likelihood of developing certain kinds of cancers is more closely linked to environmental exposures than genetics. The results of this large-scale, landmark study reveal that environmental factors are linked to twice as many cancers as genetic factors. In fact, the risk of developing only three types of cancer (albeit some of the most common cancers)—breast, colorectal, and prostate cancers—shows a significant genetic correlation.

Prostate cancer was found to have the strongest genetic link, with 42% of risk explained by genetic factors and 58% by environmental factors, while breast and colorectal cancers were found to have less than a 35% link to genetics. Lichtenstein et al. conclude that inherited genetic factors make a minor contribution to susceptibility to most types of cancer.

In "Breast Tissue Composition and Susceptibility to Breast Cancer" (Journal of the National Cancer Institute, vol. 102, no. 16, August 18, 2010), Norman F. Boyd et al. explain that breast density is associated with breast cancer risk more strongly than most other risk factors for the disease and assert that high breast density may account for a substantial fraction of breast cancer. The researchers observe that twin studies demonstrate that breast density is a highly heritable trait and posit that the risk of breast cancer may increase when women with extensive breast density are exposed to mitogens (substances that induce mitosis and cell transformation) and mutagens (substances that cause changes in DNA such as radioactive compounds, x-rays, ultraviolet radiation, and certain chemicals).

Twin and familial studies have also identified genetic factors that influence susceptibility to mental illnesses. Jerry Guintivano and Zachary A. Kaminsky of the Johns Hopkins University School of Medicine report in "Role of Epigenetic Factors in the Development of Mental Illness throughout Life" (Neuroscience Research, August 20, 2014) that twin studies and GWASs identify a range of heritable psychiatric disorders. The researchers note that psychiatric disorders are also linked to environmental exposures and that these influences are strongest during certain developmental periods, including pregnancy, shortly after birth, and major hormonal changes.

In "Meta-analysis of the Heritability of Human Traits Based on Fifty Years of Twin Studies" (Nature Genetics, vol. 47, no. 7, July 2015), Tinca J. C. Polderman et al. report the results of an analysis of all twin studies published between 1958 and 2012, which provide data for nearly 14.6 million twin pairs in 39 countries. They find that all human traits are heritable: not one trait had a heritability estimate of zero. After looking at nearly 18,000 traits, Polderman et al. determine that across all traits the reported heritability is 49%.

Epigenetic Changes Pass from One Generation to the Next

Exposure to stress or traumatic experiences may also be transmitted from one generation to the next. Termed epigenetic transgenerational inheritance, it is the germline (egg or sperm) transmission of epigenetic information between generations in the absence of any environmental exposure. Michael K. Skinner reports in "Environmental Stress and Epigenetic Transgenerational Inheritance" (BMC Medicine, vol. 12, no. 153, September 5, 2014) that multiple studies show multigenerational effects of stress. For example, when mothers provide excellent care for their infants, it promotes epigenetic programming of the brain that encourages good care of offspring, and this characteristic is then passed on to subsequent generations. By contrast, poor maternal care promotes altered epigenetic programming of the brain and is also transmitted generationally. Skinner concludes, "The observation that environmental stress can also promote transgenerational pathologies suggests ancestral stress conditions may be a significant factor in our own disease and what we pass down to our grandchildren."

Two studies—Casey Sarapas et al.'s "Genetic Markers for PTSD Risk and Resilience among Survivors of the World Trade Center Attacks" (Disease Markers, vol. 30, nos. 2–3, 2011) and Rachel Yehuda et al.'s "Gene Expression Patterns Associated with Posttraumatic Stress Disorder Following Exposure to the World Trade Center Attacks" (Biological Psychiatry, vol. 66, no. 7, October 1, 2009)—considered pregnant women who were at or near the World Trade Center on September 11, 2001, and were traumatized by witnessing the terrorist attack and its aftermath. Both studies find that women who developed post-traumatic stress disorder (PTSD; an anxiety disorder that some people get after seeing or living through a dangerous event) and had lower levels of the stress hormone cortisol passed some of this trauma on to their children. Expectant mothers who developed PTSD had significantly lower cortisol levels in their saliva than those who also witnessed the attack but did not develop PTSD. When the researchers measured cortisol levels in the children, they found that those born to the women who had developed PTSD had lower cortisol levels than the others. Interestingly, low cortisol levels were most evident in children whose mothers were in the third trimester of pregnancy when they were exposed to the attack.

Rachel Yehuda et al. report in "Holocaust Exposure Induced Intergenerational Effects on *FKBP5* Methylation" (*Biological Psychiatry*, vol. 80, no. 5, September 1, 2016) that the effects of stress can be transmitted intergenerationally through alterations in certain genes. Specifically, the FKBP5 gene has been linked to mental health disorders such as posttraumatic stress disorder (PTSD) and major depression. The researchers looked at differential methylation of FKBP5 in Holocaust survivors and their offspring, and in matched, nonexposed parents and their offspring. They also considered the offspring's history of trauma during childhood. Compared with nonexposed parents, the Holocaust survivors had a 10% increase in methylation as a result of their exposure. The offspring of Holocaust survivors showed a 7.7% downregulation of methylation compared with the nonexposed offspring, and this difference did not change when the researchers considered "presence of lifetime PTSD in offspring" or "childhood trauma." However, it did decrease when the researchers controlled for parental PTSD, indicating that the downregulation resulted from parental influence rather than from the offspring's own experiences. Yehuda et al. observe that Holocaust survivors and their offspring experienced changes in methylation in opposite directions and attribute these changes to epigenetic changes that occurred in utero (in the uterus) in response to stress.

Do Genes Govern Sexual Orientation?

Homosexuality seems to be a stable sexual phenotype in humans, which according to Andrea Burri et al., in "Genetic and Environmental Influences on Female Sexual Orientation, Childhood Gender Typicality and Adult Gender Identity" (*PLoS One*, vol. 6, no. 7, 2011), has a lifetime prevalence (the proportion of a population that at some point during their life will experience this orientation) of 2% to 4% in men and 0.5% to 1.5% in women. The role of genetics in establishing sexual orientation (sexual attraction to men or women) and its link to homosexuality have been hotly debated in the relevant scientific literature and the media. Studies of monozygotic twins reveal that sexual orientation, like the overwhelming majority of human traits and characteristics, is not exclusively governed by genetics, but is more likely the result of a gene-environment interaction. For example, if homosexuality was exclusively controlled by genes, then either both members of a set of monozygotic twins would be homosexual or neither would be. Multiple

studies show that if one twin is homosexual his or her sibling is also homosexual less than 40% of the time.

J. Michael Bailey, Michael P. Dunne, and Nicholas G. Martin systematically evaluated gender identity and sexual orientation of twins and reported their findings in "Genetic and Environmental Influences on Sexual Orientation and Its Correlates in an Australian Twin Sample" (*Journal of Personality and Social Psychology*, vol. 78, no. 3, March 2000). The researchers observe that both male and female homosexuality appears to run in families and that studies of unseparated twins suggest that this is primarily because of genetic rather than familial environmental influences. Bailey, Dunne, and Martin assessed twins from the Australian Twin Registry rather than recruiting twins especially for the purpose of their research. Using probandwise concordance (an estimate of the probability that a twin is nonheterosexual given that his or her co-twin is nonheterosexual), the researchers find lower rates of twin concordance for nonheterosexual orientation than in previous studies. Previously, the lowest concordances for single-sex monozygotic twins were 47% for women and 48% for men. This study documents concordances of just 20% for women and 24% for men, which are significantly lower than the rates reported for the two largest previous twin studies of sexual orientation. Bailey, Dunne, and Martin conclude that sexual orientation is familial; however, their study does not provide statistically significant support for the importance of genetic factors for this trait. They caution that this does not mean that their results entirely exclude heritability. In fact, they consider their findings consistent with moderate heritability for male and female sexual orientation, even though their male monozygotic concordance suggests that any major gene for homosexuality has either low penetrance or low frequency.

Bailey, Dunne, and Martin attribute their markedly different results to the observation that in previous studies twins deciding whether to participate in research that was clearly designed to study homosexuality probably considered the sexual orientation of their co-twin before agreeing to participate. In contrast, the more general focus of the Bailey, Dunne, and Martin study and its anonymous response format made such considerations less likely.

Although it remains unclear in 2017 whether concordance is closer to 30% or 50%, all researchers concur that it is not 100%, which suggests that the influence of genes on sexual orientation is indirect and influenced by environment. Tuck C. Ngun and Eric Vilain of the University of California, Los Angeles, assert in "The Biological Basis of Human Sexual Orientation: Is There a Role for Epigenetics?" (*Advances in Genetics*, vol. 86, 2014) that sexual orientation is more concordant in monozygotic twins than in dizygotic twins and that male sexual orientation is linked to several regions of the genome. The researchers suggest there may be a link between sexual orientation and epigenetic mechanisms. For example, female fetuses exposed to high levels of testosterone have significantly higher rates of nonheterosexual orientation compared with those without such exposure.

In "Genetic and Environmental Effects on Same-Sex Sexual Behavior: A Population Study of Twins in Sweden" (*Archives of Sexual Behavior*, vol. 39, no. 1, February 2010), Niklas Långström et al. report the results of a twin study of same-sex sexual behavior in men and women. Because it was a true population-based study as opposed to a small sample, it overcame some of the biases present in previous research such as self-selection. According to the researchers, this study revealed heritability estimates of 39% for any lifetime same-sex partners in men. Furthermore, the genetic influences on any lifetime same-sex partners and total number of same-sex partners were weaker in women than in men. The results of this study support Bailey, Dunne, and Martin's findings, including the premise that same-sex behavior arises not only from heritable but also from specific environmental sources. Långström et al. conclude that their results are "consistent with moderate, primarily genetic, familial effects, and moderate to large effects of the nonshared environment (social and biological) on same-sex sexual behavior."

AN EPIGENETIC MODEL OF HOMOSEXUALITY? William R. Rice, Urban Friberg, and Sergey Gavrilets describe in "Homosexuality via Canalized Sexual Development: A Testing Protocol for a New Epigenetic Model" (*Bioassays*, vol. 35, no. 9, September 2013) an epigenetic model of homosexuality. The researchers explain that twin studies initially led to the conclusion that genetic polymorphisms accounted for most of the variation in sexual orientation observed within human populations. However, because homosexuality has substantial heritability but low concordance between monozygotic twins in both sexes (about 20%) and GWASs have not found any associated genetic markers with male homosexuality, epigenetics may offer an alternative explanation to genetic polymorphisms as the biological basis for homosexuality.

In "Homosexuality as a Consequence of Epigenetically Canalized Sexual Development" (*Quarterly Review of Biology*, vol. 87, no. 4, December 2012), Rice, Friberg, and Gavrilets explain that homosexuality is linked to epigenetic markers (additional layers of information that control how certain genes are expressed). These epigenetic markers are generally, but not always, "erased" between generations. In homosexuals, the epi-marks are not erased; instead, they are passed from father to daughter or from mother to son.

GENETIC AND ENVIRONMENTAL INFLUENCES ON INTELLIGENCE

The role of genetics in determining a person's intelligence is another controversial subject. Few would deny that genes play some role, but many are uncomfortable with the idea that genes determine intelligence. For if intelligence is a genetic trait, the implication is that some people are born to be smart, others are not, and education and upbringing cannot change it. The results of many studies and contentious debate in the scientific community have produced little consensus about the relationship between genetics and intelligence. Part of the problem stems from the fact that the term *intelligence* is defined differently by different people.

This issue has been argued since the 1870s—when Galton proposed his arguably racist notions about the heritability of intelligence—however, the debate was reignited during the 1990s, when Richard Herrnstein (1930–1994) and Charles Murray (1943–) published *The Bell Curve: Intelligence and Class Structure in American Life* (1994). Herrnstein and Murray expressed their beliefs that between 40% and 80% of intelligence is determined by genetics and that the root of many of the world's social problems involves intelligence levels, rather than environmental circumstances, poverty, or lack of education. Critics argued that Herrnstein and Murray not only manipulated and misinterpreted data to support their contention that intelligence levels differ among ethnic groups but also reintroduced outdated and harmful racial stereotypes. Many observers did agree with Herrnstein and Murray's premises that intellect is spread unevenly among individuals and population subgroups; that innate intelligence is distributed through the entire population on a "bell curve," with most people near the average and fewer at the high and low ends; and even that the distribution varies by race and ethnicity. However, few have been willing to accept the idea that intelligence is entirely genetically encoded, permanently fixed, and unresponsive to environmental influences.

One traditional measure of intelligence is a standardized intelligence quotient (IQ) test, which measures an individual's ability to reason and solve problems. Nearly all studies that focus on the link between intelligence and genetics rely on results obtained from IQ tests, which generally provide an overall score along with measures of verbal ability and performance ability. Although there are several versions of IQ tests available, test takers generally perform comparably on all of them, presumably indicating that they all measure similar aspects of cognitive ability. Critics of IQ tests as measures of intelligence contend that they do not measure all abilities in the complex realm of intelligence. They observe that the tests evaluate only analytic abilities, fail to assess creative and practical abilities, and measure only a small sample of the skills that define the domain of intelligent human behavior.

Despite concern about whether IQ tests fully measure intelligence, nearly all scientific study of the contributions of genetics and environment to intelligence has focused on measuring IQ and examining individuals who differ in their familial relationships. For example, if genetics plays the predominant role in IQ, then monozygotic twins should have IQs that compare more closely than the IQs of dizygotic twins, and siblings' IQs should correlate more highly than those of cousins. Alan S. Kaufman of Yale University observes in "Genetics of Childhood Disorders, II: Genetics and Intelligence" (*Journal of the American Academy of Child and Adolescent Psychiatry*, vol. 38, no. 5, May 1999) that scientists who argue in favor of genetic determination of IQ cite these data to support their assertion:

- Monozygotic twins' IQs are more similar than those of dizygotic twins' IQs.

- IQs of siblings correlate more highly than IQs of half-siblings, which, in turn, correlate higher than IQs of cousins.

- IQ correlations between a biological parent and child living together are higher than those between an adoptive parent and child living together.

Kaufman asserts that the following results support scientists who believe that environment more strongly determines IQ:

- IQs of dizygotic twins correlate more highly than IQs of siblings of different ages despite the same degree of genetic similarity.

- Unrelated siblings reared together, such as biological and adopted children, have IQs that are more similar than biological siblings reared apart.

- Correlations between IQs of an adoptive parent and a child living together are similar to correlations of a biological parent and a child living apart.

- Siblings reared together have IQs that are more similar than siblings reared apart, and the same finding holds for parents and children, when they live together or apart.

The preponderance of evidence from twin, familial, and adoption studies supports increasing heritability of intelligence over time, ranging from 20% in infancy to 60% to 80% in adulthood, along with environmental factors estimated to contribute about 30%. Kaufman suggests that heredity is important in determining a person's IQ, but that environment is also crucial. Based on twin studies, the heritability percentage for IQ is approximately 50, which is comparable to the heritability value for body weight. Kaufman believes the genetic contributions to weight and intelligence are comparable. He observes that many overweight people have a genetic predisposition for a metabolism that promotes weight

gain, whereas naturally thin people have the opposite genetic predisposition. Nonetheless, for most people environmental factors such as diet and exercise have a substantial impact on weight. Similarly, genetics and environment interact to determine IQ, and people with genetic, familial relationships such as parents and siblings frequently share common environments.

In "Human Intelligence and Polymorphisms in the DNA Methyltransferase Genes Involved in Epigenetic Marking" (*PLoS One*, vol. 5, no. 6, June 25, 2010), Paul Haggerty et al. posit that because known genetic influences do not adequately explain the high degree of heritability of intelligence and because epigenetic mechanisms have been associated with some types of mental impairment, these same mechanisms may be involved in intelligence. To test this hypothesis, the researchers looked at SNPs in four DNA methyltransferases involved in epigenetic markers and then applied these SNPs to a test group, which consisted of 1,542 people born in Scotland in 1936. All the study subjects had taken the same IQ test at the age of 11 years and were retested at the age of 70 years. The researchers find that childhood intelligence and adult intelligence were associated with one of the SNPs. Haggerty et al. conclude that "the potential involvement of epigenetics, and imprinting in particular, raises the intriguing possibility that even the heritable component of intelligence could be modifiable by factors such as diet during early development."

Nevertheless, the precise contributions of genetics and environmental influences remain unresolved. Andrea Christoforou et al. explain in "GWAS-Based Pathway Analysis Differentiates between Fluid and Crystallized Intelligence" (*Genes, Brain and Behaviour*, vol. 13, no. 7, September 2014) that about half of the variation in a wide range of cognitive abilities is attributable to a general factor g (a factor that is correlated with performance on intelligence tests), which is highly heritable. There is also general fluid intelligence (gF; the ability to reason in new situations) and general crystallized intelligence (gC; the ability to apply acquired knowledge and learned skills). Based on twin studies, gF and gC are correlated at least 50%. GWASs have established that these cognitive traits have a large genetic component, such that 40% and 51% of the phenotypic variability in gC and gF, respectively, may be accounted for by genetic variants in common SNPs.

Scientists Confirm Genetic Contribution to Intelligence

In 2011 researchers testing people's DNA for genetic variations showed that much like characteristics such as height and weight a large number of individual genes are involved in human intelligence. Gail Davies et al. explain in "Genome-Wide Association Studies Establish That Human Intelligence Is Highly Heritable and Polygenic" (*Molecular Psychiatry*, vol. 16, no. 10, October 2011) that they used GWASs to look at more than half a million genetic markers, and although the studies did not pinpoint the specific genes or genetic variations involved in intelligence, they did confirm that about 40% to 50% of variation in intelligence was attributable to genetic differences.

In "Heritability of Structural Brain Network Topology: A DTI Study of 156 Twins" (*Human Brain Mapping*, vol. 35, no. 10, October 2014), Marc M. Bohlken et al. conducted a twin study to estimate structural differences in the brain that were considered to be under genetic influence. The researchers estimate that heritability was as high as 68% and find that the largest component of heritable genetic variance was in brain networks (the functional connectivity between the regions of brain tissue).

Robert Plomin and Ian J. Deary explain in "Genetics and Intelligence Differences: Five Special Findings" (*Molecular Psychiatry*, vol. 20, no. 1, February 2015) that the heritability of intelligence increases dramatically from about 20% in infancy to 40% in adolescence and 60% to 80% in adulthood. Because the same genes affect intelligence across the lifespan, increasing heritability despite genetic stability suggests some contribution from genetic amplification, which is the production of multiple copies of a particular gene or genes that serve to enlarge or increase the phenotype (in this instance intelligence) that the gene confers.

Genetic Influences on Depression

According to Kornel Schuebel et al. of the University of Würzburg, in "Making Sense of Epigenetics" (*International Journal of Neuropsychopharmacology*, vol. 19, no. 11, December 3, 2016), epigenome-wide association studies have identified methylation differences in seven genes, including three genes that were previously discovered in genetic studies. It is hypothesized that these DNA methylation changes cause the development of depression by mediating the expression of these genes.

In "Epigenetic Mechanisms of Depression" (*Neuroscience and Psychiatry*, vol. 71, no. 4, April 2014), Eric J. Nestler of the Icahn School of Medicine at Mount Sinai opines that epigenetics explains how environmental exposures interact with a person's genetic makeup to determine the risk for depression. For example, severe stress triggers changes in regions of the brain, that in turn cause changes in gene expression that contribute to episodes of depression. It is also likely that when stress-induced epigenetic changes occur early in life, they help influence a person's lifetime vulnerability or resistance to stressful events later in life.

HOW DO GENES INFLUENCE BEHAVIOR AND ATTITUDES?

The question of interest is no longer whether human social behavior is genetically determined; it is to what extent.

—Edward O. Wilson, *On Human Nature* (1978)

Most people accept the premise that genes at least in part influence personality and behavior. Families have long decried a characteristic "bad temper" or "wild streak" appearing in a new generation, or boasted about inherited musical or artistic talents. It seems intuitively correct to assume that some of our behaviors and attitudes, both the desirable and less desirable ones, are in part genetically mediated.

Studies of families and twins strongly suggest genetic influences on the development and expression of specific behaviors. For example, Wendy Johnson et al. observe in "Beyond Heritability: Twin Studies in Behavioral Research" (*Current Directions in Psychological Science*, vol. 18, no. 4, August 1, 2010) that estimates of heritability of general intelligence range from 50% to 80% and personality from 20% to 50%. The researchers assert that all behavior is at least partially heritable—from difficulties with emotional control to voting and television viewing.

Johnson et al. explain that genetic influence does not mean that genes cause behavior. Instead, genes predispose people to behavioral choices that in turn affect subsequent options for genetically influenced behavioral choices in what is called a "causal chain" of gene-environment correlation, or genetically influenced differences in environmental exposure. Like genetic influences, environmental influences are strong and pervasive but do not determine behaviors.

Furthermore, studies of environmental effects show that there are individual differences in response. Some individuals are severely affected and others experience few repercussions from environmental factors. This has given rise to the idea of varying degrees of resiliency—that people vary in their relative resistance to the harmful effects of psychosocial adversity—as well as to the premise that genetics may offer protective effects from certain environmental influences.

In "Scarred DNA and How It Might Heal" (Psychology Today.com, May 12, 2008), Peter D. Kramer opines that epigenetics may help explain why some people are resilient in the face of a challenge or threat and why others suffer lifelong problems. He cites a study in which a group of genetically identical small mice were exposed to and bullied by larger, aggressive mice. Several of the small mice frequently responded by displaying behaviors that are associated with "social defeat," such as anxiety and depression. However, some of the small mice responded differently to the bullying by remaining confident and seemingly unaffected. Although the mice had the exact same genes, their DNA segments revealed differing levels of activity, specifically differences in the methylation of a part of a gene that regulates production of brain-derived neurotrophic factor (a protein that stimulates growth and adaptability of some nerve cells), which means that the mice were actually epigenetically distinct from one another. Conceivably, the epigenetic changes associated with defeated outlooks or resiliency would be heritable and evident in their offspring.

Amy Abrahamson, Laura Baker, and Avshalom Caspi examine genetic influences on attitudes of adolescents and report their findings in the landmark study "Rebellious Teens? Genetic and Environmental Influences on the Social Attitudes of Adolescents" (*Journal of Personality and Social Psychology*, vol. 83, no. 6, December 2002). The purpose of their study was to investigate sources of familial influence on adolescent social attitudes in an effort to understand whether and how families exert an influence on the attitudes of adolescents. They wanted to pinpoint the age when genetic influences actually emerge and to determine the extent to which parents and siblings shape teens' views about controversial issues. Abrahamson, Baker, and Caspi explored genetic and environmental influences in social attitudes in 654 adopted and nonadopted children and their biological and adoptive relatives. Conservatism and religious attitudes were measured in the children annually from ages 12 to 15 and in the parents during the visit with 12-year-olds.

The researchers find that both conservatism and religious attitudes are strongly influenced by shared-family environmental factors throughout adolescence. Familial resemblance for conservative attitudes arises from both genetic and common environmental factors, and familial influence on religious attitudes is almost entirely in response to shared-family environmental factors. These findings are different from previous findings in twin studies, which suggest that genetic influence on social attitudes does not emerge until adulthood. In contrast, Abrahamson, Baker, and Caspi detect significant genetic influence in conservatism as early as age 12, but find no evidence of genetic influence on religious attitudes during adolescence. They conclude that genetic factors exert an influence on social attitudes much earlier than previously indicated. Abrahamson, Baker, and Caspi provide further evidence that shared environmental factors contribute significantly to individual differences in social attitudes during adolescence.

In "A Twin-Sibling Study on the Relationship between Exercise Attitudes and Exercise Behavior" (*Behavioral Genetics*, vol. 44, no. 1, January 2014), Charlotte Huppertz et al. look at twins and their siblings to determine whether attitudes about exercise are heritable. The researchers find that the perception of exercise benefits and barriers depends in part on genetics, but is also affected by environmental influences. Huppertz et al. also determine that there is a causal relationship between attitudes about exercise and exercise behavior.

Behaviors related to risk taking also appear to have a strong genetic contribution. In "Monoamine Oxidase A Gene (MAOA) Associated with Attitude towards Longshot Risks" (*PLoS One*, vol. 4, no. 12, December 31, 2009), Songfa Zhong et al. look at the heritability of risk-taking behavior. The researchers observe that although twin studies have confirmed the heritability of economic risk taking, there has not been a direct association of this behavior with a specific gene. Zhong et al. examined 350 subjects and their attitudes toward risk taking and the genotype of a specific polymorphism, monoamine oxidase A (MAOA; an enzyme involved in the degradation of dopamine and norepinephrine, which are neurotransmitters that exert a strong influence on behavior).

The researchers offered the subjects the option of choosing an insurance program, which was a relatively low economic risk, or a lottery purchase, which was considered riskier—a long-shot preference. Zhong et al. find that more subjects with the MAOA high-activity allele preferred the long-shot lottery purchase than did the subjects with the MAOA low-activity allele and conclude that the high activity of the MAOA polymorphism is significantly associated with economic risk taking and preference for the long-shot and higher-risk option.

Similarly, Elizabeth K. Do et al. find in "Genetic and Environmental Influences on Smoking Behavior across Adolescence and Young Adulthood in the Virginia Twin Study of Adolescent Behavioral Development and the Transitions to Substance Abuse Follow-Up" (*Twin Research and Human Genetics*, vol. 18, no. 1, February 2015) that genes influence the initiation of adolescents' cigarette smoking. The researchers posit that environmental influences may be more important for 14- to 15-year-olds than for older age groups because they have less access to cigarettes than adolescents of legal age who can purchase them.

In "Genetic Influences on Political Ideologies: Twin Analyses of 19 Measures of Political Ideologies from Five Democracies and Genome-Wide Findings from Three Populations" (*Behavioral Genetics*, vol. 44, no. 3, May 2014), Peter K. Hatemi et al. present the results from analyses of a combined sample of more than 12,000 twin pairs from nine studies that were conducted in five democracies over the course of four decades. The researchers find evidence that genetic factors play a role in the formation of political ideology, independent of how ideology is measured across different populations. Hatemi et al. conclude that political ideology is a fundamental aspect of an individual's genetically informed psychological disposition and that "formulation of an integrated theory of political ideology"—including the acquisition and application of social values and behavior—"requires the integration of genes and environment, embedded within a developmental framework."

CHAPTER 5
GENETIC DISORDERS

We could wish that ... life-histories were found in every family, showing the health and diseases of its different members. We might thus in time find evidence of pathological connections and morbid liabilities not now suspected.

—Sir William Gull, 1896

It has long been known that heredity affects health. Genetics (the study of single genes and their effects on the body and mind) explains how and why certain traits such as hair color and blood type run in families. Genomics (the study of more than single genes) considers the functions and interactions of all the genes in the genome. In terms of health and disease, genomics has a broader and more promising range than genetics. Genomics relies on knowledge of and access to the entire genome and applies to common conditions, such as breast and colorectal cancer, Parkinson's disease, and Alzheimer's disease. It also plays a role in infectious diseases once believed to be caused entirely by the environment, such as the human immunodeficiency virus and tuberculosis. Like most diseases, these frequently occurring disorders result from the interactions of multiple genes and environmental factors. Genetic variations may have a protective or a causative role in the expression of diseases.

It is commonly accepted that diseases fall into one of three broad categories: those that are primarily genetic in origin, those that are largely attributable to environmental causes, and those in which genetics and environmental factors make comparable, though not necessarily equal, contributions. As understanding in genomics and epigenetics (heritable changes that influence gene expression and are caused by mechanisms other than by changes in a deoxyribonucleic acid [DNA] sequence) advance and scientists identify genes and heritable changes in gene function that are linked to a growing number of diseases, the distinctions between these classes of disorders are diminishing. This chapter considers some of the disorders that are believed to

be predominantly genetic in origin and some that are the result of genes acted on by environmental factors.

There are two types of genes: dominant and recessive. When a dominant gene is passed on to offspring, the feature or trait it determines appears regardless of the characteristics of the corresponding gene on the chromosome inherited from the other parent. If the gene is recessive, the feature it determines will not show up in the offspring unless both the parents' chromosomes contain the recessive gene for that characteristic. Similarly, among diseases and conditions primarily attributable to a gene or genes, there are autosomal dominant disorders and autosomal recessive disorders (an autosomal chromosome is a chromosome that is not involved in the determination of sex).

Another way to characterize genetic disorders is by their pattern of inheritance, as single gene, multifactorial, chromosomal, or mitochondrial. Single-gene disorders (also called Mendelian or monogenic) are caused by mutations in the DNA sequence of one gene. Because genes code for proteins, when a gene is mutated so that its protein product can no longer carry out its normal function, it may produce a disorder. According to Melissa Conrad Stöppler, in "Genetic Diseases Overview" (MedicineNet.com, August 22, 2016), there are more than 6,000 known single-gene disorders that occur in about 1 out of every 200 births. Examples are cystic fibrosis, sickle-cell anemia, Huntington's disease, and hereditary hemochromatosis (a disorder in which the body absorbs too much iron from food; rather than the excess iron being excreted, it is stored throughout the body, and the iron deposits damage the pancreas, liver, skin, and other tissues). Figure 5.1 shows the cystic fibrosis gene and its location on chromosome 7, and Figure 5.2 shows the sickle-cell anemia gene found on chromosome 11. Single-gene disorders are the result of either autosomal dominant, autosomal recessive, or X-linked inheritance.

FIGURE 5.1

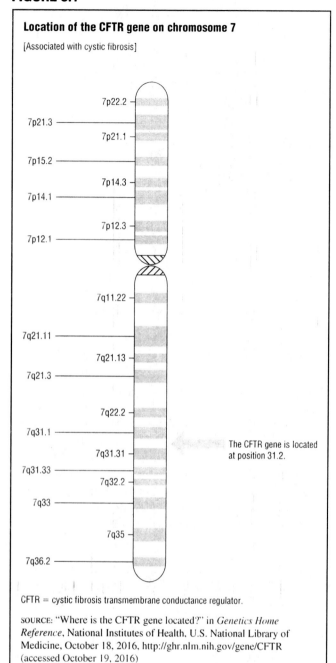

Location of the CFTR gene on chromosome 7

[Associated with cystic fibrosis]

The CFTR gene is located at position 31.2.

CFTR = cystic fibrosis transmembrane conductance regulator.

SOURCE: "Where is the CFTR gene located?" in *Genetics Home Reference*, National Institutes of Health, U.S. National Library of Medicine, October 18, 2016, http://ghr.nlm.nih.gov/gene/CFTR (accessed October 19, 2016)

FIGURE 5.2

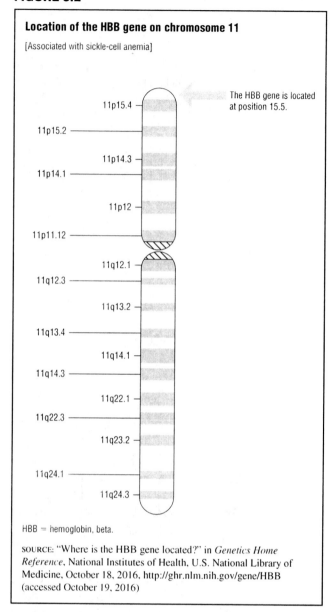

Location of the HBB gene on chromosome 11

[Associated with sickle-cell anemia]

The HBB gene is located at position 15.5.

HBB = hemoglobin, beta.

SOURCE: "Where is the HBB gene located?" in *Genetics Home Reference*, National Institutes of Health, U.S. National Library of Medicine, October 18, 2016, http://ghr.nlm.nih.gov/gene/HBB (accessed October 19, 2016)

Multifactorial or polygenic disorders result from a complex combination of environmental factors and mutations in multiple genes. For example, different genes that influence breast cancer susceptibility have been found on seven different chromosomes, rendering it more difficult to analyze than single-gene or chromosomal disorders. Some of the most common chronic diseases are multifactorial in origin. Examples include heart disease, Alzheimer's disease, arthritis, diabetes, and cancer.

Chromosomal disorders are produced by abnormalities in chromosome structure, missing or extra copies of chromosomes, or errors such as translocations (movement of a chromosome section from one chromosome

to another). Down syndrome or trisomy 21 is a chromosomal disorder that results when an individual has an extra copy, or a total of three copies, of chromosome 21. Mitochondrial disorders result from mutations in the nonchromosomal DNA of mitochondria, which are organelles involved in cellular respiration. When compared with the three other patterns of inheritance, mitochondrial disorders occur infrequently.

There are significant differences between the 19th-century germ theory of disease and the 21st-century genomic theory of disease. By the mid-20th century it became possible to improve the quality of life and to save the lives of people with some genetic diseases. Effective treatment included changes in diet to prevent or manage conditions such as phenylketonuria (PKU) and glucose galactose malabsorption (GGM). PKU is an inherited error of metabolism caused by a deficiency in the enzyme phenylalanine hydroxylase that may result in mental retardation, organ damage, and unusual posture. (See Figure 5.3.) GGM is a rare metabolic disorder caused by a lack of the enzyme that converts galactose into glucose. For people with severe cases, it is vital to avoid lactose (milk sugar), sucrose (table sugar), glucose, and galactose.

During the second decade of the 21st century the possibility of preventing and changing genetic legacies appears within reach of modern medical science. Genomic medicine predicts the risk of disease in the individual, whether highly probable, as in the case of some of the well-established single-gene disorders, or in terms of an increased susceptibility likely to be influenced by environmental factors. The promise of genomic medicine is to make preventive medicine more powerful and treatment more specific to the individual, enabling investigation and treatments that are custom-tailored to an individual's genetic susceptibilities, or to the characteristics of the specific disease or disorder.

In "From Genomic Medicine to Precision Medicine: Highlights of 2015" (*Genome Medicine*, vol. 29, no. 8, January 2016), Charles Auffray et al. report on recent advances including exome sequencing, a technique for sequencing all the expressed genes in a genome, which has led to improved diagnosis and enabled the study of specific rare variants that cause disease. Matchmaker Exchange, a web-based platform for rare gene discovery, enables families with rare diseases to connect with one another and the health professionals who care for them.

Auffray et al. describe the 2015 launch of the precision medicine initiative, which focuses on cancer genomics and on generating long-term data about health and disease by tracking more than 1 million people. Furthermore, molecular technologies that rapidly identify viral pathogens (disease causing agents) are in use worldwide. For example, pathogen genomics were used to track the Ebola virus (a rare but often deadly disease) epidemic in West Africa between 2014 and 2015.

COMMON GENETICALLY INHERITED DISEASES

Although many diseases, disorders, and conditions are termed *genetic*, classifying a disease as genetic simply means there is an identified genetic component to its origin or its expression. Many medical geneticists contend that most diseases cannot be classified as strictly genetic or environmental. The phenotype of genetic diseases can sometimes be modified, even to the point of nonexpression, by controlling environmental factors. Similarly, environmental (infectious) diseases may not be expressed because of some genetic predisposition to immunity. Each disease, in each individual, exists along a continuum between a genetic disease and an environmental disease.

Many diseases are believed to have strong genetic contributions, including:

- Heart disease—coronary atherosclerosis, hypertension (high blood pressure), and hyperlipidemia (elevated blood levels of cholesterol and other lipids)

- Diabetes—abnormally high blood glucose resulting from the body's inability to use blood glucose for energy

- Cancer—retinoblastomas, colon, stomach, ovarian, uterine, lung, bladder, breast, skin (melanoma), pancreatic, and prostate

- Neurological disorders—Alzheimer's disease, amyotrophic lateral sclerosis (a degenerative neurological condition commonly known as Lou Gehrig's disease), Gaucher's disease (a rare disorder of lipid metabolism), Huntington's disease, multiple sclerosis (a chronic progressive disorder in which the immune system destroys the myelin sheath that protects nerve fibers), narcolepsy (a disorder characterized by sudden, uncontrollable episodes of deep sleep), neurofibromatosis (a disease characterized by tumors of the fibrous covering of peripheral nerves and other developmental problems), Parkinson's disease (a disorder that affects nerve cells in the part of the brain that controls muscle movement), Tay-Sachs disease, and Tourette's syndrome (a neurological disorder

FIGURE 5.3

A normal level of the enzyme phenylalanine hydroxylase converts the amino acid phenylalanine to the amino acid tyrosine

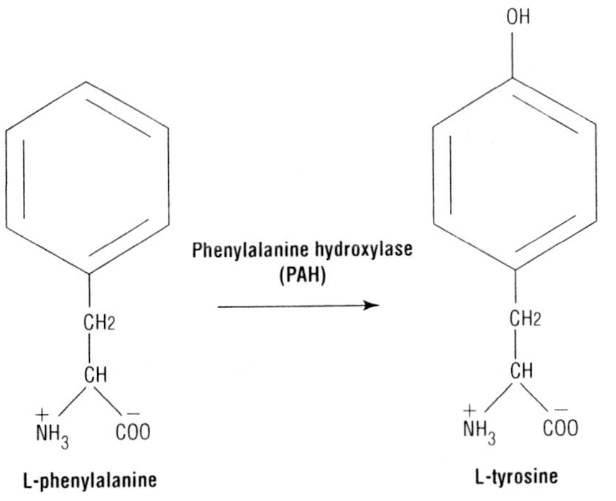

Phenylalanine hydroxylase (PAH)

L-phenylalanine

L-tyrosine

SOURCE: Adapted from "The Enzyme Phenylalanine Hydroxylase Converts the Amino Acid Phenylalanine to Tyrosine," in "Nutritional and Metabolic Diseases," *Genes and Disease*, National Institutes of Health, U.S. National Library of Medicine, National Center for Biotechnology Information, 1998–2012, http://www.ncbi.nlm.nih.gov/books/NBK22253/ (accessed October 19, 2016)

characterized by repetitive, involuntary movements and vocalizations; a syndrome is a set of symptoms or conditions that taken together suggest the presence of a specific disease or an increased risk of developing the disease)

- Mental illnesses, mental retardation, and behavioral conditions—alcoholism, anxiety disorders, attention-deficit/hyperactivity disorder, eating disorders, Lesch-Nyhan syndrome (a disorder caused by a deficiency of the enzyme hypoxanthine-guanine phosphoribosyltransferase, which causes a buildup of uric acid in all body fluids and produces symptoms such as severe gout, poor muscle control, and moderate retardation)

- Other genetic disorders—Behcet's disease (inflammation of the blood vessels; symptoms include sores on the skin, mouth, and genitals as well as swelling of the eyes and joints), cleft lip and cleft palate, clubfoot, cystic fibrosis, Duchenne muscular dystrophy, GGM, hemophilia, Hurler's syndrome (a disease of metabolism in which a person cannot break down long chains of sugar molecules called glycosaminoglycans), Marfan's syndrome (a disorder of connective tissue that affects the skeletal system, cardiovascular system, eyes, and skin), PKU, sickle-cell disease, and thalassemia (a disorder characterized by the abnormal production of hemoglobin that can lead to growth failure, bone deformities, and enlarged liver and spleen)

- Other medical conditions—including alpha-1-antitrypsin deficiency (a lack of a liver protein that blocks the destructive effects of certain enzymes that may lead to emphysema and liver disease), arthritis, asthma, baldness, congenital adrenal hyperplasia (a disorder of the adrenal gland that causes male characteristics to appear early or inappropriately), migraine headaches, obesity, periodontal disease, porphyria (a disorder in which a crucial part of hemoglobin, called heme, does not develop properly, which can lead to abdominal pain, light sensitivity, rashes, and nerve and muscle problems), and some speech disorders

ALZHEIMER'S DISEASE

Alzheimer's disease (AD) is a progressive, degenerative disease that affects the brain and results in severely impaired memory, thinking, and behavior. Figure 5.4 shows the difference in the appearance of a healthy brain and one suffering from severe AD. The Alzheimer's Association estimates in *2016 Alzheimer's Disease Facts and Figures* (https://www.alz.org/documents_custom/2016-facts-and-figures.pdf) that in 2016, one out of nine (11%) Americans aged 65 years and older had AD and about 5.4 million Americans were living with AD.

FIGURE 5.4

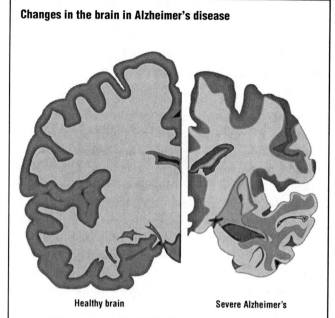

Changes in the brain in Alzheimer's disease

Healthy brain   Severe Alzheimer's

SOURCE: "Changes in the Brain," in *Alzheimer's Disease Fact Sheet*, National Institute on Aging, August 18, 2016, https://www.nia.nih.gov/alzheimers/publication/alzheimers-disease-fact-sheet#changes (accessed October 21, 2016)

AD is of particular concern in the United States because the nation's older adult population is growing rapidly. Grayson K. Vincent and Victoria A. Velkoff of the U.S. Census Bureau project in *The Next Four Decades—The Older Population in the United States: 2010 to 2050* (May 2010, http://www.census.gov/prod/2010pubs/p25-1138.pdf) that by 2050 approximately 88.5 million Americans will be over age 65. The Alzheimer's Association estimates that the annual number of new cases of AD is projected to double by 2050. Prevalence (the number of people with a disease at a given time) is partially determined by the length of time people with AD survive. Although the average survival is eight years after diagnosis, some AD patients have lived longer than 20 years with the disease. Therefore, improvements in AD care, as well as increased life spans of the older adult population in general, will increase the numbers of AD patients.

Genetic Causes of AD

AD is not a normal consequence of growing older, and scientists continue to seek its cause. Researchers have found promising genetic clues to the disease. Mutations in more than 10 genes are thought to be involved in the disease.

The formation of lesions made of fragmented brain cells surrounded by amyloid-family proteins is characteristic of the disease. Tangles of filaments largely made up of a protein associated with the cytoskeleton have also been observed in samples taken from AD brain tissue.

The first genetic breakthrough was reported by Alison Goate et al. in "Segregation of a Missense Mutation in the Amyloid Precursor Protein Gene with Familial Alzheimer's Disease" (*Nature*, vol. 349, no. 6311, February 21, 1991). The researchers discovered that a mutation in a single gene could cause a defect in the gene that directs cells to produce a substance called amyloid protein. They also discovered that low levels of the brain chemical acetylcholine contribute to the formation of hard deposits of amyloid protein that accumulate in the brain tissue of AD patients. In unaffected people the protein fragments are broken down and excreted. Amyloid protein is found in cells throughout the body, and scientists do not yet know why it is harmful in the brain cells of some people and not others.

In 1995 three more genes linked to AD were identified. The APP gene, located on chromosome 21, is related to the most devastating form of AD, which can strike people in their 30s. (See Figure 5.5.) When defective, this gene may prevent brain cells from correctly processing a substance called beta amyloid precursor protein. The PSEN1 gene, located on chromosome 14, is linked to another early-onset form of AD. (See Figure 5.6.) This gene provides instructions for making the presenelin 1 protein. The PSEN2 gene, located on chromosome 1, provides instructions for making the presenilin 2 protein, which helps process amyloid precursor protein. (See Figure 5.7.)

Another gene, known as apolipoprotein E (apoE), is located on the long arm of chromosome 19 and plays several roles. (See Figure 5.8.) It regulates lipid metabolism and helps redistribute cholesterol. In the brain apoE participates in repairing injured nerve tissue. There are three forms (alleles) of the gene: apoE-2, apoE-3, and apoE-4. ApoE-2 is a variant that may confer protection against developing AD or may delay its onset. The apoE-3 variant does not appear to influence the risk of developing AD. People with one copy of the apoE-4 variant have an increased risk of developing AD. People with two copies of the variant, one from each parent, have an even greater risk. The apoE-4 gene variant may also determine when a person develops AD and is associated with earlier memory loss and other symptoms.

Other candidate genes have been linked to the risk of developing AD. Denise Harold et al. identify in "Genome-Wide Association Study Identifies Variants at CLU and PICALM Associated with Alzheimer's Disease, and Shows Evidence for Additional Susceptibility Genes" (*Nature Genetics*, vol. 41, no. 10, October 2009) two more genes, CLU and PICALM, that have a significant association with AD. The PICALM gene may be involved in the breakdown of the synapses, the places where nerve impulses are conveyed from one nerve cell to another. In "Genetic Variation and Neuroimaging Measures in Alzheimer Disease" (*Archives of Neurology*, vol. 67, no. 6, June 2010), Alessandro Biffi et al. identify

FIGURE 5.5

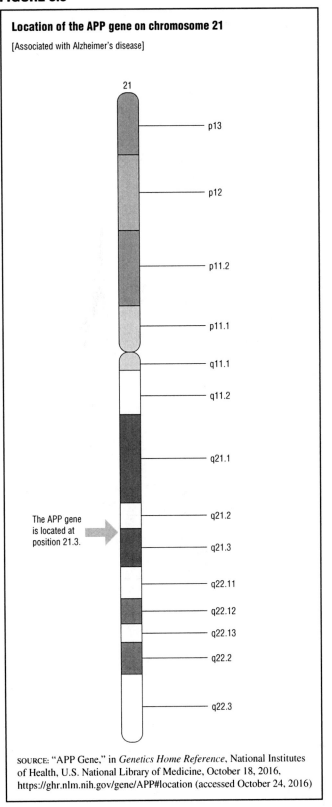

Location of the APP gene on chromosome 21

[Associated with Alzheimer's disease]

The APP gene is located at position 21.3.

SOURCE: "APP Gene," in *Genetics Home Reference*, National Institutes of Health, U.S. National Library of Medicine, October 18, 2016, https://ghr.nlm.nih.gov/gene/APP#location (accessed October 24, 2016)

two additional gene variants, BIN1 and CNTN5, that are linked to the risk for AD.

Adam C. Naj et al. report in "Dementia Revealed: Novel Chromosome 6 Locus for Late-Onset Alzheimer Disease Provides Genetic Evidence for Folate-Pathway

FIGURE 5.6

Location of the PSEN1 gene on chromosome 14

[Associated with Alzheimer's disease]

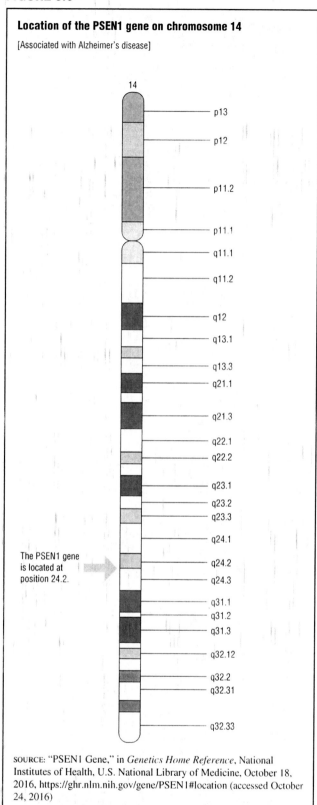

The PSEN1 gene is located at position 24.2.

SOURCE: "PSEN1 Gene," in *Genetics Home Reference*, National Institutes of Health, U.S. National Library of Medicine, October 18, 2016, https://ghr.nlm.nih.gov/gene/PSEN1#location (accessed October 24, 2016)

FIGURE 5.7

Location of the PSEN2 gene on chromosome 1

[Associated with Alzheimer's disease]

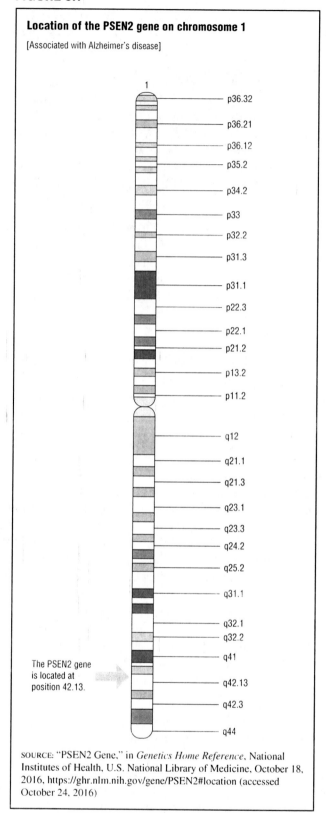

The PSEN2 gene is located at position 42.13.

SOURCE: "PSEN2 Gene," in *Genetics Home Reference*, National Institutes of Health, U.S. National Library of Medicine, October 18, 2016, https://ghr.nlm.nih.gov/gene/PSEN2#location (accessed October 24, 2016)

Abnormalities" (*PLoS Genetics*, vol. 6, no. 9, September 23, 2010) that people with a variant of the gene MTHFD1L are nearly twice as likely to develop AD as those without the gene variant. MTHFD1L on chromosome 6 plays a role in regulating levels of homocysteine, a naturally occurring amino acid that is found in blood plasma. High levels of homocysteine increase the risk of developing late-onset AD as well as heart disease, stroke, and osteoporosis.

According to Jean-Charles Lambert et al., in "The CALHM1 P86L Polymorphism Is a Genetic Modifier of

FIGURE 5.8

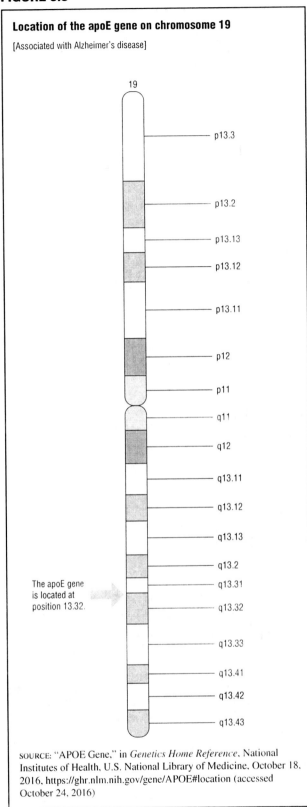

Location of the apoE gene on chromosome 19

[Associated with Alzheimer's disease]

19

p13.3

p13.2

p13.13

p13.12

p13.11

p12

p11

q11

q12

q13.11

q13.12

q13.13

q13.2

q13.31

The apoE gene is located at position 13.32.

q13.32

q13.33

q13.41

q13.42

q13.43

SOURCE: "APOE Gene," in *Genetics Home Reference*, National Institutes of Health, U.S. National Library of Medicine, October 18, 2016, https://ghr.nlm.nih.gov/gene/APOE#location (accessed October 24, 2016)

may influence when the disease begins through an interaction with apoE-4.

In "The Genetics of Alzheimer's Disease" (*Progress in Molecular Biology and Translational Science*, vol. 107, 2012), Lars Bertram and Rudolph E. Tanzi explain that although rare genetic mutations transmitted in a Mendelian fashion, such as APP, PSEN1, and PSEN2, cause AD, disease risk is largely determined by common polymorphisms that, together with nongenetic risk factors, affect the risk for AD. Epigenetic changes, whether protective or harmful, may help explain why one family member develops AD and another does not when both are at risk. Figure 5.9 shows two epigenetic mechanisms that may turn genes on or off. The first is methylation, in which methyl groups attach to the backbone of a DNA molecule. The second is when chemical tags attach to the tails of histones (spool-like proteins that wrap DNA neatly into chromosomes and change how tightly DNA is wound).

By early 2017 genome-wide association studies (GWASs) had identified several more loci (the specific location of a gene, DNA sequence, or position on a chromosome) that make some contribution to the risk for AD, and new polygenic models were under investigation to accurately predict who will develop the disease. Perry G. Ridge et al. estimate in "Assessment of the Genetic Variance of Late-Onset Alzheimer's Disease" (*Neurobiology of Aging*, vol. 41, May 2016) that genetics can explain 53.2% of phenotypic variance, but that only 30.6% of the genetic variance can be explained by known AD single nucleotide polymorphisms (SNPs). Most (59%) of the genetic variance remains unexplained and is likely accounted for by genetic variation near already identified AD-risk variants and other as yet undiscovered genetic regions.

Testing for AD

A complete physical, psychiatric, and neurological evaluation can usually produce a diagnosis of AD that is about 90% accurate. For many years the only sure way to diagnose AD was to examine brain tissue under a microscope, which was not possible while the AD patient was still alive. An autopsy of someone who has died of AD reveals a pattern that is the hallmark of the disease: tangles of fibers (neurofibrillary tangles) and clusters of degenerated nerve endings (neuritic plaques) in areas of the brain that are crucial for memory and intellect. Also, the cortex of the brain is shrunken.

Diagnostic tests for AD include analysis of blood and spinal fluid as well as the use of magnetic resonance imaging (a radiological technique that provides detailed images of organs and tissues) to measure the volume of tissue in areas of the brain that are used for memory, organizational ability, and planning to accurately identify people with AD and predict who will develop AD in the future.

Age at Onset in Alzheimer's Disease: A Meta-analysis Study" (*Journal of Alzheimer's Disease*, vol. 21, no. 1, August 30, 2010), a specific variant, P86L, of the calcium homeostasis modulator 1 gene (CALHM1) is associated with a risk of developing AD. The CALHM1 variant does not determine whether an individual will develop AD but

FIGURE 5.9

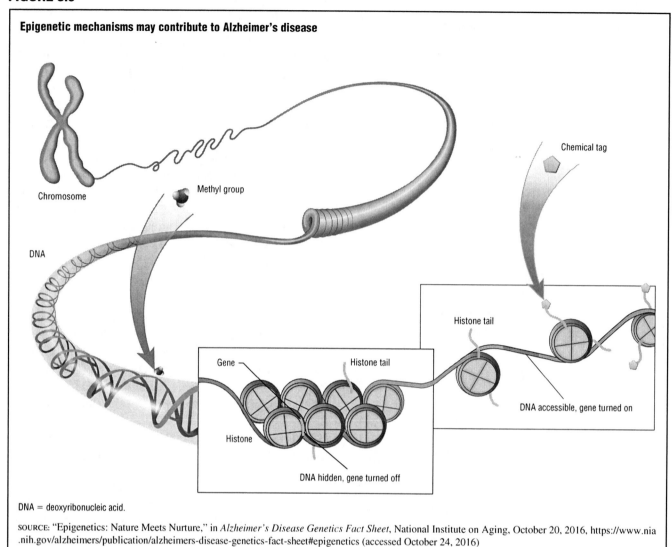

Epigenetic mechanisms may contribute to Alzheimer's disease

DNA = deoxyribonucleic acid.

SOURCE: "Epigenetics: Nature Meets Nurture," in *Alzheimer's Disease Genetics Fact Sheet*, National Institute on Aging, October 20, 2016, https://www.nia .nih.gov/alzheimers/publication/alzheimers-disease-genetics-fact-sheet#epigenetics (accessed October 24, 2016)

The presence of specific proteins called biomarkers in spinal fluid and blood also assist to diagnose people with AD and may be able to predict the disease up to a decade before symptoms appear. For example, in "Diagnostic Accuracy of ^{18}F Amyloid PET Tracers for the Diagnosis of Alzheimer's Disease: A Systematic Review and Meta-analysis" (*European Journal of Nuclear Medicine and Molecular Imaging*, vol. 43, no. 2, February 2016), Elizabeth Morris et al. report that there is increasing interest in the use of positron emission tomography (an imaging test that uses a radioactive drug to show this activity) to detect the presence of amyloid plaques in the brain. The researchers note that this testing is able to distinguish between people with and without AD.

Blood biomarkers are preferable to spinal fluid biomarkers because it is easier to obtain a blood sample than a spinal fluid sample, which requires lumbar puncture (also known as a spinal tap), a procedure in which a needle is inserted into the spinal canal. Julia Muenchhoff et al. report in "Plasma Protein Profiling of Mild Cognitive Impairment

and Alzheimer's Disease across Two Independent Cohorts" (*Journal of Alzheimer's Disease*, vol. 43, no. 4, January 1, 2015) that decreases in two blood plasma proteins—fibronectin and C1 inhibitor—identify the mild cognitive impairment characteristic of early-onset AD. According to Vincent Chourak et al., in "Plasma Amyloid-β and Risk of Alzheimer's Disease in the Framingham Heart Study" (*Alzheimer's & Dementia*, September 9, 2014), risk for AD may be assessed using the blood biomarker plasma amyloid beta peptide. Lower plasma amyloid beta levels are associated with risk for AD.

The National Institute on Aging and the Alzheimer's Association provide diagnostic criteria that identify three stages of AD:

• Preclinical AD—there are measurable changes in the brain, cerebrospinal fluid, and/or blood (biomarkers that show the level of beta amyloid accumulation in the brain and/or those that show nerve cells in the brain are injured or degenerating), but there are no symptoms such as memory loss. This presymptomatic stage may be as long as 20 years because brain changes are thought to precede the appearance of symptoms.

- Mild cognitive impairment (MCI) due to AD—people with MCI have mild but measurable changes in thinking abilities that are noticeable to the person affected and to family members and friends, but the person remains able to perform everyday activities. Research suggests that as many as 20% of people aged 65 years and older have MCI and nearly half of all people who visit a doctor about MCI symptoms develop dementia in three or four years.

- Dementia due to AD—includes all stages of symptomatic AD from mild/early stage to severe/late stage.

Treatments for AD

There is still no cure or prevention for AD, and treatment focuses on managing symptoms. Medication can lessen some of the symptoms, such as agitation, anxiety, unpredictable behavior, and depression. Physical exercise and good nutrition are important, as is a calm and highly structured environment. The objective is to help the AD patient maintain as much comfort, normalcy, and dignity for as long as possible.

In 2017 prescription drugs were available to treat people who suffer from AD. These drugs may prevent or delay the deterioration of memory and other cognitive skills in some patients; however, they offer mild benefits at best and may lose their effectiveness over time. They may also produce undesirable side effects. Table 5.1 lists medications that are used to treat AD, how they work, and the common side effects associated with their use.

CANCER

Cancer is a group of diseases that are characterized by uncontrolled cell division and the growth and spread

TABLE 5.1

Medications to treat Alzheimer's disease

Drug name	Drug type and use	How it works	Common side effects	Manufacturer's recommended dosage
Aricept® (donepezil)	Cholinesterase inhibitor prescribed to treat symptoms of mild, moderate, and severe Alzheimer's	Prevents the breakdown of acetylcholine in the brain	Nausea, vomiting, diarrhea, muscle cramps, fatigue, weight loss	• Tablet*: Initial dose of 5 mg once a day • May increase dose to 10 mg/day after 4–6 weeks if well tolerated, then to 23 mg/day after at least 3 months • Orally disintegrating tablet*: Same dosage as above • 23-mg dose available as brand-name tablet only
Exelon® (rivastigmine)	Cholinesterase inhibitor prescribed to treat symptoms of mild to moderate Alzheimer's (patch is also for severe Alzheimer's)	Prevents the breakdown of acetylcholine and butyrylcholine (a brain chemical similar to acetylcholine) in the brain	Nausea, vomiting, diarrhea, weight loss, indigestion, muscle weakness	• Capsule*: Initial dose of 3 mg/day (1.5 mg twice a day) • May increase dose to 6 mg/day (3 mg twice a day), 9 mg (4.5 mg twice a day), and 12 mg/day (6 mg twice a day) at minimum 2-week intervals if well tolerated • Patch*: Initial dose of 4.6 mg once a day; may increase dose to 9.5 mg once a day and 13.3 mg once a day at minimum 4-week intervals if well tolerated
Namenda® (memantine)	N-methyl D-aspartate (NMDA) antagonist prescribed to treat symptoms of moderate to severe Alzheimer's	Blocks the toxic effects associated with excess glutamate and regulates glutamate activation	Dizziness, headache, diarrhea, constipation, confusion	• Tablet*: Initial dose of 5 mg once a day • May increase dose to 10 mg/day (5 mg twice a day), 15 mg/day (5 mg and 10 mg as separate doses), and 20 mg/day (10 mg twice a day) at minimum 1-week intervals if well tolerated • Oral solution*: Same dosage as above • Extended-release capsule: Initial dose of 7 mg once a day; may increase dose to 14 mg/day, 21 mg/day, and 28 mg/day at minimum 1-week intervals if well tolerated
Namzaric® (memantine extended-release and donepezil)	NMDA antagonist and cholinesterase inhibitor prescribed to treat symptoms of moderate to severe Alzheimer's (for patients stabilized on both memantine and donepezil taken separately)	Blocks the toxic effects associated with excess glutamate and prevents the breakdown of acetylcholine in the brain	Headache, nausea, vomiting, diarrhea, dizziness	• Capsule: 28 mg memantine extended-release + 10 mg donepezil once a day • 14 mg memantine extended-release + 10 mg donepezil once a day (for patients with severe renal impairment)
Razadyne® (galantamine)	Cholinesterase inhibitor prescribed to treat symptoms of mild to moderate Alzheimer's	Prevents the breakdown of acetylcholine and stimulates nicotinic receptors to release more acetylcholine in the brain	Nausea, vomiting, diarrhea, decreased appetite, dizziness, headache	• Tablet*: Initial dose of 8 mg/day (4 mg twice a day) • May increase dose to 16 mg/day (8 mg twice a day) and 24 mg/day (12 mg twice a day) at minimum 4-week intervals if well tolerated • Extended-release capsule*: Same dosage as above but taken once a day

*Available as a generic drug.

SOURCE: "Medication to Treat Alzheimer's Disease," in *Alzheimer's Disease Medications Fact Sheet*, National Institutes of Health, August 2016, https://www.nia.nih.gov/espanol/node/1843 (accessed October 24, 2016)

of abnormal cells. These cells may grow into masses of tissue called tumors. Tumors composed of cells that are not cancerous are called benign tumors. Tumors consisting of cancer cells are called malignant tumors. Cancer cells invade and destroy normal tissue.

The mechanisms of action that disrupt the cell cycle are impairment of a DNA repair pathway, transformation of a normal gene into an oncogene (a hyperactive gene that stimulates cell growth), and the malfunction of a tumor-suppressor gene (a gene that inhibits cell division). Figure 5.10 shows the multiple systems that interact to control the cell cycle.

The spread of cancer cells occurs either by local growth of the tumor or by some of the cells detaching and traveling through the blood and lymphatic system to seed additional tumors in other parts of the body. Metastasis (the spread of cancer cells) may be confined to a local region of the body, but if left untreated (and often despite treatment), cancer cells can spread throughout the entire body, eventually causing death. It is perhaps the rapid, invasive, and destructive nature of cancer that makes it, arguably, the most feared of all diseases, even though it is second to heart disease as the leading cause of death in the United States.

Cancer can be caused by both external environmental influences (chemicals, radiation, and viruses) and internal factors (hormones, immune conditions, and inherited mutations). These factors may act together or in sequence to begin or promote cancer. There is consensus in the scientific community that several cancer-promoting influences accrue and interact before an individual develops a malignant growth. With only a few exceptions, no single factor or risk alone is sufficient to cause cancer. As with other disorders that arise in response to multiple factors, susceptibility to certain cancers is often attributed to a mutated gene. Figure 5.11 shows how the inheritance of a mutated gene increases susceptibility for retinoblastoma (cancer of the eye). According to the American Cancer Society, in "What Are the Key Statistics about Retinoblastoma?" (2017, https://www.cancer.org/cancer/retinoblastoma/about/key-statistics.html), retinoblastoma affects between 200 and 300 children in the United States each year.

FIGURE 5.10

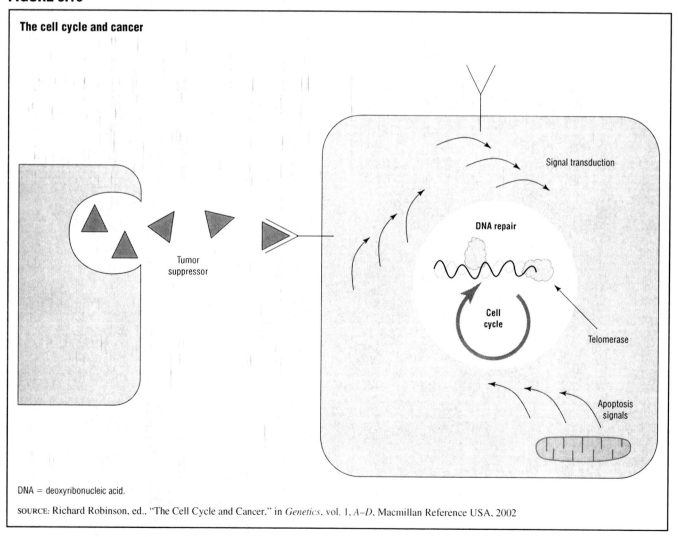

The cell cycle and cancer

Tumor suppressor

Signal transduction

DNA repair

Cell cycle

Telomerase

Apoptosis signals

DNA = deoxyribonucleic acid.

SOURCE: Richard Robinson, ed., "The Cell Cycle and Cancer," in *Genetics*, vol. 1, *A–D*, Macmillan Reference USA, 2002

FIGURE 5.11

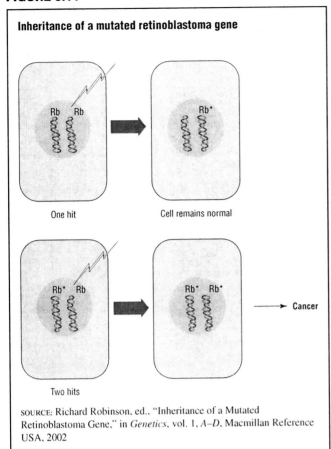

Inheritance of a mutated retinoblastoma gene

One hit

Cell remains normal

Two hits

→ Cancer

SOURCE: Richard Robinson, ed., "Inheritance of a Mutated Retinoblastoma Gene," in *Genetics*, vol. 1, A–D, Macmillan Reference USA, 2002

Because DNA loops, genes that would otherwise be widely separated come in close proximity to one another. In "Insulator Dysfunction and Oncogene Activation in IDH Mutant Gliomas" (*Nature*, vol. 529, no. 7584, January 7, 2016), William A. Flavahan et al. of the Harvard Medical School explain that when a specific gene associated with brain tumors comes close to a gene activating switch, it is switched on and becomes a cancer-growth gene. Hannah Naughton of the National Human Genome Research Institute (NHGRI) describes in "The 'Bunny Ear' Hypothesis: How Defective DNA Looping May Contribute to Cancer" (February 3, 2016, https://www.genome.gov/27563797/the-bunny-ear-hypothesis-how-defective-dna-looping-may-contribute-to-cancer/) these DNA loops as "bunny ears" that bring normally distant DNA "neighborhoods" closer together. Flavahan et al. find that the when two loops merge, a gene from one neighborhood that causes cells to grow, or increase in number, but is rarely turned on, can come under the control of another gene that turns it on. The researchers find that drug treatment can reform the DNA loops, separate the DNA neighborhoods, and turn the cancer causing gene off.

Genetic Research for Breast Cancer

Scientists and physicians have known for some time that predisposition to some forms of breast cancer are inherited and have been searching for the gene or genes responsible so that they can test patients and identify those at increased risk. In 1994 researchers identified the BRCA1 gene, and in late 1995 they isolated the BRCA2 gene. The National Library of Medicine's Genetics Home Reference notes in "Breast Cancer" (February 14, 2017, https://ghr.nlm.nih.gov/condition/breast-cancer) that breast cancer is associated with mutations in dozens of other genes and that women are at risk of developing breast cancer if they have variations of the BRCA1, BRCA2, CDH1, STK11, and TP53 genes. Some genes that have been isolated from breast tumors, such as ERBB2, DIRAS3, and TP53, are somatic mutations, which means they are not inherited.

When a woman with a family history of breast cancer inherits a defective form of either BRCA1 or BRCA2, she has an estimated 80% to 90% chance of developing breast cancer. Figure 5.12 shows the location of the BRCA1 gene on the long arm of chromosome 17 at position 21. Figure 5.13 shows the location of the BRCA2 gene on the long arm of chromosome 13 at position 12.3. These two genes are also linked to ovarian, prostate, and colon cancer and the BRCA2 gene likely plays some role in breast cancer in men. These and other genes may also participate in the development of breast cancer in women with no family history of the disease.

Fergus J. Couch et al. report in "Risks of Triple Negative Breast Cancer Associated with Cancer Predisposition Gene Mutations" (*Journal of Clinical Oncology*, vol. 34, suppl. abstr. 1513, May 20, 2016) the results of multigene testing for hereditary cancer and identify the cancer risks associated with mutations in specific genes. Mutations in the PALB2 gene, which encodes a protein that may suppress tumors, were associated with a high risk of developing breast cancer. Figure 5.14 shows the location of the PALB2 gene on chromosome 16.

Mutations in the ATM, BRIP1, CHEK2, MSH2, MSH6, NBN, RAD51C, and RAD51D genes are associated with a risk of developing breast cancer. Mutations in the PALB2, RAD51D, and BARD1 genes are associated with a high risk of developing triple-negative breast cancer (the three most common types of receptors known to fuel most breast cancer growth—estrogen, progesterone, and the HER2/neu gene—are not present in the cancer tumor). Figure 5.15 shows the location of the RAD51D gene on chromosome 17 and Figure 5.16 shows the BARD1 gene, which is located on chromosome 2. Like some of the other known breast cancer genes such as BRCA1 and BRCA2, BRIP1, RAD51C, and NBN are DNA repair genes, so women with faulty versions of these genes cannot repair damaged DNA correctly, which increases their risk of cancer because their healthy cells are more likely to accumulate genetic damage that can trigger the cell to replicate uncontrollably. The BRIP1

FIGURE 5.12

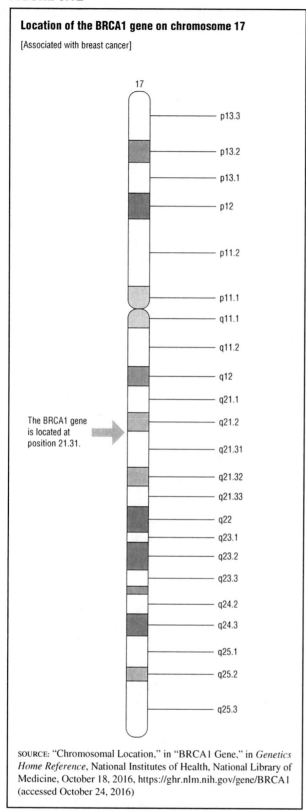

Location of the BRCA1 gene on chromosome 17

[Associated with breast cancer]

17

p13.3
p13.2
p13.1
p12
p11.2
p11.1
q11.1
q11.2
q12
q21.1
q21.2
q21.31
q21.32
q21.33
q22
q23.1
q23.2
q23.3
q24.2
q24.3
q25.1
q25.2
q25.3

The BRCA1 gene is located at position 21.31.

SOURCE: "Chromosomal Location," in "BRCA1 Gene," in *Genetics Home Reference*, National Institutes of Health, National Library of Medicine, October 18, 2016, https://ghr.nlm.nih.gov/gene/BRCA1 (accessed October 24, 2016)

FIGURE 5.13

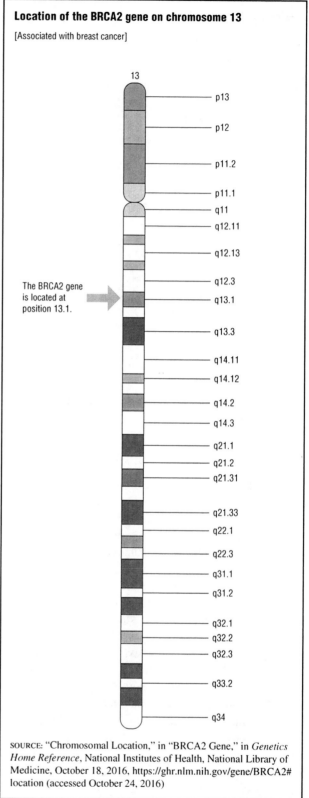

Location of the BRCA2 gene on chromosome 13

[Associated with breast cancer]

13

p13
p12
p11.2
p11.1
q11
q12.11
q12.13
q12.3
q13.1
q13.3
q14.11
q14.12
q14.2
q14.3
q21.1
q21.2
q21.31
q21.33
q22.1
q22.3
q31.1
q31.2
q32.1
q32.2
q32.3
q33.2
q34

The BRCA2 gene is located at position 13.1.

SOURCE: "Chromosomal Location," in "BRCA2 Gene," in *Genetics Home Reference*, National Institutes of Health, National Library of Medicine, October 18, 2016, https://ghr.nlm.nih.gov/gene/BRCA2#location (accessed October 24, 2016)

and RAD51C genes are also located on chromosome 17. (See Figure 5.17 and Figure 5.18.) The NBN gene, which provides instructions for making nibrin, a protein involved in critical cell functions including repair of damaged DNA, is located on chromosome 8. (See Figure 5.19.)

In "The Landscape of Cancer Genes and Mutational Processes in Breast Cancer" (*Nature*, vol. 486, no. 7403, June 21, 2012), Philip J. Stephens et al. note that by examining the genomes of 100 cases of breast cancer they identified nine new genes—AKT2, ARID1B, CASP8, CDKN1B, MAP3K1, MAP3K13, NCOR1, SMARCD1,

FIGURE 5.14

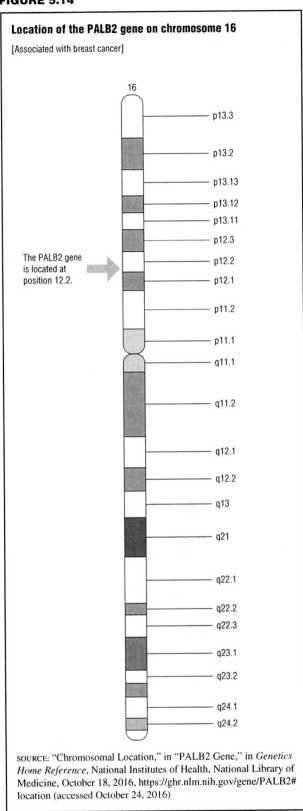

Location of the PALB2 gene on chromosome 16

[Associated with breast cancer]

The PALB2 gene is located at position 12.2.

SOURCE: "Chromosomal Location," in "PALB2 Gene," in *Genetics Home Reference*, National Institutes of Health, National Library of Medicine, October 18, 2016, https://ghr.nlm.nih.gov/gene/PALB2#location (accessed October 24, 2016)

FIGURE 5.15

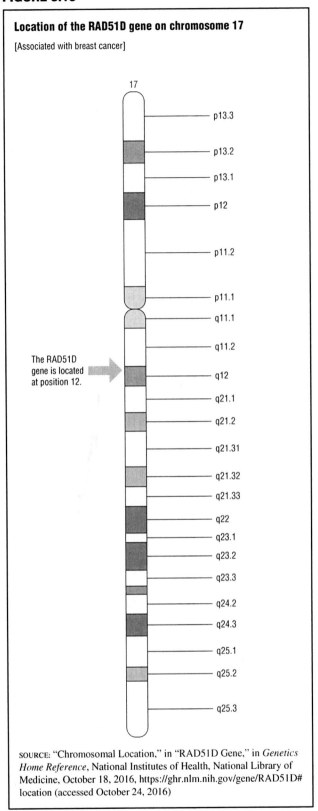

Location of the RAD51D gene on chromosome 17

[Associated with breast cancer]

The RAD51D gene is located at position 12.

SOURCE: "Chromosomal Location," in "RAD51D Gene," in *Genetics Home Reference*, National Institutes of Health, National Library of Medicine, October 18, 2016, https://ghr.nlm.nih.gov/gene/RAD51D#location (accessed October 24, 2016)

and TBX3—that lead to its development. The researchers observe that different combinations of the mutated genes were in different breast cancer samples, underscoring the fact that the disease is highly genetically diverse. Figure 5.20 shows the AKT2 gene located on chromosome 19.

Figure 5.21 shows the position of the CASP8 gene on the long arm of chromosome 2. Figure 5.22 shows the location of the TBX3 gene on chromosome 12.

Additional genes associated with breast cancer risk have been identified. Klesia Pirola Madeira et al. find in

FIGURE 5.16

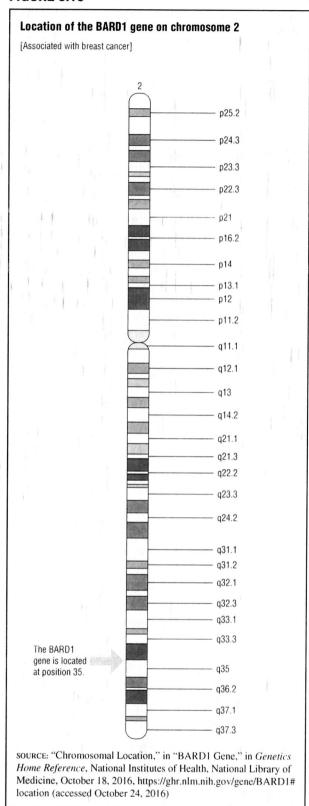

Location of the BARD1 gene on chromosome 2

[Associated with breast cancer]

The BARD1 gene is located at position 35.

SOURCE: "Chromosomal Location," in "BARD1 Gene," in *Genetics Home Reference*, National Institutes of Health, National Library of Medicine, October 18, 2016, https://ghr.nlm.nih.gov/gene/BARD1#location (accessed October 24, 2016)

FIGURE 5.17

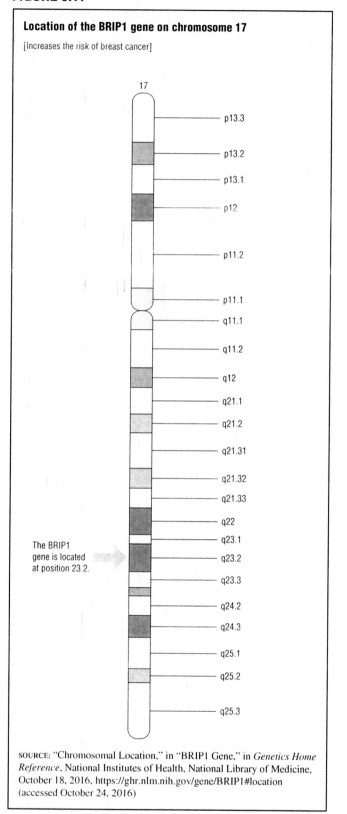

Location of the BRIP1 gene on chromosome 17

[Increases the risk of breast cancer]

The BRIP1 gene is located at position 23.2.

SOURCE: "Chromosomal Location," in "BRIP1 Gene," in *Genetics Home Reference*, National Institutes of Health, National Library of Medicine, October 18, 2016, https://ghr.nlm.nih.gov/gene/BRIP1#location (accessed October 24, 2016)

"Estrogen Receptor Alpha (ERS1) SNPs c454-397T>C (PvuII) and c454-351A>G (XbaI) Are Risk Biomarkers for Breast Cancer Development" (*Molecular Biology Reports*, vol. 41, no. 8, August 2014) that SNPs on the ESR1 gene may be associated with increased risk for breast cancer. Figure 5.23 shows the location of the ESR1 gene on chromosome 6. Other research has determined that DP103, a gene that is activated in metastatic breast cancer, may predict the effectiveness of chemotherapy. Eun Myoung Shin et al. indicate in "DEAD-Box

FIGURE 5.18

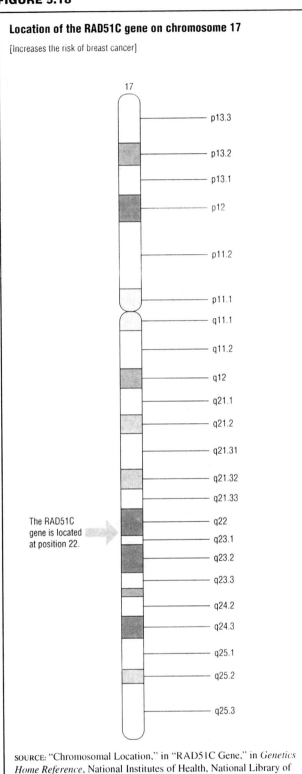

Location of the RAD51C gene on chromosome 17

[Increases the risk of breast cancer]

The RAD51C gene is located at position 22.

SOURCE: "Chromosomal Location," in "RAD51C Gene," in *Genetics Home Reference*, National Institutes of Health, National Library of Medicine, October 18, 2016, https://ghr.nlm.nih.gov/gene/RAD51C# location (accessed October 24, 2016)

FIGURE 5.19

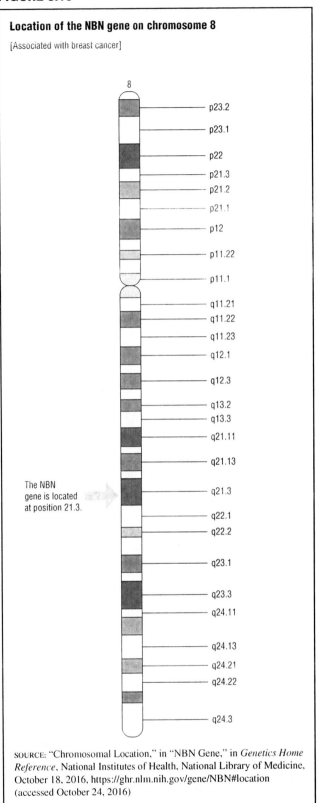

Location of the NBN gene on chromosome 8

[Associated with breast cancer]

The NBN gene is located at position 21.3.

SOURCE: "Chromosomal Location," in "NBN Gene," in *Genetics Home Reference*, National Institutes of Health, National Library of Medicine, October 18, 2016, https://ghr.nlm.nih.gov/gene/NBN#location (accessed October 24, 2016)

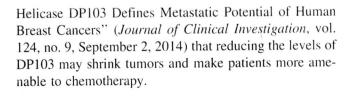

Helicase DP103 Defines Metastatic Potential of Human Breast Cancers" (*Journal of Clinical Investigation*, vol. 124, no. 9, September 2, 2014) that reducing the levels of DP103 may shrink tumors and make patients more amenable to chemotherapy.

CYSTIC FIBROSIS

Cystic fibrosis (CF) is the most common inherited fatal disease of children and young adults in the United States. According to Genetics Home Reference, in "Cystic Fibrosis" (February 14, 2017, https://ghr.nlm

FIGURE 5.20

Location of the AKT2 gene on chromosome 19

[Increases the risk of breast cancer]

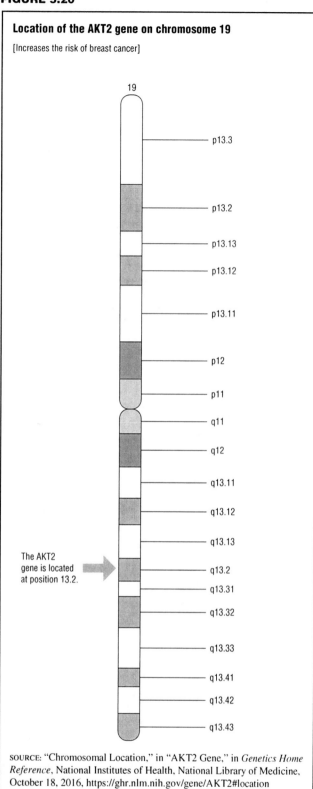

SOURCE: "Chromosomal Location," in "AKT2 Gene," in *Genetics Home Reference*, National Institutes of Health, National Library of Medicine, October 18, 2016, https://ghr.nlm.nih.gov/gene/AKT2#location (accessed October 24, 2016)

FIGURE 5.21

Location of the CASP8 gene on chromosome 2

[Increases the risk of breast cancer]

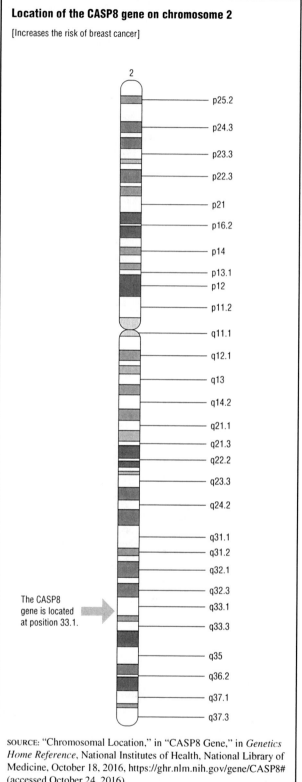

SOURCE: "Chromosomal Location," in "CASP8 Gene," in *Genetics Home Reference*, National Institutes of Health, National Library of Medicine, October 18, 2016, https://ghr.nlm.nih.gov/gene/CASP8# (accessed October 24, 2016)

.nih.gov/condition/cystic-fibrosis), CF occurs in about 1 out of 2,500 to 3,500 whites, 1 out of 17,000 African Americans, and 1 out of 31,000 Asian Americans. In "Learning about Cystic Fibrosis" (December 27, 2013, https://www.genome.gov/10001213/), the NHGRI notes that more than 10 million Americans, almost all of whom are white, are symptomless carriers of the CF gene. Like sickle-cell disease, it is a recessive genetic disorder—to inherit this disease, a child must receive the CF gene from both parents.

FIGURE 5.22

FIGURE 5.23

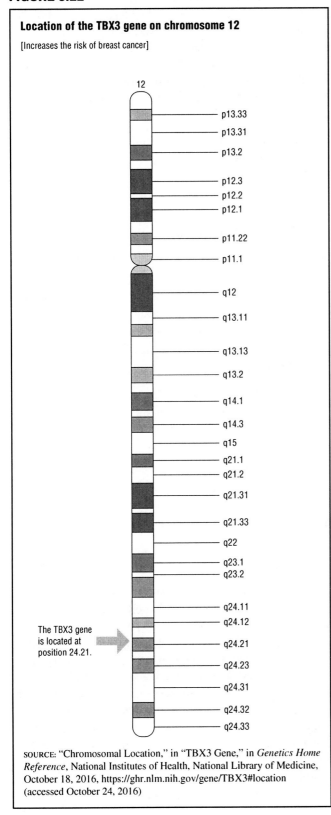

Location of the TBX3 gene on chromosome 12

[Increases the risk of breast cancer]

The TBX3 gene is located at position 24.21.

SOURCE: "Chromosomal Location," in "TBX3 Gene," in *Genetics Home Reference*, National Institutes of Health, National Library of Medicine, October 18, 2016, https://ghr.nlm.nih.gov/gene/TBX3#location (accessed October 24, 2016)

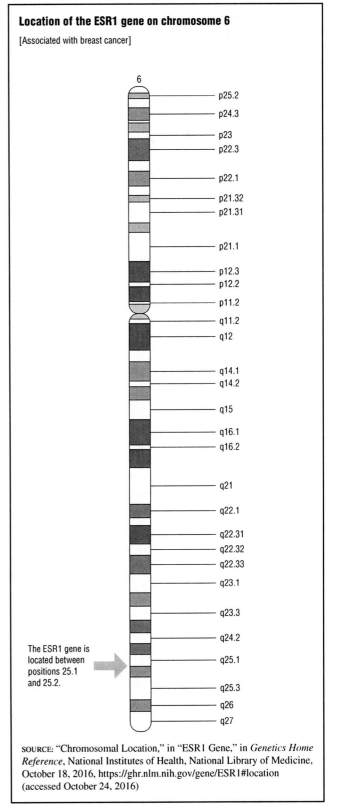

Location of the ESR1 gene on chromosome 6

[Associated with breast cancer]

The ESR1 gene is located between positions 25.1 and 25.2.

SOURCE: "Chromosomal Location," in "ESR1 Gene," in *Genetics Home Reference*, National Institutes of Health, National Library of Medicine, October 18, 2016, https://ghr.nlm.nih.gov/gene/ESR1#location (accessed October 24, 2016)

In 1989 the CF gene was identified, and in 1991 it was cloned and sequenced. It is located on the long arm of chromosome 7 at position 31.2. (See Figure 5.24.) The gene was called cystic fibrosis transmembrane conductance regulator (CFTR) because it was discovered to encode a membrane protein that controls the transit of chloride ions across the plasma membrane of cells. Nearly 1,000 mutations of the large gene have been identified. Although most are extremely rare, several account for more than two-thirds of all mutations.

FIGURE 5.24

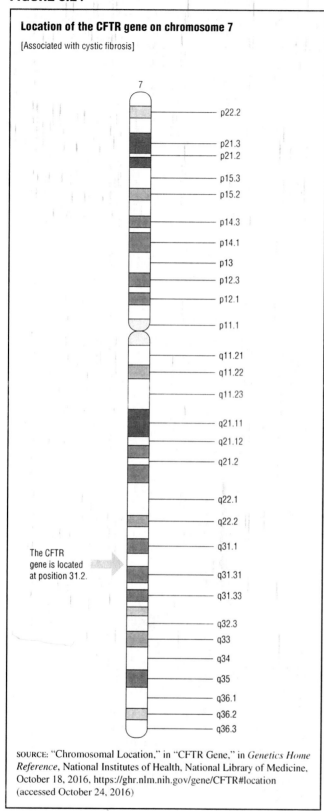

Location of the CFTR gene on chromosome 7

[Associated with cystic fibrosis]

The CFTR gene is located at position 31.2.

SOURCE: "Chromosomal Location," in "CFTR Gene," in *Genetics Home Reference*, National Institutes of Health, National Library of Medicine, October 18, 2016, https://ghr.nlm.nih.gov/gene/CFTR#location (accessed October 24, 2016)

The mutated versions of the gene found in people with CF cause relatively modest impairment of chloride transport in cells. However, this seemingly minor defect can result in a multisystem disease that affects organs and tissues throughout the body, provoking abnormal, thick secretions from glands and epithelial cells. Ultimately, these secretions fill the lungs and cause affected children to die of respiratory failure.

The progression from the defective gene and the protein it encodes to life-threatening illness follows a complex path:

1. The defect in chloride passage across the cell membrane indirectly produces an accumulation of thick mucus secretions in the lungs.

2. The bacteria *Pseudomonas aeruginosa* grows in the mucus.

3. To combat the bacterial invasion, the body's immune system is activated but is unable to access the bacteria because the thick mucus protects it.

4. The immune reaction persists and becomes chronic, resulting in inflammation that harms the lungs.

5. Ultimately, it is the affected individual's own immune response, rather than the defective CF gene, the protein, or the bacterial infection, that produces the often fatal damage to the lungs.

A simple sweat test is currently the standard diagnostic test for CF. The test measures the amount of salt in the sweat; abnormally high levels are the hallmark of cystic fibrosis. Newborns may also be screened for CF using a blood test.

CF Gene-Screening Falters

In 1989 researchers isolated the specific gene that causes CF. The mutation of this gene accounts for about 70% of the cases of the disease. In 1990 scientists successfully corrected the biochemical defect by inserting a healthy gene into diseased cells grown in the laboratory—this was a major step toward developing new therapies for the disease. In 1992 they injected healthy genes into laboratory rats with a deactivated common cold virus as the vector, or delivery agent. The rats began manufacturing the missing protein, which regulates the chloride and sodium in the tissues, preventing the deadly buildup of mucus. Scientists were hopeful that within a few years CF would be eliminated as a fatal disease, giving many children the chance for a healthy and normal life.

Optimism faded in 1993, when it was discovered that the CF gene was more complicated than expected and that many people who have inherited mutated genes from both parents do not have CF. Alan W. Cuthbert of the University of Cambridge notes in "New Horizons in the Treatment of Cystic Fibrosis" (*British Journal of*

Figure 5.25 shows three types of defects in the CFTR gene that can cause CF. The most frequently occurring mutation causes faulty processing of the protein such that the protein is degraded before it reaches the cell membrane.

FIGURE 5.25

Genetic defects that can cause cystic fibrosis

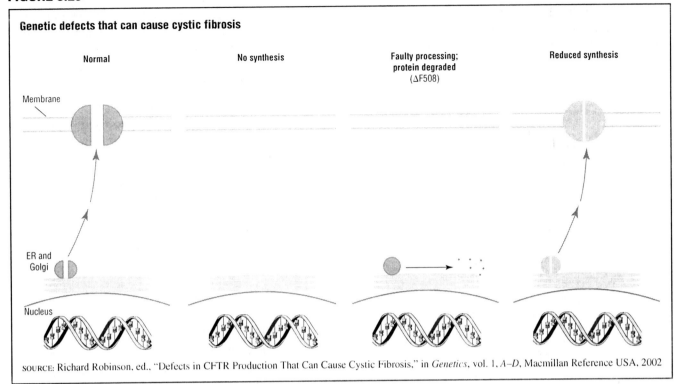

SOURCE: Richard Robinson, ed., "Defects in CFTR Production That Can Cause Cystic Fibrosis," in *Genetics*, vol. 1, *A–D*, Macmillan Reference USA, 2002

Pharmacology, vol. 163, no. 1, May 2011) that by 2010 scientists found that the gene can be mutated at more than 1,700 points. With so many possible mutations, the potential combinations in a person who inherits one gene from each parent are immeasurable. It was also discovered that CF mutations are much more common in the population than previously thought.

The combinations of different mutations create different effects. Some may result in crippling and fatal CF, whereas others may cause less serious disorders, such as infertility, asthma, or chronic bronchitis. To further complicate the picture, other genes can alter the way different mutations of the CF gene affect the body.

Because only a few dozen of the mutations occur with a frequency that is greater than 1 out of 1,000 in the general population and approximately 20 mutations are identified in more than 80% of carriers (e.g., just seven mutations account for nearly all CF cases among east European [Ashkenazi] Jews), the American College of Obstetricians and Gynecologists (ACOG) recommended in 2001 routine screening for the most common mutations. Nearly three-quarters of CF cases among whites are caused by the DF508 mutation, and the W1282X mutation is most common among Ashkenazi Jews. In 2011 the ACOG updated its guidelines in "Update on Carrier Screening for Cystic Fibrosis" (http://www.acog .org/Resources-And-Publications/Committee-Opinions/ Committee-on-Genetics/Update-on-Carrier-Screening-for -Cystic-Fibrosis), which recommended screening all

women of reproductive age for 23 of the most common mutations to greatly reduce the risk of having a child with CF.

In 2012 researchers identified why females with CF do not fare as well as males. In "Effect of Estrogen on Pseudomonas Mucoidy and Exacerbations in Cystic Fibrosis" (*New England Journal of Medicine*, vol. 366, no. 21, May 24, 2012), Sanjay H. Chotirmall et al. find that the female hormone estrogen promotes a specific strain of bacteria that produces more severe symptoms for female CF patients. Interestingly, CF patients taking the oral contraceptive pill, which reduces the amount of naturally occurring estrogen in their body, had lower levels of the troublesome bacteria.

Michael C. Paul-Smith et al. observe in "Gene Therapy for Cystic Fibrosis: Recent Progress and Current Aims" (*Expert Opinion on Orphan Drugs*, vol. 4, no. 6, 2016) that nonviral gene transfer agents can slow the decline of lung function in CF patients. They also note that recent advances in specially engineered viral vectors may overcome problems related to loss of efficacy with repeated administration.

According to the Cystic Fibrosis Foundation, in "Drug Development Pipeline" (2017, https://www .cff.org/trials/pipeline), there are two CFTR modulators that have been approved by the U.S. Food and Drug Administration (FDA): Kalydeco, for people with 10 specific mutations of CF; and Orkambi, for people with two copies of the F508del mutation, which is the most common CF mutation. Kalydeco helps the defective CFTR protein work at the surface of the cell, enabling salt and fluid to move into the airways and thin the thick, sticky mucus so it is easier to cough out.

DIABETES

Diabetes is a disease that affects the body's use of food, causing blood glucose (sugar levels in the blood) to become dangerously high. Normally, the body converts sugars, fats, starches, and proteins into a form of sugar called glucose. The blood then carries glucose to all the cells throughout the body. In the cells, with the help of the hormone insulin, which facilitates the entry of glucose into the cells, the glucose is either converted into energy for use immediately or stored for the future. Beta cells of the pancreas (an organ located behind the stomach) manufacture the insulin. The process of turning food into energy via glucose is important because the body depends on glucose as its energy source.

In a person with diabetes, food is converted to glucose, but there is a problem with insulin. In one type of diabetes the pancreas does not manufacture enough insulin, and in another type the body has insulin but cannot use it effectively. When insulin is either absent or ineffective, glucose cannot get into the cells to be converted into energy. Instead, the unused glucose accumulates in the blood. If a person's blood-glucose level rises high enough, the excess glucose is excreted from the body via urine, causing frequent urination. This, in turn, leads to an increased feeling of thirst as the body tries to compensate for the fluid lost through urination.

Types of Diabetes

There are two distinct types of diabetes. Type 1 diabetes occurs most often in children and young adults.

The pancreas stops manufacturing insulin, so the hormone must be injected daily. Type 2 diabetes is most often seen in adults. In this type the pancreas produces insulin, but it is not used effectively, and the body resists its effects. Table 5.2 compares the phenotype (presentation) and genotype of type 1 and type 2 diabetes.

According to the Centers for Disease Control and Prevention (CDC), in "Diabetes" (July 25, 2016, https:// www.cdc.gov/chronicdisease/resources/publications/aag/ diabetes.htm), more than 29 million Americans had diabetes in 2016. Figure 5.26 shows that the percentage of Americans diagnosed with diabetes has risen sharply since the late 1990s. The CDC notes that in 2013 diabetes was the seventh-leading cause of death. People at risk for type 2 diabetes are usually overweight, over 40 years old, and have a family history of diabetes. Type 2 represents about 95% of adult diabetes patients, whereas type 1 accounts for roughly 5% of cases.

Causes of Diabetes

The causes of both type 1 and type 2 diabetes are unknown, but a family history of the disease increases the risk for both types, leading researchers to believe there is a genetic component. A flaw in the body's immune system may also be a factor in type 1 diabetes.

Mutations in many genes contribute to the origin and onset of type 1 diabetes. For example, an insulin-dependent diabetes mellitus (IDDM1) site on chromosome 6 may harbor at least one susceptibility gene for type 1 diabetes. The role of this mutation in increasing susceptibility is not yet known; however, because chromosome 6 also contains genes for antigens (the molecules that normally tell the immune system not to attack itself), there may be some interaction between immunity and diabetes. In type 1 diabetes the body's immune system mounts an

TABLE 5.2

Comparison of type 1 and type 2 diabetes

	Type 1 diabetes	Type 2 diabetes
Phenotype (observable characteristics)	Onset primarily in childhood and adolescence	Onset predominantly after 40 years of age*
	Often thin or normal weight	Often obese
	Prone to ketoacidosis	No ketoacidosis
	Insulin administration required for survival	Insulin administration not required for survival
	Pancreas is damaged by an autoimmune attack	Pancreas is not damaged by an autoimmune attack
	Absolute insulin deficiency	Relative insulin deficiency and/or insulin resistance
	Treatment: insulin injections	Treatment: (1) healthy diet and increased exercise; (2) hypoglycemic tablets; (3) insulin injections
Genotype (genetic makeup)	Increased prevalence in relatives	Increased prevalence in relatives
	Identical twin studies: <50% concordance	Identical twin studies: usually above 70% concordance
	HLA association: yes	HLA association: no

*Type 2 diabetes is increasingly diagnosed in younger patients.
Note: HLA is human leukocyte antigen.

SOURCE: Adapted from Laura Dean and Johanna McEntyre, "Table 1. Comparison of Type 1 and Type 2 Diabetes," in *The Genetic Landscape of Diabetes*, U.S. Department of Health and Human Services, National Institutes of Health, U.S. National Library of Medicine, National Center for Biotechnology Information, 2004, http://www.ncbi.nlm.nih.gov/bookshelf/br.fcgi?book=diabetes&part=A3 (accessed October 24, 2016)

Genetics and Genetic Engineering

FIGURE 5.26

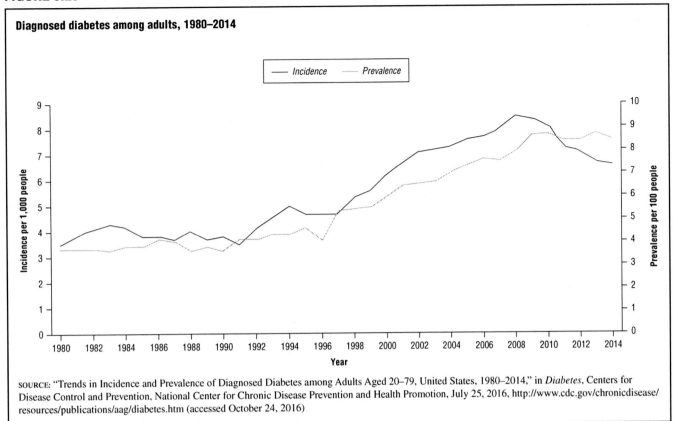

Diagnosed diabetes among adults, 1980–2014

SOURCE: "Trends in Incidence and Prevalence of Diagnosed Diabetes among Adults Aged 20–79, United States, 1980–2014," in *Diabetes*, Centers for Disease Control and Prevention, National Center for Chronic Disease Prevention and Health Promotion, July 25, 2016, http://www.cdc.gov/chronicdisease/resources/publications/aag/diabetes.htm (accessed October 24, 2016)

immunological assault on its own insulin and the pancreatic cells that manufacture it. Some 10 sites in the human genome, including a gene at the locus IDDM2 on chromosome 11 and the gene for glucokinase (an enzyme that is crucial for glucose metabolism) on chromosome 7, appear to increase susceptibility to type 1 diabetes.

The American Diabetes Association indicates in "Genetics of Diabetes" (January 27, 2017, http://www.diabetes.org/diabetes-basics/genetics-of-diabetes.html) that whites with the HLA-DR3 and HLA-DR4 genes are at increased risk for type 1 diabetes. The HLA-DR7 gene may increase risk among African Americans, and the HLA-DR9 gene may increase risk among Asian Americans. Genetics Home Reference explains in "Type 1 Diabetes" (February 21, 2017, https://ghr.nlm.nih.gov/condition/type-1-diabetes) that combinations of variations in the HLA-DRB1 and other HLA genes affect risk of type 1 diabetes and that diabetes risk is greatly increased by two combinations of variations of the HLA-DRB1 gene and other HLA genes called HLA-DQA1 and HLA-DQB1. Figure 5.27 shows the location of the HLA-DRB1 gene on the short arm of chromosome 6.

In type 2 diabetes heredity plays a role, but because the pancreas continues to produce insulin, the disease is considered to be a problem of insulin resistance, in which the body is not using the hormone efficiently. In people prone to type 2 diabetes, overweight and obesity may be environmental triggers for the disease because excess fat prevents insulin from working correctly. Maintaining a healthy weight and keeping physically fit can usually prevent noninsulin-dependent diabetes. To date, insulin-dependent diabetes (type 1) cannot be prevented.

In "Genetic Screening for the Risk of Type 2 Diabetes: Worthless or Valuable?" (*Diabetes Care*, vol. 36, suppl. 2, August 2013), Valeriya Lyssenko and Markku Laakso report that the genetic contribution to diabetes is strong. In monozygotic (identical) twins the concordance rate (the rate of agreement, when both members of the pair of twins have the same trait) of type 2 diabetes is about 70%, whereas in dizygotic (fraternal, meaning nonidentical) twins it is 20% to 30%. The researchers note that more than 65 genetic variants have been found that increase the risk for type 2 diabetes between 10% and 30%. The contribution that each of the known variants alone makes to increasing risk varies, and each variant is assumed to increase only slightly the risk of developing the disease. In combination, however, they likely exert a greater effect.

Complications of Diabetes

Because diabetes deprives body cells of the glucose needed to function properly, it can cause life-threatening complications. These include higher risk and rates of heart disease; circulatory problems, especially in the legs,

FIGURE 5.27

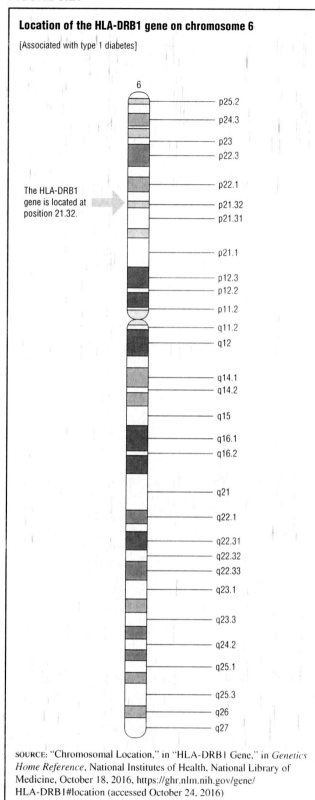

Location of the HLA-DRB1 gene on chromosome 6

[Associated with type 1 diabetes]

6

The HLA-DRB1 gene is located at position 21.32.

p25.2
p24.3
p23
p22.3
p22.1
p21.32
p21.31
p21.1
p12.3
p12.2
p11.2
q11.2
q12
q14.1
q14.2
q15
q16.1
q16.2
q21
q22.1
q22.31
q22.32
q22.33
q23.1
q23.3
q24.2
q25.1
q25.3
q26
q27

SOURCE: "Chromosomal Location," in "HLA-DRB1 Gene," in *Genetics Home Reference*, National Institutes of Health, National Library of Medicine, October 18, 2016, https://ghr.nlm.nih.gov/gene/HLA-DRB1#location (accessed October 24, 2016)

who pay close attention to the roles of diet, exercise, weight management, and proper use of insulin and other medications to manage their disease suffer the fewest complications.

Stem Cell Therapy for Type 1 Diabetes

Ahmed El-Badawy and Nagwa El-Badri of the Zewail City of Science and Technology in Cairo, Egypt, review in "Clinical Efficacy of Stem Cell Therapy for Diabetes Mellitus: A Meta-analysis" (*PLoS One*, vol. 11, no. 4, April 13, 2016) the results of 22 clinical trials of stem cell therapy as treatment for diabetes to determine its efficacy. They find that stem cell therapy is safe and effective and can induce remission of diabetes. Furthermore, they note that the type of cells is a key determinant of the outcome of therapy. The best results were achieved from intravenous administration of CD34+ human stem cells. CD34 is an antigen found on the surface of blood-forming stem cells that regulates cell-to-cell adhesion. El-Badawy and El-Badri also observe that early stem cell therapy soon after diagnosis is more effective than treatment at later stages.

HUNTINGTON'S DISEASE

Huntington's disease (HD) is an inherited, progressive brain disorder. It causes the degeneration of cells in the basal ganglia, a pair of nerve clusters deep in the brain that affect both the body and the mind. HD is caused by a single dominant gene that affects men and women of all races and ethnic groups. Figure 5.28 shows the inheritance pattern of HD.

Gene Responsible for HD Found

The gene mutation that produces HD was mapped to chromosome 4 in 1983 and cloned in 1993. (See Figure 5.29.) In the huntingtin gene, the mutation involves a triplet of nucleotides: cytosine (C), adenine (A), and guanine (G). The mutation is an expansion of a nucleotide triplet repeat in the DNA that codes for the protein huntingtin. In unaffected people the gene has 30 or fewer of these triplets, but HD patients have 40 or more. These increased multiples either destroy the gene's ability to make the necessary protein or cause it to produce a misshapen and malfunctioning protein. Either way, the defect results in the death of brain cells.

The number of repeated triplets is inversely related to the age when the individual first experiences symptoms—the more repeated triplets, the younger the age of onset of the disease. Like myotonic dystrophy, in which the symptoms of the disease often increase in severity from one generation to the next, the unstable triplet repeat sequence can lengthen from one generation to the next, with a resultant decrease in the age when symptoms first appear.

often severe enough to require surgery or even amputation; diabetic retinopathy, a condition that can cause blindness; kidney disease that may require dialysis; dental problems; impaired healing and increased risk of infection; and problems during pregnancy. People

FIGURE 5.28

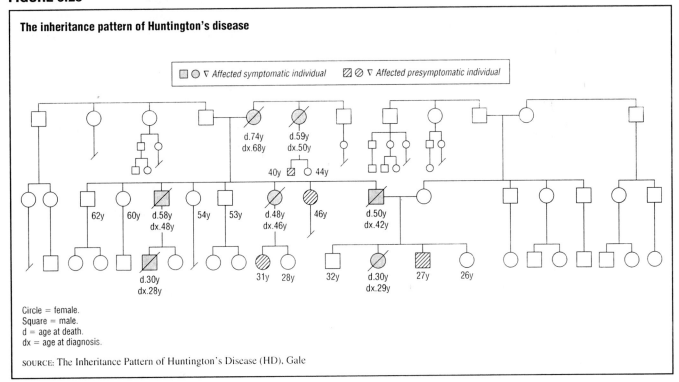

The inheritance pattern of Huntington's disease

☐ ○ ▽ Affected symptomatic individual   ⊠ ⊘ ▽ Affected presymptomatic individual

Circle = female.
Square = male.
d = age at death.
dx = age at diagnosis.

SOURCE: The Inheritance Pattern of Huntington's Disease (HD), Gale

HD does not usually strike until mid-adulthood, between the ages of 30 and 50, although there is a juvenile form that can affect children and adolescents. Early symptoms, such as forgetfulness, a lack of muscle coordination, or a loss of balance, are often ignored, delaying the diagnosis. The disease gradually takes its toll over a 10- to 25-year period.

Within a few years characteristic involuntary movement (chorea) of the body, limbs, and facial muscles appears. As HD progresses, speech becomes slurred and swallowing becomes difficult. The patients' cognitive abilities decline and there are distinct personality changes—depression and withdrawal, sometimes countered with euphoria. There is no cure for HD and no way to stop the disease from progressing. Treatment, which may include medication to control abnormal movements, aims to reduce symptoms and help patients function for as long as possible. Eventually, however, nearly all patients must be institutionalized, and they usually die as a result of choking or infections.

Prevalence of HD

HD, once considered rare, is now recognized as one of the more common hereditary diseases. Genetics Home Reference indicates in "Huntington Disease" (February 14, 2017, https://ghr.nlm.nih.gov/condition/huntington -disease) that HD affects 3 to 7 per 100,000 people of European ancestry. HD appears to be less common in other populations, including people of African, Chinese, and Japanese descent. In "Learning about Huntington's

Disease" (November 17, 2011, https://www.genome .gov/10001215/), the NHGRI reports that in the United States about 30,000 people have HD, an additional 35,000 have some of the symptoms of HD, and 75,000 are carriers of the huntingtin gene.

Prediction Test

In 1983 researchers identified a DNA marker that made it possible to determine whether an individual has inherited the huntingtin gene before symptoms appear. It is also possible to make a prenatal diagnosis on an unborn child. Many people, however, prefer not to know whether they carry the defective gene. Researchers are trying to determine if the exact number of excess triplets indicates when in life a person will be affected by the disease. Some scientists fear that the ability to tell people that they are going to develop an incurable disease and pinpoint when they will develop it will make genetic testing, which is already a difficult decision, even more complicated.

MUSCULAR DYSTROPHY

Muscular dystrophy (MD) is a term that applies to a group of more than 30 types of hereditary muscle-destroying disorders. More than a million Americans are affected by one of the forms of MD. Each variant of the disease is caused by defects in the genes that play important roles in the growth and development of muscles. Duchenne muscular dystrophy (DMD) is one of the most frequently occurring types of MD and is characterized by rapid progression of muscle degeneration that

FIGURE 5.29

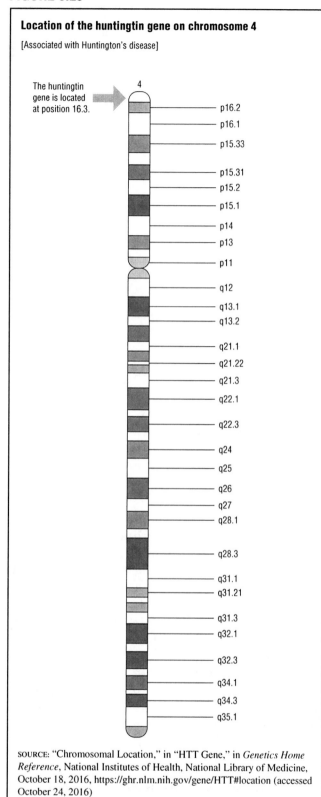

Location of the huntingtin gene on chromosome 4

[Associated with Huntington's disease]

The huntingtin gene is located at position 16.3.

4

- p16.2
- p16.1
- p15.33
- p15.31
- p15.2
- p15.1
- p14
- p13
- p11
- q12
- q13.1
- q13.2
- q21.1
- q21.22
- q21.3
- q22.1
- q22.3
- q24
- q25
- q26
- q27
- q28.1
- q28.3
- q31.1
- q31.21
- q31.3
- q32.1
- q32.3
- q34.1
- q34.3
- q35.1

SOURCE: "Chromosomal Location," in "HTT Gene," in *Genetics Home Reference*, National Institutes of Health, National Library of Medicine, October 18, 2016, https://ghr.nlm.nih.gov/gene/HTT#location (accessed October 24, 2016)

function properly, the muscle cells die and are replaced by fat and connective tissue. Symptoms of MD may not be noticed until as much as 50% of the muscle tissue has been affected.

DMD is X-linked, affects mostly males, and, according to the NHGRI, in "Learning about Duchenne Muscular Dystrophy" (April 18, 2013, https://www.genome.gov/19518854/), strikes 1 out of 3,500 males worldwide. The gene for DMD is located on the short arm of the X chromosome between positions 21.2 and 21.1 and encodes a large protein called dystrophin. (See Figure 5.30.) Dystrophin provides structural support for muscle cells, and without it the cell membrane becomes penetrable, allowing extracellular components into the cell. These additional components increase the intracellular pressure, causing the muscle cell to die.

With myotonic dystrophy the muscles contract but have diminishing ability to relax, and there is muscle weakening and wasting. Typically, the initial complaints are the loss of hand strength or tripping while walking or climbing stairs. Along with decreased muscle strength, myotonic dystrophy may cause mental deficiency, hair loss, and cataracts. The myotonic dystrophy gene is a protein kinase gene, called DMPK, found on the long arm of chromosome 19. (See Figure 5.31.) The defect is a repeated set of three nucleotides—cytosine (C), thymine (T), and guanine (G), known as CTG—in the gene. The symptoms of myotonic dystrophy frequently become more severe with each generation because mistakes in copying the gene from one generation to the next result in amplification of a genomic AGC/CTG triplet repeat, which is similar to the process that is observed in HD. Unaffected individuals have CTG repeats of between three and 37 iterations (repetitions) of the triplet. By contrast, people with the mild phenotype of myotonic dystrophy have between 40 and 170 iterations, and those with more serious forms of the disease have between 100 and 1,000 iterations.

All the various disorders labeled MD cause progressive weakening and wasting of muscle tissue. They vary, however, in terms of the usual age at the onset of symptoms, the rate of progression, and the initial group of muscles affected. The most common type, DMD, affects young boys, who show symptoms in early childhood and usually die from respiratory weakness or heart damage before adulthood. The gene is passed from the mother to her children. Females who inherit the defective gene generally do not manifest symptoms—they become carriers of the defective gene, and their children have a 50% chance of inheriting the disease. Other forms of MD appear later in life and are usually not fatal.

In 1992 scientists discovered the defect in the gene that causes myotonic dystrophy. In people with this

occurs early in life. DMD may also cause enlargement of some muscles, such as those in the calves, which is caused by fat and connective tissue replacing muscle. In all forms of MD the proteins produced by the defective genes are abnormal, causing the muscles to waste away. Unable to

FIGURE 5.30

Location of the dystrophin gene on the X chromosome

[Associated with Duchenne muscular dystrophy]

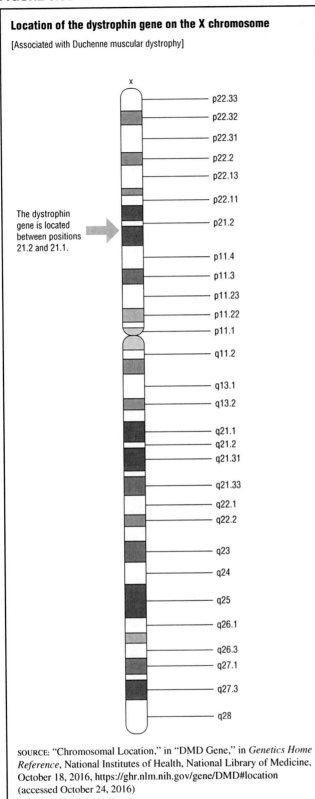

The dystrophin gene is located between positions 21.2 and 21.1.

SOURCE: "Chromosomal Location," in "DMD Gene," in *Genetics Home Reference*, National Institutes of Health, National Library of Medicine, October 18, 2016, https://ghr.nlm.nih.gov/gene/DMD#location (accessed October 24, 2016)

FIGURE 5.31

Location of the DMPK gene on chromosome 19

[Associated with myotonic dystrophy]

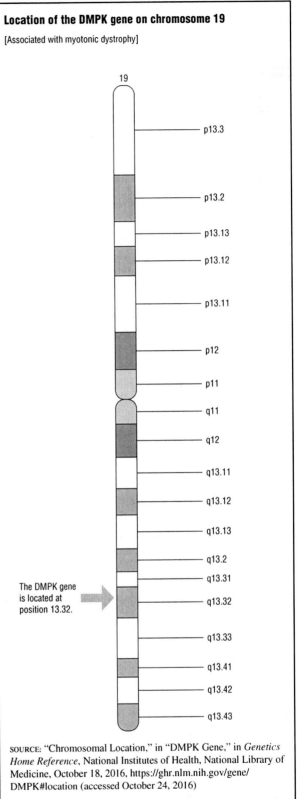

The DMPK gene is located at position 13.32.

SOURCE: "Chromosomal Location," in "DMPK Gene," in *Genetics Home Reference*, National Institutes of Health, National Library of Medicine, October 18, 2016, https://ghr.nlm.nih.gov/gene/DMPK#location (accessed October 24, 2016)

disorder, a segment of the gene is enlarged and unstable. This finding helps physicians diagnose myotonic dystrophy more accurately. Researchers have since identified genes that are linked to other types of MD, including DMD, Becker MD, limb-girdle MD, and Emery-Dreifuss MD.

Treatment and Hope

There is no known cure for MD, but patients can be made more comfortable and functional by a combination of physical therapy, exercise programs, and orthopedic devices (special shoes, braces, or powered wheelchairs)

FIGURE 5.32

Dystrophin and utrophin

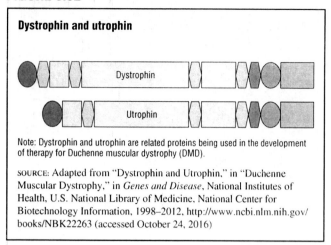

Note: Dystrophin and utrophin are related proteins being used in the development of therapy for Duchenne muscular dystrophy (DMD).

SOURCE: Adapted from "Dystrophin and Utrophin," in "Duchenne Muscular Dystrophy," in *Genes and Disease*, National Institutes of Health, U.S. National Library of Medicine, National Center for Biotechnology Information, 1998–2012, http://www.ncbi.nlm.nih.gov/books/NBK22263 (accessed October 24, 2016)

that help them maintain mobility and independence as long as possible.

Genetic research via gene therapy offers hope of finding effective treatments, and even cures, for these diseases. Research teams have identified the crucial proteins produced by these genes, such as dystrophin, beta sarcoglycan, gamma sarcoglycan, and adhalin. One experimental treatment approach involves substituting a protein of comparable size, such as utrophin, for dystrophin to compensate for the loss of dystrophin. (See Figure 5.32.)

PHENYLKETONURIA

Phenylketonuria (PKU) is an example of a rare, inherited metabolic disorder caused by a gene-environment interaction. As a result of the defect, the affected individual is unable to convert phenylalanine into tyrosine. Phenylalanine in the body accumulates in the blood and can reach toxic levels. (See Figure 5.3.) This toxicity may impair brain and nerve development and result in mental retardation, organ damage, and unusual posture. When it occurs during pregnancy, it may jeopardize the health and viability of the unborn child.

Originally, PKU was considered to be an autosomal recessive inherited error of metabolism that occurred when an individual received two defective copies, caused by mutations in both alleles of the phenylalanine hydroxylase (PAH) gene found in the long arm of chromosome 12 between positions 22 and 24.2. (See Figure 5.33.) More than 500 different genetic mutations have been identified that impair functioning of the PAH enzyme and result in elevated phenylalanine levels. The environmental trigger—dietary phenylalanine—was not identified at first because phenylalanine is so prevalent in the diet, occurring in common foods such as milk and eggs and in the artificial sweetener aspartame. Recognition of dietary phenylalanine as a critical environmental trigger has enabled children born with PKU to lead normal lives

FIGURE 5.33

Location of the phenylalanine hydroxylase gene on chromosome 12

[Associated with phenylketonuria]

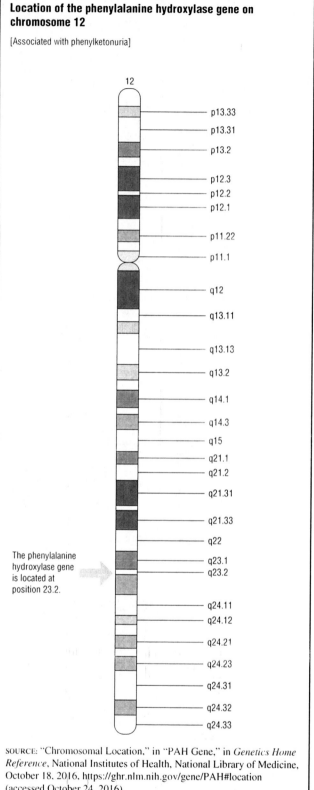

The phenylalanine hydroxylase gene is located at position 23.2.

SOURCE: "Chromosomal Location," in "PAH Gene," in *Genetics Home Reference*, National Institutes of Health, National Library of Medicine, October 18, 2016, https://ghr.nlm.nih.gov/gene/PAH#location (accessed October 24, 2016)

when they are placed on low-phenylalanine diets, and mothers with the disease can bear healthy children.

In 2007 the FDA approved the drug Kuvan, which helps reduce blood phenylalanine levels in people with

PKU by increasing the activity of the PAH enzyme. Kuvan only works in people who have some PAH activity. Also, patients treated with Kuvan must still maintain a phenylalanine-restricted diet and have regular blood tests to measure their phenylalanine levels. The most common side effects associated with use of Kuvan are headaches, diarrhea, abdominal pain, upper respiratory tract infections (cold symptoms), throat pain, vomiting, and nausea.

All newborns in the United States are screened at birth for high levels of phenylalanine in their blood. In "Overview of Phenylketonuria" (UptoDate.com, June 21, 2016, http://www.uptodate.com/contents/overview-of-phenylketonuria), Olaf A. Bodamer reports that 1 out of 13,500 to 19,000 newborns is diagnosed with PKU every year and, with proper diet, is likely to lead a healthy, normal life. PKU occurs less frequently among African Americans, with just 1 out of 50,000 based on newborn screening, and is rare in Finland and Japan.

SCHIZOPHRENIA

Schizophrenia is a severe mental disorder characterized by delusions, hallucinations, incoherence, and physical agitation that occurs in less than 1% of the population and has long been thought to be heritable given that 10% of people with a close relative are also affected. Symptoms generally appear during the teen years and early 20s.

Treatment of Schizophrenia

Medication and psychotherapy are used to control symptoms, but there is no cure for this disorder. Because medication can produce serious side effects, some people with schizophrenia are reluctant to take their medication. Many people with schizophrenia benefit from help preparing for and holding down jobs and obtaining housing and other social services.

Genes Linked to Schizophrenia

Thorsten M. Kranz et al. identify in "Phenotypically Distinct Subtypes of Psychosis Accompany Novel or Rare Variants in Four Different Signaling Genes" (EBioMedicine, April 6, 2016) four genes—PTPRG, SLC39SA13, ARMS/KIDINS220, and TGM5—involved in the growth or regulation of nerve circuits that are associated with different types of schizophrenia. Identification of these genes may lead to targeted treatment.

In "Schizophrenia Risk from Complex Variation of Complement Component 4" (Nature, vol. 530, no. 7589, February 11, 2016), Aswin Sekar et al. observe that more than 100 loci in the human genome contain SNP haplotypes (a set of alleles or markers on one of a pair of homologous chromosomes) associated with the risk of schizophrenia. In an effort to identify the gene or genes that actually cause schizophrenia, the researchers analyzed genome data from healthy and affected people and found that those with higher numbers of copies of the C4 gene, specifically a version called C4A, were diagnosed with schizophrenia more frequently. The C4 gene seems to trigger too much "pruning" of adolescents still-growing and developing brain functions. Although pruning is a normal function and helps rid the brain of excess connections, too much pruning can harm mental function. Treatment aimed at suppressing excessive pruning could prevent the development of this debilitating disorder.

SICKLE-CELL DISEASE

Sickle-cell disease (SCD) is a group of hereditary diseases, including sickle-cell anemia (SCA) and sickle B-thalassemia, in which the red blood cells contain an abnormal hemoglobin, called hemoglobin S (HbS). HbS is responsible for hemolysis (premature destruction of red blood cells). In addition, it causes the red blood cells to become deformed, actually taking on a sickle shape, particularly in parts of the body where the amount of oxygen is relatively low. These abnormally shaped cells cannot travel smoothly through the smaller blood vessels and capillaries. They tend to clog the vessels and prevent blood from reaching vital tissues. This blockage produces anoxia (lack of oxygen), which in turn causes more sickling and more damage.

SCA is an autosomal recessive disease caused by a point mutation in the hemoglobin beta gene found on the short arm of chromosome 11 at position 15.5. (See Figure 5.34.) A mutation in the hemoglobin beta gene results in the production of hemoglobin with an abnormal structure. Figure 5.35 shows how a point mutation in SCA causes the amino acid glutamine to be replaced by valine to produce HbS. It also shows how when red blood cells with HbS are oxygen-deprived they become sickle shaped and may cause blockages that result in tissue death.

Symptoms of SCA

People with SCA have symptoms of anemia, including fatigue, weakness, fainting, and palpitations or an increased awareness of their heartbeat. These palpitations result from the heart's attempts to compensate for the anemia by pumping blood faster than normal.

In addition, patients experience occasional sickle-cell crises—attacks of pain in the bones and abdomen. Blood clots may also develop in the lungs, kidneys, brain, and other organs. A severe crisis or several acute crises can permanently damage various organs of the body. This damage can lead to death from heart failure, kidney failure, or stroke. The frequency of these crises varies from patient to patient. A sickle-cell crisis, however, occurs more often during infections and after an accident or injury.

FIGURE 5.34

Location of the hemoglobin beta gene on chromosome 11

[Associated with sickle-cell anemia]

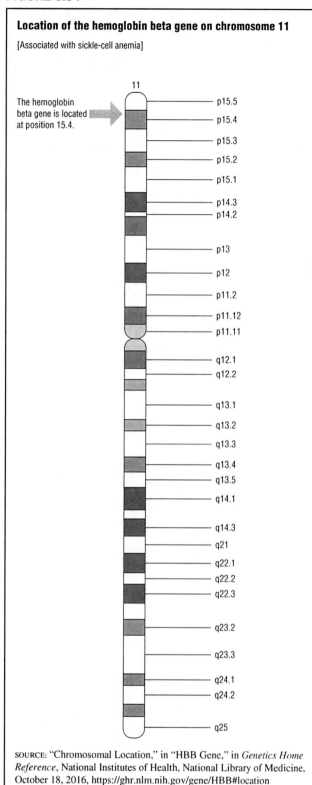

The hemoglobin beta gene is located at position 15.4.

11

p15.5
p15.4
p15.3
p15.2
p15.1
p14.3
p14.2
p13
p12
p11.2
p11.12
p11.11
q12.1
q12.2
q13.1
q13.2
q13.3
q13.4
q13.5
q14.1
q14.3
q21
q22.1
q22.2
q22.3
q23.2
q23.3
q24.1
q24.2
q25

SOURCE: "Chromosomal Location," in "HBB Gene," in *Genetics Home Reference*, National Institutes of Health, National Library of Medicine, October 18, 2016, https://ghr.nlm.nih.gov/gene/HBB#location (accessed October 24, 2016)

FIGURE 5.35

Sickle-cell anemia

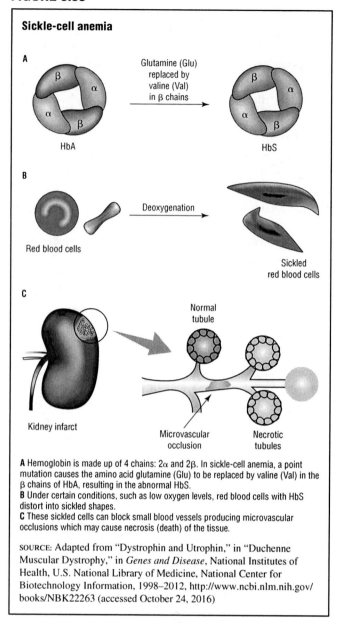

A Hemoglobin is made up of 4 chains: 2α and 2β. In sickle-cell anemia, a point mutation causes the amino acid glutamine (Glu) to be replaced by valine (Val) in the β chains of HbA, resulting in the abnormal HbS.
B Under certain conditions, such as low oxygen levels, red blood cells with HbS distort into sickled shapes.
C These sickled cells can block small blood vessels producing microvascular occlusions which may cause necrosis (death) of the tissue.

SOURCE: Adapted from "Dystrophin and Utrophin," in "Duchenne Muscular Dystrophy," in *Genes and Disease*, National Institutes of Health, U.S. National Library of Medicine, National Center for Biotechnology Information, 1998–2012, http://www.ncbi.nlm.nih.gov/books/NBK22263 (accessed October 24, 2016)

Who Has SCD?

Both the sickle-cell trait and the disease exist almost exclusively in people of African, Hispanic, and Native American descent and in those from India, Italy, Greece, and the Middle East. If one parent has the sickle-cell gene, then the couple's offspring will carry the trait; if both the mother and the father have the trait, then their children may be born with SCA. This trait is relatively common among African Americans. People of African descent are advised to seek genetic counseling and testing for the trait before starting a family. Genetics Home Reference notes in "Sickle Cell Disease" (February 21, 2017, https://ghr.nlm.nih.gov/condition/sickle-cell-disease) that SCD is the most common inherited blood disorder in the United States, affecting between 70,000 and 80,000 Americans, most of whom have African ancestry. SCA occurs in approximately 1 out of 500 African American births and in 1 out of 1,000 to 1,400 Hispanic births.

Treatment of SCD

There is no universal cure for SCD, but the symptoms can be treated. Crises accompanied by extreme pain are the most common problems and can usually be treated with painkillers. Maintaining a healthy diet and receiving prompt treatment for any infection or injury is important. Because a drop in body fluids promotes and sustains sickling, maintaining hydration (adequate fluid intake) is vitally important. Special precautions are often necessary before any type of surgery, and for major surgery some patients receive a transfusion to boost their level of hemoglobin (the oxygen-bearing, iron-containing protein in red blood cells). In 1995 hydroxyurea, a medication that causes the body to produce red blood cells that resist sickling, was approved by the FDA.

USING STEM CELLS TO REPAIR GENES AND CURE SCD. Throughout the first decades of the 21st century scientists attempted to use gene therapy to replace the faulty genes that produce hemoglobin. The article "Chicago Woman Cured of Sickle Cell Disease" (ScienceDaily.com, June 18, 2012) reports that in 2012 a Chicago woman was among the first SCD patients to be successfully treated using stem cell therapy. Suppressing the patient's immune system before the transplant enabled the transplanted stem cells to survive. Over time, the transplanted stem cells assumed the bone marrow's role of producing red blood cells and began manufacturing normal, healthy red blood cells. Just six months after the stem cell transplant, the patient was declared disease free and no longer required red blood cell transfusions.

In "β-Globin Gene Transfer to Human Bone Marrow for Sickle Cell Disease" (*Journal of Clinical Investigation*, July 2013), Zulema Romero et al. describe how an antisickling gene was inserted into hematopoietic stem cells (blood-producing stem cells located in the core of bones), enabling them to produce healthy red blood cells that do not sickle. Shanmuganathan Chandrakasan and Punam Malik endorse this approach in "Gene Therapy for Hemoglobinopathies: The State of the Field and the Future" (*Hematology/Oncology Clinics of North America*, vol. 28, no. 2, April 2014). The researchers explain that although stem cell transplant is curative, it is limited by the availability of donors. By contrast, genetic modification of SCD patients' own hematopoietic stem cells overcomes the problem of donors and the immunological consequences of transplant, which includes rejection of the transplant.

Despite these promising treatments, Megan D. Hoban, Stuart H. Orkin, and Daniel E. Bauer of the Harvard Medical School assert in "Genetic Treatment of a Molecular Disorder: Gene Therapy Approaches to Sickle Cell Disease" (*Blood*, vol. 127, no. 7, February 18, 2016) that gene therapy researchers must establish safe, effective gene transfer, addition, and correction therapies. The researchers report that clinical trials using lentiviral vectors have met with some success. Another approach, genome engineering, which does not permanently insert foreign DNA into the genome, has also shown promise. Hoban, Orkin, and Bauer conclude, "After many years of preclinical laboratory investigation, gene therapy options are now on the horizon for patients with SCD."

TAY-SACHS DISEASE

Tay-Sachs disease (TSD) is caused by mutations in the HEXA gene, which is located on the long arm of chromosome 15 at position 23. (See Figure 5.36.) It is a fatal genetic disorder in children caused by the absence of

FIGURE 5.36

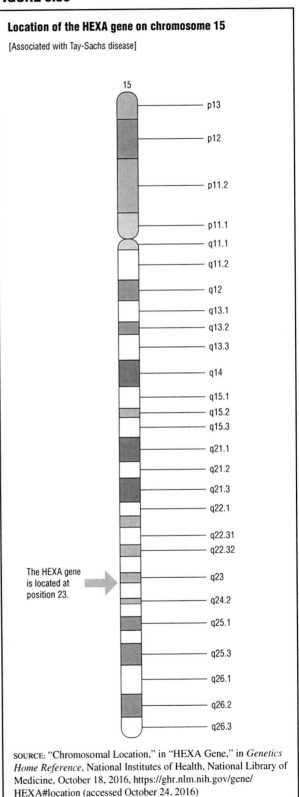

Location of the HEXA gene on chromosome 15

[Associated with Tay-Sachs disease]

The HEXA gene is located at position 23.

SOURCE: "Chromosomal Location," in "HEXA Gene," in *Genetics Home Reference*, National Institutes of Health, National Library of Medicine, October 18, 2016, https://ghr.nlm.nih.gov/gene/HEXA#location (accessed October 24, 2016)

an important enzyme, hexosaminidase A (hex-A), that leads to the progressive destruction of the central nervous system. Without hex-A, a fatty substance called GM2 ganglioside builds up abnormally in the cells, particularly in the brain's nerve cells. (Figure 5.37 shows how the absence of, or defect in, the hex-A protein prevents complete processing of GM2 ganglioside.) Eventually, these cells degenerate and die. This destructive process begins early in the development of a fetus, but the disease is not usually diagnosed until the baby is several months old. By the time a child with TSD is four or five years old, the nervous system is so badly damaged that the child dies.

How Is TSD Inherited?

TSD is an autosomal recessive genetic disorder caused by mutations in both alleles of the HEXA gene on chromosome 15. Both the mother and the father must be carriers of the defective HEXA gene to produce a child with the disease.

People who carry the gene for TSD are entirely unaffected and usually unaware that they have the potential to pass this disease onto their offspring. A blood test distinguishes Tay-Sachs carriers from noncarriers. Blood samples may be analyzed by enzyme assay or DNA studies. Enzyme assay measures the level of hex-A in the blood. Carriers have less hex-A than noncarriers. When only one parent is a carrier, the couple will not have a child with TSD. When both parents carry the recessive HEXA gene, they have a one in four chance in every pregnancy of having a child with the disease. They also have a 50% chance of bearing a child who is a carrier. Prenatal diagnosis early in pregnancy can predict

FIGURE 5.37

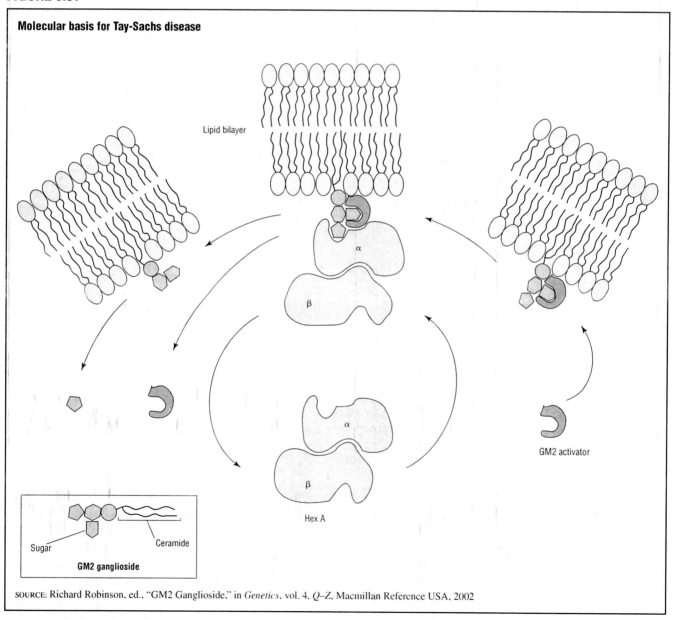

Molecular basis for Tay-Sachs disease

Lipid bilayer

α

β

α

β

Hex A

GM2 activator

Sugar

Ceramide

GM2 ganglioside

SOURCE: Richard Robinson, ed., "GM2 Ganglioside," in *Genetics*, vol. 4, *Q–Z*, Macmillan Reference USA, 2002

if the unborn child has TSD. If the fetus has the disease, the couple may choose to terminate the pregnancy.

Who Is at Risk?

Some genetic diseases, such as TSD, occur most frequently in a specific population. As the American neurologist Bernard Sachs (1858–1944) observed, individuals of Ashkenazi Jewish descent have the highest risk of being carriers of TSD. The National Tay-Sachs and Allied Diseases Association explains in "Tay-Sachs Disease" (2017, http://www.tay-sachs.org/taysachs_disease.php) that approximately 1 out of 27 Jews in the United States is a carrier of the HEXA gene. French Canadians and Cajuns also have the same carrier rate as Ashkenazi Jews. Among Irish Americans, the carrier rate is approximately 1 out of 50. In the general population the carrier rate is 1 out of 250.

CHAPTER 6
GENETIC TESTING

New genomic technologies mean increasing numbers of clinically significant genetic results are being generated through research studies. ... A broad ethical framework underpinning return of research results is emerging. It is recognized that a duty to 'maximize benefit and minimize harm' should remain an overarching aim.

— Kate A. McBride et al., in "Timing and Context: Important Considerations in the Return of Genetic Results to Research Participants" (*Journal of Community Genetics*, vol. 7, no. 1, January 2016)

Advances in genetics have changed the definitions and determination of health and disease. Genetic testing enables researchers and clinicians to detect inherited traits, diagnose heritable conditions, determine and quantify the likelihood that a heritable disease will develop, and identify genetic susceptibility to familial disorders. Many of the strides made in genetic diagnostics are direct results of the Human Genome Project, an international effort begun in 1990 by the U.S. Department of Energy and the National Institutes of Health (NIH), which mapped and sequenced the human genome in its entirety. The increasing availability of genetic testing has been one of the most immediate applications of this groundbreaking research.

A genetic test is the analysis of human deoxyribonucleic acid (DNA), ribonucleic acid (RNA), chromosomes, and proteins to detect heritable disease-related genotypes, mutations, phenotypes, or karyotypes (standard pictures of the chromosomes in a cell) for the purposes of diagnosis, treatment, and other clinical decision making. Most genetic testing involves drawing a blood sample and extracting DNA from white blood cells. Genetic tests may detect mutations at the chromosomal level, such as additional, absent, or rearranged chromosomal material, or even subtler abnormalities, such as a substitution in one of the bases that make up the DNA. There is a broad range of techniques that can be used for genetic testing. Genetic tests have diverse purposes, including screening for and diagnosis of genetic disease in newborns, children, and adults; the identification of future health risks; the prediction of drug responses; and the assessment of risks to future children.

There is a difference between genetic testing to establish a diagnosis and testing to screen for a disease. Diagnostic tests are intended to definitively determine whether a patient has a particular problem. They are generally complex tests and require sophisticated analysis and interpretation. They may be expensive and are generally performed only on people believed to be at risk, such as patients who already have symptoms of a specific disease.

By contrast, screening is performed on healthy people with no signs of disease and often on the entire relevant population. A good screening test is relatively inexpensive, easy to use and interpret, and helps identify which individuals in the population are at higher risk of developing a specific disease. By definition, screening tests identify people who need further testing or those who should take special preventive measures or precautions. Examples of genetic tests used to screen relevant populations include those that screen people of Ashkenazi Jewish heritage (the east European Jewish population primarily from Germany, Poland, and Russia, as opposed to the Sephardic Jewish population primarily from Spain, France, Italy, and North Africa) for Tay-Sachs disease, African Americans for sickle-cell disease, and the fetuses of expectant mothers over the age of 35 years for Down syndrome.

QUALITY AND UTILITY OF GENETIC TESTS

Like all diagnostic and screening tests, the quality and utility of genetic tests depend on their reliability, validity, sensitivity, specificity, positive predictive value, and negative predictive value. Reliability of testing refers to the test's ability to be repeated and to produce equivalent results in comparable circumstances. A reliable test is consistent and measures the same way each time it is used with the same patients in the same circumstances. For example, a well-calibrated balance scale is a reliable instrument for measuring body weight.

Validity is the accuracy of the test. It is the degree to which the test correctly identifies the presence of disease, blood level, or other quality or characteristic it is intended to detect. For example, if a person puts an object he or she knows weighs 10 pounds (4.5 kg) on a scale and the scale states it weighs 10 pounds, then the scale's results are valid. There are two components of validity: sensitivity and specificity.

Sensitivity is the test's ability to identify people who have the disease. Mathematically speaking, it is the percentage of people with the disease who test positive for the disease. Specificity is the test's ability to identify people who do not have the disease—it is the percentage of people without the disease who test negative for the disease. Ideally, diagnostic and screening tests should be highly sensitive and highly specific, thereby accurately classifying all people tested as either positive or negative. In practice, however, sensitivity and specificity are frequently inversely related—most tests with high levels of sensitivity have low specificity, and the reverse is also true.

The likelihood that a test result will be incorrect can be gauged based on the sensitivity and specificity of the test. For example, if a test's sensitivity is 95%, then when 100 patients with the disease are tested, 95 will test positive and five will test false negative—they have the disease but the test has failed to detect it. For example, disorders such as Charcot-Marie-Tooth disease (a group of inherited, slowly developing disorders that result from progressive damage to the nerves) can arise from mutations in one of many different genes, and because some of these genes have not yet been identified, they will not be detected, so a false negative result might be reported. By contrast, if a test is 90% specific, when 100 healthy, disease-free people are tested, 90 will receive negative test results and 10 will be given false-positive results, meaning they do not have the disease but the test has inaccurately classified them as being positive.

The positive predictive value is the percentage of people who actually have the disease of all those with positive test results. The negative predictive value measures the percentage of all the people with negative test results who do not have the disease.

PREGNANCY, CHILDBIRTH, AND GENETIC TESTING

There are thousands of genetic diseases, such as sickle-cell anemia, cystic fibrosis, and Tay-Sachs disease, that may be passed from one generation to the next.

Many tests have been developed to help screen parents who are at risk of passing on genetic diseases to their children as well as to identify embryos, fetuses, and newborns who suffer from genetic diseases.

Carrier Identification

Carrier identification is the term for genetic testing to determine whether a healthy individual has a gene that may cause disease if passed on to his or her offspring. It is usually performed on people considered to be at higher than average risk, such as those of Ashkenazi Jewish descent, who have a 1 out of 27 chance of being Tay-Sachs carriers (in other populations the risk is 1 out of 250), according to the National Tay-Sachs and Allied Diseases Association, in "Tay-Sachs Disease" (2017, http://www.tay-sachs.org/taysachs_disease.php). Testing is necessary because many carriers have just one copy of a gene for an autosomal recessive trait and are unaffected by the trait or disorder. Only someone with two copies of the gene will actually have the disorder. So although it is assumed that everyone is an unaffected carrier of at least one autosomal recessive gene, it only presents a problem in terms of inheritance when two parents have the same recessive disorder gene (or both are carriers). In this instance, the offspring would each have a one in four chance of receiving a defective copy of the gene from each parent and developing the disorder. Figure 6.1 shows the probability of transmitting an autosomal recessive gene mutation to offspring.

Carrier testing is offered to individuals who have family members with a genetic condition, people with family members who are identified carriers, and members of racial and ethnic groups known to be at high risk. Figure 6.2 indicates how carrier testing would be used for a family that is affected by cystic fibrosis and among African Americans who may carry the gene for sickle-cell anemia.

Preimplantation Genetic Diagnosis

Preimplantation genetic diagnosis enables parents undergoing in vitro fertilization (fertilization that takes place outside the body) to screen an embryo for specific genetic mutations when it is no larger than six or eight cells and before it is implanted in the uterus to grow and develop.

Prenatal Genetic Testing

Prenatal genetic testing enables physicians to diagnose diseases in the fetus. Most genetic tests examine blood or other tissue to detect abnormalities. An example of a blood test is the triple-marker screen. This test measures levels of alpha fetoprotein (AFP), human chorionic gonadotropin (hCG), and unconjugated estriol and can identify some birth defects such as Down syndrome and neural tube defects. (Two of the most common neural

FIGURE 6.1

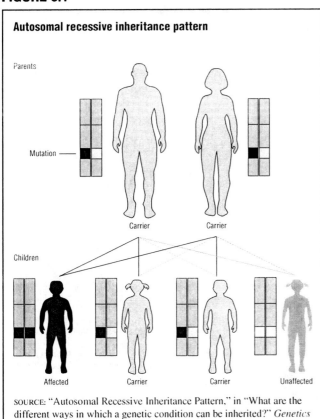

Autosomal recessive inheritance pattern

Parents

Mutation

Carrier　　　Carrier

Children

Affected　　Carrier　　Carrier　　Unaffected

SOURCE: "Autosomal Recessive Inheritance Pattern," in "What are the different ways in which a genetic condition can be inherited?" *Genetics Home Reference*, Lister Hill National Center for Biomedical Communications, U.S. National Library of Medicine, National Institute of Health, January 24, 2017, https://ghr.nlm.nih.gov/primer/illustrations/autorecessive.jpg (accessed January 30, 2017)

FIGURE 6.2

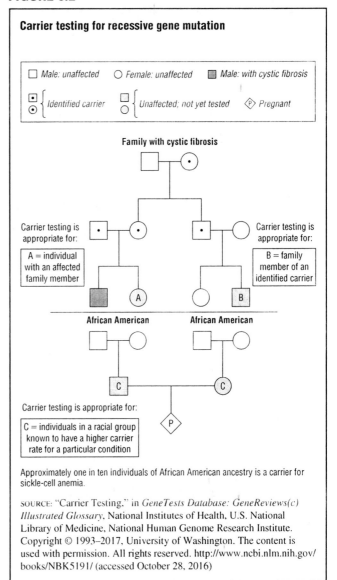

Carrier testing for recessive gene mutation

☐ Male: unaffected　　○ Female: unaffected　　■ Male: with cystic fibrosis

⊡ ⊙ } Identified carrier　　☐ ○ } Unaffected; not yet tested　　⟨P⟩ Pregnant

Family with cystic fibrosis

Carrier testing is appropriate for:

A = individual with an affected family member

Carrier testing is appropriate for:

B = family member of an identified carrier

African American　　African American

Carrier testing is appropriate for:

C = individuals in a racial group known to have a higher carrier rate for a particular condition

Approximately one in ten individuals of African American ancestry is a carrier for sickle-cell anemia.

SOURCE: "Carrier Testing," in *GeneTests Database: GeneReviews(c) Illustrated Glossary*, National Institutes of Health, U.S. National Library of Medicine, National Human Genome Research Institute. Copyright © 1993–2017, University of Washington. The content is used with permission. All rights reserved. http://www.ncbi.nlm.nih.gov/books/NBK5191/ (accessed October 28, 2016)

tube defects are anencephaly [absence of most of the brain] and spina bifida [incomplete development of the back and spine].)

The fetal yolk sac and the fetal liver make AFP, which is continuously processed by the fetus and excreted into the amniotic fluid. A small amount crosses the placenta and enters maternal blood. Screening for AFP levels is based on maternal age, fetal gestation, and the number of fetuses the mother is carrying. Elevated levels of AFP are associated with conditions such as spina bifida and low levels are found with Down syndrome. Because AFP levels alone may not always adequately detect disorders, two other blood serum tests have been developed. The glycoprotein hCG is produced by the placenta. Normally, hCG is elevated at the time of implantation, but decreases at about eight weeks of gestation, and then drops again at approximately 12 weeks of gestation. Elevated levels of hCG are found with Down syndrome. The placenta also produces unconjugated estriol. As with AFP, lower unconjugated estriol maternal serum levels are also found with Down syndrome. Triple-marker screen results are usually available within several days and women with abnormal results are often advised to undergo additional diagnostic testing such as chorionic villus sampling (CVS), amniocentesis, or periumbilical blood sampling.

CVS enables obstetricians and perinatologists (physicians specializing in the evaluation and care of high-risk expectant mothers and infants) to assess the progress of pregnancy during the first trimester (the first three months). A physician passes a small, flexible tube called a catheter through the cervix to extract chorionic villi tissue (cells that will become the placenta). Because the chorion and the fetus develop from trophoblasts, they are genetically identical (both contain the same DNA and chromosomes). The cells obtained via CVS are examined in the laboratory for indications of genetic disorders, such as cystic fibrosis, Down syndrome, Tay-Sachs disease, and thalassemia (a disorder characterized by the abnormal production of hemoglobin that can lead to growth failure, bone deformities, and enlarged liver and spleen), and the results of testing are available within seven to 14 days. Table 6.1 describes CVS and other prenatal diagnostic tests and specifies when they are performed during pregnancy.

TABLE 6.1

Common prenatal tests

Test	What it is	How it is done
Amniocentesis (AM-nee-oh-sen-TEE-suhss)	This test can diagnosis certain birth defects, including: • Down syndrome • Cystic fibrosis • Spina bifida It is performed at 14 to 20 weeks. It may be suggested for couples at higher risk for genetic disorders. It also provides DNA for paternity testing.	A thin needle is used to draw out a small amount of amniotic fluid and cells from the sac surrounding the fetus. The sample is sent to a lab for testing.
Biophysical profile (BPP)	This test is used in the third trimester to monitor the overall health of the baby and to help decide if the baby should be delivered early.	BPP involves an ultrasound exam along with a nonstress test. The BPP looks at the baby's breathing, movement, muscle tone, heart rate, and the amount of amniotic fluid.
Chorionic villus (KOR-ee-ON-ihk VIL-uhss) sampling (CVS)	A test done at 10 to 13 weeks to diagnose certain birth defects, including: • Chromosomal disorders, including Down syndrome • Genetic disorders, such as cystic fibrosis CVS may be suggested for couples at higher risk for genetic disorders. It also provides DNA for paternity testing.	A needle removes a small sample of cells from the placenta to be tested.
First trimester screen	A screening test done at 11 to 14 weeks to detect higher risk of: • Chromosomal disorders, including Down syndrome and trisomy 18 • Other problems, such as heart defects It also can reveal multiple births. Based on test results, your doctor may suggest other tests to diagnose a disorder.	This test involves both a blood test and an ultrasound exam called nuchal translucency (NOO-kuhl trans-LOO-sent-see) screening. The blood test measures the levels of certain substances in the mother's blood. The ultrasound exam measures the thickness at the back of the baby's neck. This information, combined with the mother's age, help doctors determine risk to the fetus.
Glucose challenge screening	A screening test done at 26 to 28 weeks to determine the mother's risk of gestational diabetes. Based on test results, your doctor may suggest a glucose tolerance test.	First, you consume a special sugary drink from your doctor. A blood sample is taken one hour later to look for high blood sugar levels.
Glucose tolerance test	This test is done at 26 to 28 weeks to diagnose gestational diabetes.	Your doctor will tell you what to eat a few days before the test. Then, you cannot eat or drink anything but sips of water for 14 hours before the test. Your blood is drawn to test your "fasting blood glucose level." Then, you will consume a sugary drink. Your blood will be tested every hour for three hours to see how well your body processes sugar.
Group B streptococcus (STREP-tuh-KOK-uhss) infection	This test is done at 36 to 37 weeks to look for bacteria that can cause pneumonia or serious infection in newborn.	A swab is used to take cells from your vagina and rectum to be tested.
Maternal serum screen (also called quad screen, triple test, triple screen, multiple marker screen, or AFP)	A screening test done at 15 to 20 weeks to detect higher risk of: • Chromosomal disorders, including Down syndrome and trisomy 18 • Neural tube defects, such as spina bifida Based on test results, your doctor may suggest other tests to diagnose a disorder.	Blood is drawn to measure the levels of certain substances in the mother's blood.
Nonstress test (NST)	This test is performed after 28 weeks to monitor your baby's health. It can show signs of fetal distress, such as your baby not getting enough oxygen.	A belt is placed around the mother's belly to measure the baby's heart rate in response to its own movements.
Ultrasound exam	An ultrasound exam can be performed at any point during the pregnancy. Ultrasound exams are not routine. But it is not uncommon for women to have a standard ultrasound exam between 18 and 20 weeks to looks for signs of problems with the baby's organs and body systems and confirm the age of the fetus and proper growth. It also might be able to tell the sex of your baby. Ultrasound exam is also used as part of the first trimester screen and biophysical profile (BPP). Based on exam results, your doctor may suggest other tests or other types of ultrasound to help detect a problem.	Ultrasound uses sound waves to create a "picture" of your baby on a monitor. With a standard ultrasound, a gel is spread on your abdomen. A special tool is moved over your abdomen, which allows your doctor and you to view the baby on a monitor.
Urine test	A urine sample can look for signs of health problems, such as: • Urinary tract infection • Diabetes • Preeclampsia If your doctor suspects a problem, the sample might be sent to a lab for more in-depth testing.	You will collect a small sample of clean, midstream urine in a sterile plastic cup. Testing strips that look for certain substances in your urine are dipped in the sample. The sample also can be looked at under a microscope.

SOURCE: "Common Prenatal Tests," in "Prenatal Care and Tests," *Pregnancy*, Office on Women's Health, U.S. Department of Health and Human Services, January 27, 2017, https://www.womenshealth.gov/pregnancy/youre-pregnant-now-what/prenatal-care-and-tests?no_redirect=true (accessed January 30, 2017)

Amniocentesis involves taking a sample of the fluid that surrounds the fetus in the uterus for chromosome analysis. An amniocentesis is usually performed at 14 to 20 weeks of gestation, although it can be done as early as

12 weeks. (See Table 6.1.) The physician inserts a hollow needle through the abdominal wall and the wall of the uterus to obtain approximately 0.7 ounces (20 ml) of amniotic fluid. Fetal karyotyping, DNA analysis, and biochemical testing may be performed on the isolated fetal cells. Like CVS, amniocentesis samples and analyzes cells that are derived from the baby to enable parents to learn of chromosomal abnormalities, as well as the sex of the unborn child, about two weeks after the test is performed.

Using samples of genetic material obtained from amniocentesis or CVS, physicians can detect disease in an unborn child. Down syndrome (also known as trisomy 21, because it is caused by an extra copy of chromosome 21) is the genetic disease most often identified using this technique. Down syndrome is rarely inherited; most cases result from an error in the formation of the ovum (egg) or sperm, leading to the inclusion of an extra chromosome 21 at conception. Like prenatal diagnosis for most inherited genetic diseases, this use of genetic testing is focused on reproductive decision making.

The most invasive prenatal procedure for genetic testing is periumbilical blood sampling. This procedure poses the greatest risk to an unborn child—1 out of 50 miscarriages occurs as a result of it. It is used when a diagnosis must be made quickly. For example, when an expectant mother is exposed to an infectious agent with the potential to produce birth defects, it may be used to examine fetal blood for the presence of infection.

In "Modeled Fetal Risk of Genetic Diseases Identified by Expanded Carrier Screening" (*Journal of the American Medical Association*, vol. 316, no. 7, August 16, 2016), Imran S. Haque et al. report that an analysis of nearly 347,000 adults finds that expanded carrier screening for as many as 94 severe conditions may increase the detection of carrier status for a variety of potentially serious genetic conditions compared with current recommendations. The researchers note that current guidelines emphasize diseases common in people of European descent such as cystic fibrosis but do not adequately identify risk for conditions prevalent in other racial and ethnic groups. Haque et al. conclude, "Expanded carrier screening revealed that many non-European racial/ethnic categories have a risk of a profound or severe genetic disease that may not be detected by the guidelines in place at the time of this analysis."

Genetic Testing of Newborns

The most common form of genetic testing is the screening of blood taken from newborn infants for genetic abnormalities. The March of Dimes notes in "Newborn Screening Tests for Your Baby" (February 2016, http://www.marchofdimes.org/baby/newborn-screen ing-tests-for-your-baby.aspx) that although all states require newborn screening, each state determines which tests to offer. The March of Dimes asserts that all babies in all states should be screened for at least 34 health conditions. The Health Resources and Services Administration's Advisory Committee on Heritable Disorders in Newborns and Children lists in "Recommended Uniform Screening Panel" (November 2016, https://www.hrsa.gov/advisorycommittees/mchbadvisory/heritabledis orders/recommendedpanel/index.html) the recommended conditions.

Laboratory Techniques for Genetic Prenatal Testing

Genetic testing is performed on chromosomes, genes, or gene products to determine whether a mutation is causing or may cause a specific condition. Direct testing examines the DNA or RNA that makes up a gene. Linkage testing looks for disease-causing gene markers in family members from at least two generations. Biochemical testing assays certain enzymes or proteins, which are the products of genes. Cytogenetic testing examines the chromosomes.

Generally, a blood sample or buccal smear (cells from the mouth) is used for genetic tests. Other tissues used include skin cells from a biopsy, fetal cells, or stored tissue samples. (Table 6.1 lists the tissue samples used in each procedure.) Testing requires highly trained, certified technicians and laboratories because the procedures are complex and varied, the technology is evolving, and hereditary conditions are often rare. In the United States laboratories performing clinical genetic tests must be approved under the Clinical Laboratory Improvement Amendments (CLIA), passed by Congress in 1988 to establish the standards with which all laboratories that test human specimens must comply to receive certification. CLIA standards determine the qualifications of laboratory personnel, categorize the complexity of various tests, and oversee quality improvement and assurance.

The polymerase chain reaction (PCR) technique permits rapid cloning and DNA analysis and allows selective amplification of specific DNA sequences. PCR can be performed in hours and is a sensitive test that may be used to screen for altered genes, but it is limited by the size and length of the DNA sequences that can be cloned. Figure 6.3 explains how PCR is performed.

Fluorescent in situ hybridization (FISH), in which a fluorescent label is attached to a DNA probe that will bind to the complementary DNA strands, provides a unique opportunity to view specific genetic codes. FISH may be used on cells and fluid obtained by CVS and amniocentesis as well as on maternal blood. With this technique, a single strand of DNA is used to create a probe that attaches at the specific gene location. (See Figure 6.4.) To separate the double-stranded DNA, heat

FIGURE 6.3

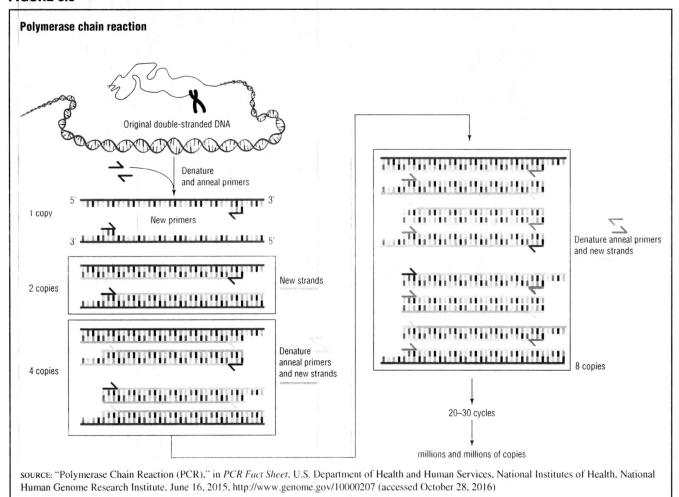

Polymerase chain reaction

SOURCE: "Polymerase Chain Reaction (PCR)," in *PCR Fact Sheet*, U.S. Department of Health and Human Services, National Institutes of Health, National Human Genome Research Institute, June 16, 2015, http://www.genome.gov/10000207 (accessed October 28, 2016)

or chemicals are used to break the chemical bonds of the DNA and obtain a single strand.

There are three kinds of chromosome-specific probes: repetitive probes, painting probes, and locus-specific probes. Repetitive probes produce intense signals by creating tandem repeats of base pairs. Painting probes are collections of specific DNA sequences that may extend along either part or all of an individual chromosome. These probe labels are most useful for identifying complex rearrangements of genetic material in structurally abnormal chromosomes. Probes that can hybridize to a single gene locus are called locus-specific probes. They can be used to identify a gene in a particular region of a chromosome. Locus-specific probes are used to identify a deletion or duplication of genetic material. FISH allows a signal to be visualized that indicates the presence or absence of DNA. The entire process can be completed in less than eight hours using five to seven probes. For this reason, FISH is frequently used as a rapid screen for trisomies and genetic disorders.

Although FISH is the most commonly used and most readily available prenatal diagnostic cytogenetic technique,

it has limited ability to detect translocations, deletions, and inversions. Microdissection FISH (also used for prenatal diagnosis) is another method that is more sensitive to these alterations. Microdissection FISH constructs probes to define specific regions of the human chromosome. Figure 6.5 shows how the microdissection technique is used to identify structurally abnormal chromosomes.

Additional technologies allow the identification of all human chromosomes, thereby expanding the FISH application to the entire genome. Multiplex FISH applies combinations of probes to color components of fluorescent dye during metaphase to visualize each chromosome. This type of FISH procedure uses a combination of fluorochromes to identify each chromosome. Just five fluorophores are needed to decode the entire complement of human chromosomes. Multicolor spectral karyotyping uses computer imaging and Fourier spectroscopy to increase the analysis of genetic markers. (See Figure 6.6.) Although FISH is considered highly reliable, there are limitations to multiplex and multicolor FISH, in that inversion or subtle deletions may be overlooked. Spectral karyotyping techniques can detect cryptic unbalanced translocations that cannot be identified by other techniques.

FIGURE 6.4

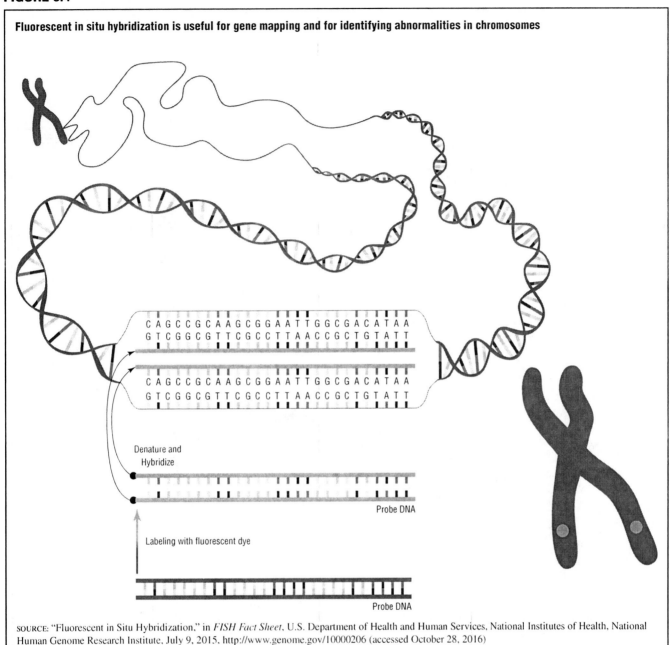

Fluorescent in situ hybridization is useful for gene mapping and for identifying abnormalities in chromosomes

CAGCCGCAAGCGGAATTGGCGACATAA
GTCGGCGTTCGCCTTAACCGCTGTATT

CAGCCGCAAGCGGAATTGGCGACATAA
GTCGGCGTTCGCCTTAACCGCTGTATT

Denature and
Hybridize

Probe DNA

Labeling with fluorescent dye

Probe DNA

SOURCE: "Fluorescent in Situ Hybridization," in *FISH Fact Sheet*, U.S. Department of Health and Human Services, National Institutes of Health, National Human Genome Research Institute, July 9, 2015, http://www.genome.gov/10000206 (accessed October 28, 2016)

GENETIC TESTING FOR SICKLE-CELL ANEMIA. Sickle-cell anemia is an autosomal recessive disease that results when hemoglobin S is inherited from both parents. (When hemoglobin S is inherited from only one parent, the individual is a sickle-cell carrier.) Because normal and sickle hemoglobins differ at only one amino acid in the hemoglobin gene, a test called hemoglobin electrophoresis is used to establish the diagnosis.

Genetic testing for sickle-cell anemia involves restriction fragment length polymorphism (DNA sequence variant) that uses specific enzymes to cut the DNA. (See Figure 6.7.) These enzymes cut the DNA at a specific base sequence on the normal gene but not on a gene in which a mutation is present. As a result of this technique, there are longer fragments of sickle hemoglobin. Another technique known as gel electrophoresis sorts the DNA fragments by size. Autoradiography renders the DNA fragments by generating an image after radioactive probes have labeled the DNA fragments that contain the specific gene sequence. The location of the fragments distinguishes carrier status (heterozygous) from sickle-cell anemia (homozygous), or normal blood.

Advanced Techniques Detect Fetal Gene Mutations

Ronald Wapner et al. report in "Chromosomal Microarray versus Karyotyping for Prenatal Diagnosis" (*New England Journal of Medicine*, vol. 367, no. 23, December 2012) that a large study comparing the use of

FIGURE 6.5

Microdissection

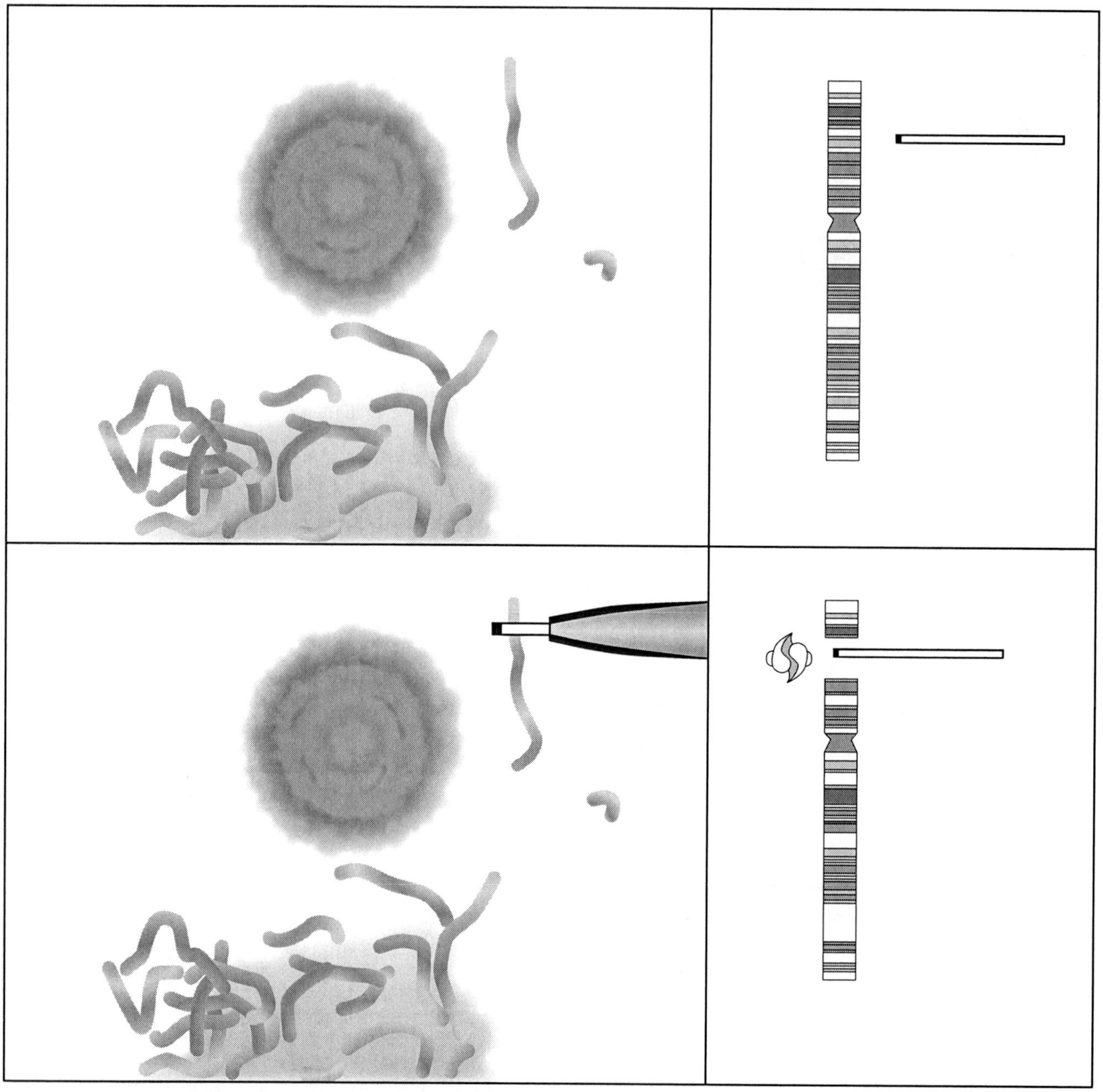

SOURCE: "How Does Microdissection Work?" in *Chromosome Microdissection Fact Sheet*, U.S. Department of Health and Human Services, National Institutes of Health, National Human Genome Research Institute, July 11, 2013, http://www.genome.gov/10000204 (accessed October 28, 2016)

chromosomal microarray (CMA; this technique determines whether the right amount of genetic material is present at many different locations in the fetus's genome) and karyotyping to test a developing fetus's DNA finds that CMA provides more information about potential disorders than karyotyping, which is the visual examination of the chromosomes. Both tests identify conditions that can be life threatening to a newborn baby or that can signal a possible problem that might be treatable, but

CMA identifies more abnormalities. (Figure 6.8 shows how microarray technology is performed to distinguish normal tissue from a tumor. It is used to look at many genes simultaneously and allows detection of RNA and DNA molecules that are labeled with a dye such as biotin.)

Another type of prenatal genetic test, array comparative genomic hybridization, which can provide a much

FIGURE 6.6

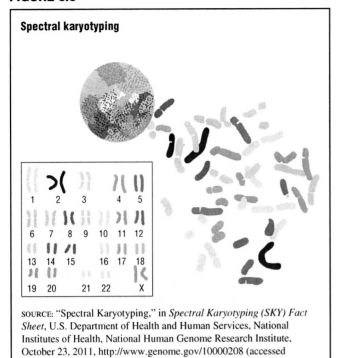

Spectral karyotyping

SOURCE: "Spectral Karyotyping," in *Spectral Karyotyping (SKY) Fact Sheet*, U.S. Department of Health and Human Services, National Institutes of Health, National Human Genome Research Institute, October 23, 2011, http://www.genome.gov/10000208 (accessed October 28, 2016)

higher resolution than conventional karyotyping, is currently the genetic test of choice for investigating causes of intellectual disability and/or congenital anomalies in infants. Its use in prenatal diagnosis is described by Isabel Filges et al. in "Array Comparative Genomic Hybridization in Prenatal Diagnosis of First Trimester Pregnancies at High Risk for Chromosomal Anomalies" (*Molecular Cytogenetics*, vol. 5, no. 1, September 17, 2012). The researchers find that its detection rate is no better than conventional chromosome analysis. They also caution that its 2% detection rate of copy number variants of unknown significance may hinder its routine use in prenatal diagnosis.

Perhaps the most promising development in prenatal diagnosis is analysis of fetal DNA in maternal plasma, a technique that has been used to successfully identify trisomies and paternally inherited mutations. In "Noninvasive Whole-Genome Sequencing of a Human Fetus" (*Science Translational Medicine*, vol. 4, no. 137, June 2012), Jacob O. Kitzman et al. report that by combining genome sequencing of two parents and genome-wide maternal haplotyping (an examination of all the genes on a chromosome that are inherited from the mother), they are able to determine the genome sequence of a fetus at 18.5 weeks' gestation without the use of any invasive testing such as CVS or amniocentesis.

However, Jean Gekas et al. caution in "Non-invasive Prenatal Testing for Fetal Chromosome Abnormalities: Review of Clinical and Ethical Issues" (*Applied Clinical*

FIGURE 6.7

Cutting deoxyribonucleic acid with restrictive enzymes

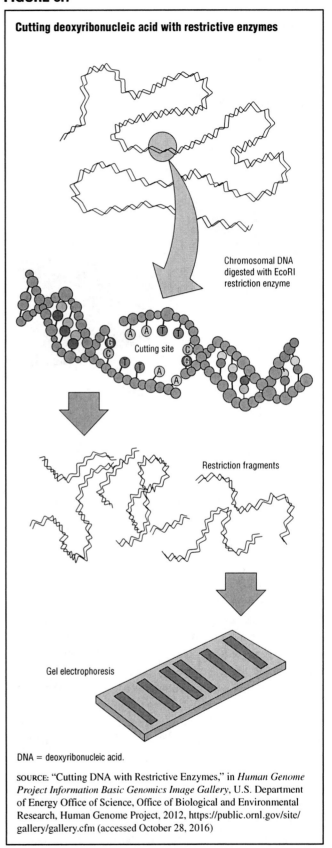

Chromosomal DNA digested with EcoRI restriction enzyme

Cutting site

Restriction fragments

Gel electrophoresis

DNA = deoxyribonucleic acid.

SOURCE: "Cutting DNA with Restrictive Enzymes," in *Human Genome Project Information Basic Genomics Image Gallery*, U.S. Department of Energy Office of Science, Office of Biological and Environmental Research, Human Genome Project, 2012, https://public.ornl.gov/site/gallery/gallery.cfm (accessed October 28, 2016)

Genetics, no. 9, February 4, 2016) that although genomics-based noninvasive prenatal screening using cell-free DNA (cfDNA; fetal DNA that circulates in the

FIGURE 6.8

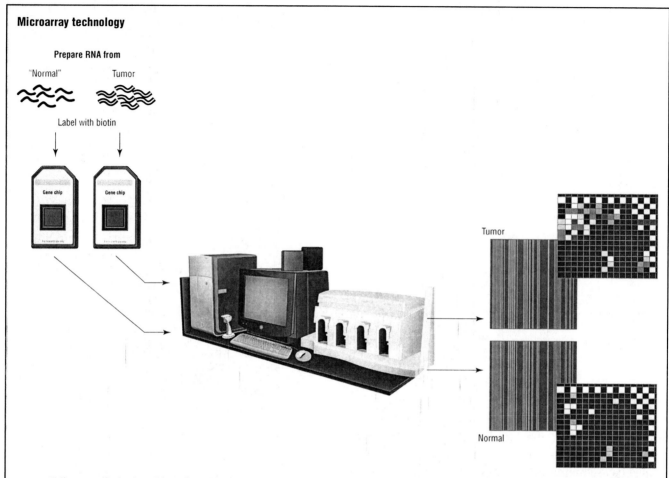

Microarray technology

SOURCE: "Microarray Technology," in *Talking Glossary of Genetic Terms*, U.S. Department of Health and Human Services, National Institutes of Health, National Human Genome Research Institute, Division of Intramural Research, undated, http://www.genome.gov/Glossary/index.cfm?id=125 (accessed October 28, 2016)

maternal blood stream) offers earlier results and reduces the number of invasive procedures, important clinical and ethical questions about its use remain unanswered. For example, one study found that since its introduction fewer patients were referred for genetic counseling, which might result in misdiagnosis of disorders that are not detectable by this form of screening. The researchers also observe that asking women to pay for this screening may limit its access only to those who can afford to pay for it. Furthermore, Gekas et al. note that because cfDNA is a screening test, positive tests must be confirmed by invasive tests such as amniocentesis or CVS.

GENETIC DIAGNOSIS IN CHILDREN AND ADULTS

Genetic testing can also be performed postnatally (after birth) to determine which children and adults are at increased risk of developing specific diseases. By 2017 scientists could perform predictive genetic testing to identify which individuals were at risk for cystic fibrosis, Tay-Sachs disease, Huntington's disease, amyotrophic lateral sclerosis (a degenerative neurological condition commonly known as Lou Gehrig's disease), and several types of cancers, including some cases of breast, colon, and ovarian cancer.

As of 2017, more than 2,500 genetic tests were available. Most of these tests are used only for research purposes because public health professionals do not consider it practical to screen for conditions that are extremely rare, have only minor health consequences, or have no effective treatment. The most frequently performed genetic tests are those considered to be the most useful in terms of their potential to screen populations for diseases that occur relatively often, have serious medical consequences (including death) if untreated, and have effective treatment. Table 6.2 lists those tests for which there is substantial evidence to support testing.

In 2012 the NIH launched the Genetic Testing Registry (https://www.ncbi.nlm.nih.gov/gtr/), a web-based database that provides information about genetic tests for physicians, patients, and researchers. Operated by the NIH's National Center for Biotechnology Information,

TABLE 6.2

Genetic tests with strong evidence to support their use, 2016

Disease/disorder	Test to be assessed	Intended use
Chronic gout	G6PD	Pharmacogenomic for pegloticase
Chorea associated with Huntington's disease	CYP2D6	Pharmacogenomic for tetrabenazine
Tourettes disorder	CYP2D6	Pharmacogenomic for pimozide
Epilepsy, trigeminal neuralgia	HLA-B*1502	Pharmacogenomic for carbamazepine, pretreatment screening for those with ancestry in populations genetically at-risk for certain serious dermatologic reactions
Cystic fibrosis	CFTR (G551D)	Pharmacogenomic for ivacaftor
HIV	HLA-B*5701	Pharmacogenomic for abacavir
Unresectable or metastatic melanoma	BRAF V600E	Pharmacogenomic for vemurafenib
Unresectable or metastatic melanoma	BRAF V600E	Pharmacogenomic for dabrafenib
Unresectable or metastatic melanoma	BRAF V600E/K	Pharmacogenomic for trametinib
Non-small cell lung cancer	ALK gene rearrangement	Pharmacogenomic for crizotinib
Locally advanced or metastatic non-small-cell lung cancer	EGFR (exon 19 deletions and exon 21 (L858R) substitution mutations)	Pharmacogenomic for erlotinib
Metastatic non-small-cell lung cancer	EGFR (exon 19 deletions and exon 21 (L858R) substitution mutations)	Pharmacogenomic for afatinib
Leukemia, lymphoma, solid tumor malignancies	G6PD	Pharmacogenomic for rasburicase
Non-Hodgkin's lymphoma	CD20	Pharmacogenomic for tositumomab
Persistent or recurrent cutaneous T-cell lymphoma	CD25	Pharmacogenomic for denileukin diftitox
Myelodysplastic/myeloproliferative diseases	PDGFRB	Pharmacogenomic for imatinib
Gastrointestinal stromal tumors	c-Kit protein (CD 117)	Pharmacogenomic for imatinib
Gastric or gastroesophageal junction adenocarcinoma	HER2	Pharmacogenomic for trastuzumab
Invasive colorectal cancer	Carcinoembryonic antigen-related cell adhesion molecule 5 (CEACAM5 or CEA)	Prognostic
Metastatic colorectal cancer	KRAS	Pharmacogenomic for cetuximab, panitumumab
Lynch syndrome	Various strategies	Diagnostic, screening
Lynch syndrome	Various strategies	Screening, cascade testing of relatives
Invasive breast cancer, breast cancer recurrences	ER and PgR	Pharmacogenomic
Early invasive breast cancer (ER+)	ER	Pharmacogenomic for anastrozole or letrozole
Early breast cancer (ER+)	ER	Pharmacogenomic for exemestane
Metastatic breast cancer	ER	Pharmacogenomic for fulvestrant
Invasive breast cancer	HER2	Pharmacogenomic
Advanced or metastatic breast cancer	HER2 testing	Pharmacogenomic - lapatinib (in combination with capecitabine or letrozole)
Advanced HR+ HER2− breast cancer	Human epidermal growth factor 2 (HER2) testing	Pharmacogenomic; inform use of everolimus
Metastatic breast cancer	Human epidermal growth factor receptor (HER2) testing	Pharmacogenomic predictor of adotrastuzumab emtansine response
Invasive breast cancer	Human epidermal growth factor receptor 2 (HER2) testing	Pharmacogenomic predictor of pertuzumab response
Invasive breast cancer	Human epidermal growth factor receptor 2 (HER2) testing	Pharmacogenomic predictor of trastuzumab response
Hereditary breast and ovarian cancer	Family history of known breast/ovarian cancer with deleterious BRCA mutation	Risk prediction; referral to counseling for BRCA genetic testing
BRCA-related cancer; hereditary breast and ovarian cancer	Family history	Risk prediction for referral for BRCA genetic counseling

SOURCE: Adapted from "Database Filtered by Tier," in *Public Health Genomics Knowledge Base (v1.2)*, Centers for Disease Control and Prevention, Office of Public Health Genomics, Center for Surveillance, Epidemiology, and Laboratory Services (CSELS), September 9, 2016, https://phgkb.cdc.gov/GAPPKB/phgHome.do?action=home (accessed October 28, 2016)

the database may be searched by condition, test, gene, or lab and includes information such as whether the test sequences the entire gene for mutations or looks for specific errors.

A positive test result (the presence of a mutation) from predictive genetic testing does not guarantee that the individual will develop the disease; it simply identifies the individual as being genetically susceptible and at an increased risk for developing the disease. For example, a woman who tests positive for the BRCA1 gene has an 80% chance of developing breast cancer before the age of 65 years. It is also important to note that, like other types of diagnostic medical testing, genetic tests are not 100% predictive—the results rely on the quality of laboratory procedures and the accuracy of interpretations.

Furthermore, because tests vary in their sensitivity and specificity, there is always the possibility of false-positive and false-negative test results.

Researchers hope that positive test results will encourage people who are at higher than average risk of developing a disease to be vigilant about disease prevention and screening for early detection, when many diseases are most successfully treated. There is an expectation that genetic information will increasingly be used in routine population screening to determine individual susceptibility to common disorders such as heart disease, diabetes, and cancer. This type of screening will identify groups at risk so that primary prevention efforts such as diet and exercise or secondary prevention efforts such as early detection can be initiated.

Another example of genetic testing that may prompt prevention or intervention efforts is screening smokers and others at risk for developing esophageal cancer and lung cancer for the gene RGS17, which may lead to earlier diagnosis and identify people who are at risk before the disease progresses. It may also help researchers develop personalized smoking cessation programs. Furthermore, identifying this gene in patients who have been diagnosed with lung cancer may help researchers develop more effective treatments.

Diagnostic Genetic Testing

Most genetic testing is performed on people who are asymptomatic. The objective of screening is to determine if people are carriers of a genetic disease or to identify their susceptibility or risk of developing a specific disease or disorder. There is, however, some testing performed on people with symptoms of a disease to establish the diagnosis and calculate the risk of developing the disease for other family members. This type of testing is known as diagnostic genetic testing or symptomatic genetic testing. It may also assist in directing treatment for symptomatic patients in whom a mutation in a single gene (or in a gene pair) accounts for a disorder. Cystic fibrosis and myotonic dystrophy are examples of disorders that may be confirmed or ruled out by diagnostic genetic testing and other methods (such as the sweat test for cystic fibrosis or a neurological evaluation for myotonic dystrophy).

One consideration in diagnostic genetic testing is the appropriate frequency of testing in view of rapidly expanding genetic knowledge and identification of genes that are linked to disease. Physicians frequently see symptomatic patients for whom there is neither a definitive diagnosis nor a genetic test. The as-yet-unanswered question is: Should such people be recalled for genetic testing each time a new test becomes available? Although clinics and physicians who perform genetic testing counsel patients to maintain regular contact so they may learn about the availability of new tests, there is no uniform guideline or recommendation about the frequency of testing.

Genetic Tests Help Patients Choose Optimal Treatment

In 2005 the U.S. Food and Drug Administration (FDA) approved marketing of a genetic test to help physicians make personalized drug treatment decisions for some patients. The Invader UGT1A1 Molecular Assay detects variations in a gene that affects how certain drugs are broken down and cleared by the body. Using this information, physicians can determine the optimal drug dosage for each patient and minimize the harsh and potentially life-threatening side effects of drug treatment. This test is one of many pharmacogenetic or pharmacogenomic tests that are used to customize treatment decisions. Examples of these tests are the Roche AmpliChip,

which is used to personalize the dosage of antidepressants, antipsychotics, beta-blockers, and some chemotherapy drugs, and the TRUGENE HIV-1 Genotyping Kit, which is used to detect variations in the genome of the human immunodeficiency virus (HIV) that make the virus resistant to some antiretroviral drugs.

A test that helps determine which patients are more sensitive to warfarin, a commonly used anticoagulant that prevents dangerous blood clots, received FDA approval in 2007. The Nanosphere Verigene Warfarin Metabolism Nucleic Acid Test detects variants of two genes: CYP2C9, which helps the body metabolize warfarin, and VKORC1, which helps regulate the ability of the drug to prevent blood clots. People who have polymorphisms of these genes are generally more sensitive to warfarin and may require lower doses of the drug to achieve the same therapeutic benefit. Knowing which patients will metabolize the drug differently can help prevent the risk of a dose that will produce bleeding instead of the desired anticoagulation.

In 2007 the FDA approved Mammaprint, a genetic test that aims to predict whether a breast cancer patient will suffer a relapse. The test can help distinguish patients who would benefit from chemotherapy from those who would benefit from surgery that removes the tumor. Two other tests intended to help guide cancer treatment choices were approved in 2008. The SPOT-Light HER2 CISH measures the number of copies of the HER2 gene in breast tumor tissue to identify patients who may benefit from treatment with the anticancer drug Herceptin, which targets HER2 protein production. The TOP2A FISH pharmDx determines whether patients have the normal type of or variants of the TOP2A gene. It helps assess the risk of recurrence and long-term survival for patients with relatively high-risk breast cancers.

In 2012 Millennium Laboratories debuted a saliva-based test that detects genetic variations in enzymes associated with the metabolism of medications commonly prescribed to patients suffering from debilitating chronic pain. The test helps health care practitioners identify patients who may benefit from specific pain relief drugs or doses of certain drugs such as methadone, benzodiazepines, tricyclic antidepressants, selective serotonin reuptake inhibitors, and serotonin norepinephrine reuptake inhibitors.

Other research finds that patients with a specific genetic variant—the nucleotide polymorphism rs2294008 within the PSCA gene—develop bladder cancer tumors that express a prostate stem cell antigen (PSCA) that is also expressed in many pancreatic and prostate tumors. Indu Kohaar et al. report in "Genetic Variant as a Selection Marker for Anti-prostate Stem Cell Antigen Immunotherapy of Bladder Cancer" (*Journal of the National Cancer Institute*, vol. 105, no. 1, January 2,

2013) the results of one of the first studies to show direct implications of a genetic variant identified through genome-wide association studies for common cancers. The researchers observe that PSCA antibody-based immunotherapy is currently being used in clinical trials for prostate and pancreatic cancers and assert that it may also be a good drug target for bladder cancer.

In "Direct PCR: A New Pharmacogenetic Approach for the Inexpensive Testing of HLA-B*57:01" (*Pharmacogenomics Journal*, September 9, 2014), Raffaella Cascella et al. describe direct PCR, a novel way to screen for HLA-B*57:01, which is associated with an increased risk for a serious reaction in HIV-positive patients following administration of Ziagen (an antiretroviral drug used to treat HIV infection). Because direct PCR combines DNA extraction and amplification into a single step that is performed directly on the sample without the need for extraction and sequencing, it is less expensive than other tests that may be used to assess a patient's response to a specific treatment.

The FDA notes in "Table of Pharmacogenomic Biomarkers in Drug Labeling" (https://www.fda.gov/Drugs/ScienceResearch/ResearchAreas/Pharmacogenetics/ucm083378.htm) that as of November 2016 it had identified more than 200 pharmacogenomic biomarkers. These tests may be performed to identify patients who will and will not respond to a particular drug, to prevent drug toxicity, and to guide drug dosing to optimize efficacy (the ability of an intervention to produce the intended diagnostic or therapeutic effect in optimal circumstances) and safety. For example, people with the CYP2D6 variant are called "ultra-rapid metabolizers of codeine" and may experience overdose symptoms from the normally prescribed dosages of this pain medication.

POINT-OF-CARE GENETIC TESTS OFFER RAPID RESULTS. According to Peter B. Luppa et al., in "Clinically Relevant Analytical Techniques, Organizational Concepts for Application and Future Perspectives of Point-of-Care Testing" (*Biotechnology Advances*, vol. 34, no. 3, May–June 2016), applications of point-of-care testing have been rapidly increasing during the last two decades. Point-of-care testing is now performed in various settings, from self-testing to outpatient clinics to intensive care units. Luppa et al. assert that point-of-care testing has the potential to improve patient outcomes and predict that tests will be conducted using smartphone-based devices.

Population Screening

Population screening for heritable diseases is one potentially lifesaving application of molecular genetics technology. Prenatal screening has demonstrated benefits and gained widespread use; however, genetic screening has not yet become part of routine medical practice for adults. Geneticists have identified at least eight genes that might be candidates for use as population screening tests in adults in the United States. The genes are:

- HFE, for hereditary hemochromatosis (a disorder in which the body absorbs too much iron from food; rather than the excess iron being excreted, it is stored throughout the body, and the iron deposits damage the pancreas, liver, skin, and other tissues)

- Apolipoprotein E-4, which is linked to Alzheimer's disease

- CYP2D6, which is linked to ankylosing spondylitis (arthritis of the spine)

- BRCA1 and BRCA2, the genes for hereditary breast and ovarian cancer

- APC and MUTYH, the genes that cause familial adenomatous polyposis, which is associated with pre-cancerous growths in the colon

- Factor V Leiden, a genetic mutation in the factor V gene that causes the most common hereditary blood-clotting disorder in the United States

As of February 2017, screening for variants in these genes had not entered into routine medical practice because there was considerable controversy about the predictive value of testing for these genes and how to monitor and care for people who test positive for them. For example, Mary-Claire King, Ephrat Levy-Lahad, and Amnon Lahad call in "Population-Based Screening for BRCA1 and BRCA2" (*Journal of the American Medical Association*, vol. 312, no. 11, September 17, 2014) for population-based screening of women over the age of 30 years for BRCA1 and BRCA2 as a routine part of clinical practice. This recommendation prompted heated discussion. The National Cancer Institute (https://www.cancer.gov/about-cancer/causes-prevention/genetics/brca-fact-sheet) responded in April 2015 by advising this screening "only when the person's individual or family history suggests the possible presence of a harmful mutation in BRCA1 or BRCA2."

Genetic Susceptibility

Susceptibility testing, also known as predictive testing, determines the likelihood that a healthy person with a family history of a disorder will develop the disease. Testing positive for a specific genetic mutation indicates an increased susceptibility to the disorder but does not establish a diagnosis. For example, a woman may choose to undergo testing to find out whether she has genetic mutations that would indicate a likelihood of developing hereditary cancer of the breast or ovary. If she tests positive for a genetic mutation, she may then decide to undergo some form of preventive treatment. Preventive measures may include increased surveillance, such as more frequent mammography; chemoprevention (prescription drug therapy that is intended to reduce risk); or surgical prophylaxis, such as mastectomy and/or oophorectomy (surgical removal of the breasts and ovaries, respectively).

Testing Children for Adult-Onset Disorders

In 2000 the American Academy of Pediatrics Committee on Genetics recommended genetic testing for people under the age of 18 years only when testing would offer immediate medical benefits or when there is a benefit to another family member and there is no anticipated harm to the person being tested. The committee considered genetic counseling before and after testing as essential components of the process.

That same year the American Academy of Pediatrics Committee on Bioethics and Newborn Screening Task Force recommended the inclusion of tests in the newborn-screening battery based on scientific evidence. The committee advocated informed consent for newborn screening. (As of February 2017, most states did not require informed consent.) However, it did not endorse carrier screening in people under 18 years of age, except in the case of a pregnant teenager. It also recommended against predictive testing for adult-onset disorders in people under the age of 18 years.

The American College of Medical Genetics, the American Society of Human Genetics (ASHG), and the World Health Organization have also weighed in on genetic testing of asymptomatic children, asserting that decisions about testing should emphasize the child's well-being. One issue involves the value of testing asymptomatic children for genetic mutations that are associated with adult-onset conditions such as Huntington's disease. Because no treatment can begin until the onset of the disease, and at present there is no treatment to alter the course of the disease, it may be ill advised to test for it. Another concern is testing for carrier status of autosomal recessive or X-linked conditions such as cystic fibrosis or Duchenne muscular dystrophy. Experts caution that children might confuse carrier status with actually having the condition, which in turn might provoke needless anxiety.

There are, however, circumstances in which genetic testing of children may be appropriate and useful. Examples are children with symptoms of suspected hereditary disorders or those at risk for cancers in which inheritance plays a primary role.

ETHICAL CONSIDERATIONS: CHOICES AND CHALLENGES

Rapid advances in genetic research have challenged scientists, health care professionals, ethicists, government regulators, legislators, and consumers to stay abreast of new developments. Understanding the scientific advances and their implications is critical for everyone involved in making informed decisions about the ways in which genetic research and information will affect the lives of current and future generations. U.S. citizens, scientists, ethicists, legislators, and regulators share this responsibility. According to Human Genome Project Information, in "Ethical, Legal, and Social Issues" (September 16, 2014, http://web.ornl.gov/sci/techresources/Human_Genome/elsi/index.shtml), the pivotal importance of these societal decisions was underscored by the allocation of 3% to 5% of the Human Genome Project budget for the study of ethical, legal, and social issues related to genetic research. As of February 2017, consideration of these issues had not produced simple or universally applicable answers to the many questions posed by the increasing availability of genetic information. Ongoing public discussion and debate is intended to inform, educate, and help people in every walk of life make personal decisions about their health and participate in decisions that concern others.

As researchers learn more about the genes that are responsible for a variety of illnesses, they can design more tests with increased accuracy and reliability to predict whether an individual is at risk of developing specific diseases. The ethical issues involved in genetic testing have turned out to be far more complicated than originally anticipated. Initially, physicians and researchers believed that a test to determine in advance who would develop or escape a disease would be welcomed by at risk families, who would be able to plan more realistically about having children and going about their daily lives. Nevertheless, many people with family histories of a genetic disease have decided that not knowing is better than anticipating a grim future and an agonizing, slow death. They prefer to live with the hope that they will not develop the disease rather than having the certain knowledge that they will.

The discovery of genetic links and the development of tests to predict the likelihood or certainty of developing a disease raise ethical questions for people who carry a defective gene. Should women who are carriers of Huntington's disease or cystic fibrosis have children? Should a fetus with a defective gene be carried to term or aborted?

Historically, there have also been concerns about privacy and the confidentiality of medical records and the results of genetic testing and possible stigmatization. Some people were reluctant to be tested because they feared losing their health, life, and disability insurances, or even their job if they were found to be at risk for a disease. Genetic tests were often costly, and some insurers agreed to reimburse for testing only if they were informed of the results. The insurance companies felt they could not risk selling policies to people they knew will become disabled or die prematurely.

The fear of discrimination by insurance companies or employers who learn the results of genetic testing was often justified. Insurance carriers could charge people higher rates or disqualify them based on test results, and employers could choose not to hire or to deny affected individuals advancement or promotions. Until 2008, such fears forced people to choose between having a genetic test that could provide lifesaving information and avoiding a test to retain health insurance coverage or to save a job.

Legislation Bans Genetic Discrimination

Until 2008, when the Genetic Information Nondiscrimination Act (GINA) became law, Americans' fears of recrimination or undesired consequences of genetic testing were justified. Dissemination of genetic information was protected by an uneven array of state and federal regulations. Heralded as the first major civil rights bill of the new century, GINA prevents health insurers from denying coverage, adjusting premiums on the basis of genetic test results, or requesting that an individual undergo genetic testing. It prohibits employers from using genetic information to make hiring, firing, or promotion decisions. The law also sharply restricts an employer's right to request, require, or purchase workers' genetic information.

In 2010 the Genetics and Public Policy Center at Johns Hopkins University, the National Coalition for Health Professional Education in Genetics, and the Genetic Alliance launched the interactive website GINA-help.org to help consumers, clinicians, and educators understand the law that protects against the misuse of genetic information in health insurance and employment.

Unanticipated Information and Results of Genetic Testing

Sometimes genetic testing yields unanticipated information, such as paternity, or other unexpected results, such as the presence of a disorder that was not sought directly. Many health professionals consider disclosure of such information, particularly when it does not influence health or medical treatment decision making, as counterproductive and even potentially harmful.

The ASHG explains in "Mission & Vision" (2017 http://www.ashg.org/pages/about_mission.shtml) that it is the primary professional organization for human geneticists in the United States. Its nearly 8,000 members include researchers, academicians, clinicians, laboratory practice professionals, genetic counselors, nurses, and others involved in or with a special interest in human genetics. The ASHG recommends that family members not be informed of misattributed paternity unless the test

requested was determination of paternity. The ASHG encourages all health professionals to educate, counsel, and obtain informed consents that include cautions regarding unexpected findings before performing genetic testing.

Advertising Genetic Testing Directly to Consumers

Pharmaceutical drugs have been advertised directly to consumers for more than two decades, but direct-to-consumer (DTC) advertising of genetic tests is a relatively recent phenomenon. The DTC genetic tests are marketed via print and electronic advertising and sold online. According to the National Library of Medicine's Genetics Home Reference, in "What Is Direct-to-Consumer Genetic Testing?" (February 21, 2017, https://ghr.nlm.nih.gov/primer/testing/directtoconsumer), some home genetic test companies obtain a DNA sample by having their clients swab the inside of their cheek, whereas others require clients to visit a laboratory to have a blood sample drawn. Test results are generally available by telephone or online, and in some instances a genetic counselor or other health educator is available to discuss the test results.

Supporters of home genetic tests assert that the tests enhance public awareness of genetic diseases, enable consumers to assume an active role in managing their health, and offer a convenient way to explore genealogy and family history. By contrast, many health professionals and government agencies are concerned about how these tests are marketed and used.

The U.S. Government Accountability Office (GAO) explains in *Direct-to-Consumer Genetic Tests: Misleading Test Results Are Further Complicated by Deceptive Marketing and Other Questionable Practices* (July 22, 2010, http://www.gao.gov/new.items/d10847t.pdf) that in 2006 it looked at the marketing, advertising, and efficacy of genetic tests by purchasing 10 tests from four companies. The tests ranged from $299 to $999. The GAO chose five subjects and sent two DNA samples from each subject to each company: one sample included factual information about the subject and one included incorrect information, such as the wrong age, race, or ethnicity. The GAO compared the subjects' results and sought health advice by making undercover calls to the companies. Genetics experts were asked to determine whether the tests provided medically useful information.

According to the GAO:

- The subjects' test results were "misleading and of little or no practical use."

- The disease predictions that were made conflicted with the subjects' actual medical conditions.

- Three of the companies failed to provide the expert follow-up they had promised.

To evaluate advertising practices, the GAO contacted 15 companies, including the four that were tested, to inquire about supplement sales, test reliability, and privacy policies. The GAO notes 10 glaring instances of deceptive marketing, including:

- Four companies claimed that a consumer's DNA could be used to create a personalized supplement to cure diseases.

- Two companies asserted that their supplements could "repair damaged DNA" or "cure disease," even though there is no scientific evidence for such claims.

- Two companies asserted that their DNA analyses could predict the sports in which children would excel, even though there is no scientific basis for such claims.

- One company advised a subject that her test result meant she was "in the high risk of pretty much getting" breast cancer, a statement that alarmed health professionals because it suggests the test is diagnostic.

The FDA Moves to Regulate Genetic Testing

Figure 6.9 lists three federal agencies that are involved in the regulation of genetic tests: the Centers for Medicare and Medicaid Services, the FDA, and the Federal Trade Commission. It also details how this federal oversight and regulation occurs. The chart was developed by the Johns Hopkins University's Genetics and Public Policy Center with support from the Pew Charitable Trusts.

In view of the explosive growth in the number of DTC genetic tests and the frequency of questionable claims that companies make about them, such as determining cancer risk or response to a particular drug, Jeffrey Shuren (July 22, 2010, https://www.fda.gov/News Events/Testimony/ucm219925.htm), the director of the FDA Center for Devices and Radiological Health, went before the U.S. House of Representatives' Subcommittee on Oversight and Investigations to discuss the FDA's activities regarding DTC genetic tests. He stated that in 2010 the FDA had notified 20 diagnostic test manufacturing firms that offered DTC genetic tests that their tests should meet the "statutory definition of a medical device" on the basis of their claims about the test results and thus may require FDA approval.

Shuren explained that although the FDA has the authority to regulate genetic tests because they are considered to be medical devices, in the past the agency had not exercised enforcement because there were not many tests offered and testing was marketed for the purposes of genealogy research as opposed to assessing disease risk. He noted that "none of the genetic tests now offered directly to consumers has undergone premarket review by FDA to ensure that the test results being provided to patients are accurate, reliable, and clinically meaningful."

In October 2014 the FDA issued *Draft Guidance for Industry, Food and Drug Administration Staff, and Clinical Laboratories: Framework for Regulatory Oversight of Laboratory Developed Tests* (https://www.fda.gov/down loads/MedicalDevices/DeviceRegulationandGuidance/

FIGURE 6.9

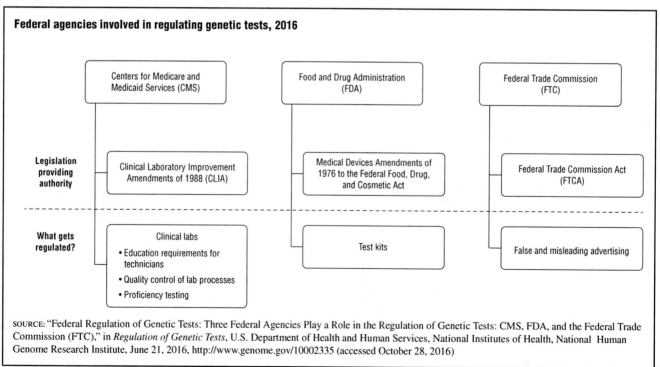

Federal agencies involved in regulating genetic tests, 2016

SOURCE: "Federal Regulation of Genetic Tests: Three Federal Agencies Play a Role in the Regulation of Genetic Tests: CMS, FDA, and the Federal Trade Commission (FTC)," in *Regulation of Genetic Tests*, U.S. Department of Health and Human Services, National Institutes of Health, National Human Genome Research Institute, June 21, 2016, http://www.genome.gov/10002335 (accessed October 28, 2016)

GuidanceDocuments/UCM416685.pdf) and *FDA Notification and Medical Device Reporting for Laboratory Developed Tests* (https://www.fda.gov/downloads/MedicalDevices/DeviceRegulationandGuidance/GuidanceDocuments/UCM416684.pdf). The first draft guidance report explained that the FDA intends to exercise its authority over laboratory developed tests (LDTs) as medical devices by classifying them by risk and category. Furthermore, the agency indicates that it will evaluate the potential for severe consequences if consumers decide, based on the results from LDTs, to initiate unnecessary treatments or to delay or forgo treatment altogether for a condition. The second draft guidance report detailed the process that laboratories offering LDTs must use to notify the FDA of all LDTs being offered. Also, it indicated that all LDTs must be registered. In early 2015 the FDA hosted public meetings to discuss the test components and labeling; clinical validity and intended use; categories for continued enforcement discretion; notification and adverse event reporting; classification and prioritization; and quality system regulation.

In "Laboratory Developed Tests Emerging in FDA Regulation" (*National Law Review*, January 19, 2016), Lisamarie A. Collins et al. observe that besides registering and meeting reporting requirements for low-risk LDTs, moderate- and high-risk LDTs will be subject to stricter regulatory oversight than low-risk LDTs. Moderate-risk LDTs must complete premarket review within five to nine years after the FDA issues its final guidance, while high-risk LDTs will be subject to premarket review just one year after the guidance is final. The most significant change is that providers of moderate- and high-risk LDTs, who marketed their tests without FDA scrutiny, will now be required to obtain FDA approval or clearance.

According to Turna Ray, in "FDA to Finalize LDT Guidance amid Uncertainty on Number of Genetic Tests Impacted" (GenomeWeb.com, February 4, 2016), by 2016 there were about 60,000 genetic test products on the market, of which 7,600 may be considered high risk by the FDA. The requirement for intensified or additional oversight would significantly increase the FDA's workload, and the agency would require additional staff to conduct premarket reviews.

OPPOSITION TO FDA OVERSIGHT OF GENETIC TESTING. The crux of the debate over LDTs is whether these tests pose a risk to users. The FDA contends that many do. For example, in November 2015 it released the report *The Public Health Evidence for FDA Oversight of Laboratory Developed Tests: 20 Case Studies* (https://www.fda.gov/downloads/AboutFDA/ReportsManualsForms/Reports/UCM472777.pdf), which cited 20 cases of LDT use that the agency believed may have caused or could have caused harm. Nonetheless, as the FDA moved to regulate LDTs some industry groups, including the American Association for Clinical Chemists, the Association for Molecular Pathology, and the College of American Pathologists, proposed alternatives that would give the oversight authority to the Centers for Medicare and Medicaid Services under an expanded CLIA program. In "Future of LDT Oversight Still Uncertain: Stakeholders Propose Alternatives to FDA Regulation" (April 1, 2016, https://www.aacc.org/publications/cln/articles/2016/april/future-of-ldt-oversight-still-uncertain), Kimberly Scott notes that the American Association for Clinical Chemists recommends that CLIA retain oversight of clinical laboratories and supports updating the regulations to ensure that they keep pace with developing technologies and safeguard consumers.

THE FDA TAKES A STANCE AGAINST 23ANDME. In November 2013 the FDA ordered 23andMe, a personal genome testing company, to stop selling and marketing its DNA testing service. After obtaining DNA from saliva samples, the company provided ancestry information and advised consumers about their relative risk of developing certain diseases. Andrew Pollack reports in "F.D.A. Orders Genetic Testing Firm to Stop Selling DNA Analysis Service" (NYTimes.com, November 25, 2013) that the FDA was concerned about the consequences of inaccurate results and felt that the company was providing interpretations of the test results.

Along with the FDA action against 23andMe, a class-action suit was filed against the company in November 2013. According to the article "Class Action Law Suit Filed against 23andMe" (Forbes.com, December 2, 2013), besides questioning the validity of the health reports, the suit charged, "Defendant uses the information it collects from the DNA tests consumers pay to take to generate databases and statistical information that it then markets to other sources and the scientific community in general, even though the test results are meaningless." The company stopped offering health-related genetic tests in December 2013, but it resumed sales in October 2015 after revising the tests to meet FDA requirements. In "Disconnect between Silicon Valley and Regulators over Health Technologies, 23andMe CEO Says" (WSJ.com, October 26, 2016), Greg Bensinger reports that Anne Wojcicki, the chief executive officer of 23andMe, suggested that "there is a disconnect between Silicon Valley and the U.S. regulatory system over how to view new health-related technologies." Wojcicki credited her company's ability to resume testing to the strong support of scientists and academic publications that support the credibility of the tests.

In "23andMe and the FDA" (*New England Journal of Medicine*, vol. 370, no. 11, March 13, 2014), George J. Annas and Sherman Elias describe the debate about home DNA tests as a struggle between medical or government control and individuals' right to information about themselves. They also predict that in the future, "Health plans will make it easy for their members to have their entire genomes sequenced and linked to their electronic health records and will provide software to help people interrogate their own genomes, with or without the help of their physicians or a genetic counselor supplied by the health plan."

Caution Is Advised When Choosing Home Genetic Tests

Although home genetic tests are available, the FDA has not authorized any DTC tests for marketing, and research reveals that they vary widely in terms of accuracy. For example, Heather Hatfield cautions buyers in "Home DNA Tests: Buyer Beware" (WebMD.com, 2007) to only purchase tests that will be analyzed by CLIA-certified laboratories and to have all test results reviewed by a physician. Hatfield also advises consumers to view with skepticism companies that sell genetic tests and products such as vitamins to address the results of these tests.

In "Direct-to-Consumer Genetic Tests" (January 2014, https://www.consumer.ftc.gov/articles/0166-direct-consumer-genetic-tests), the Federal Trade Commission adds additional warnings. It cautions against making any decisions based on the results of genetic tests alone because the results may be incomplete or inaccurate. It also advises consumers to scrutinize the company's privacy policy to determine "how the company secures the information it collects, how it will use your information, and whether it will share your information with third parties."

Personal Choices and Psychological Consequences of Genetic Testing

The results of genetic tests may be used to make decisions such as whether to have children or to end a pregnancy. Results that predict the likelihood that an individual will develop a disease may affect decisions about education, marriage, family, or career choices. The decision to undergo genetic testing and the results of the tests affect not only the individuals tested but also their family members. For example, when an unaffected person requests a genetic susceptibility test, another test of an affected relative may be required to accurately calculate probability. Family members may vary in their willingness to share genetic information and their desire to know about genetic risks.

There are psychological consequences of genetic testing and coming to terms with the results. People may be relieved or distressed when they learn the results of a genetic test. The results can change the way they feel about themselves and can influence their relationships with relatives. For example, family members who discover they are carriers for cystic fibrosis may feel isolated or estranged from siblings who have opted not to be tested. By contrast, those who find out they do not have a genetic mutation for Huntington's disease may feel guilty because they have been spared and other family members have not, or they may worry about assuming responsibility for family members who develop the disease.

The complexity of genetic testing and the uncertainty of many results pose an additional psychological challenge. The results of predictive genetic tests are often expressed in probabilities rather than in certainties, and, even for people with a high probability of developing a disease, there are often conflicting opinions about the most appropriate course of action. For example, a woman who tests positive for the BRCA1 or BRCA2 gene has an increased lifetime risk of developing breast or ovarian cancer or both, but it does not mean that her risk of developing either or both is 100%. Depending on her personal circumstances and the medical advice she receives, she may opt to intensify screening to detect disease; use prescription medication that is intended to reduce her risk, such as Soltamox; or undergo a preventive surgical procedure, such as a mastectomy or oophorectomy.

Even a negative test result can be stressful, creating nearly as many questions as it does answers about disease risk. A negative test result for BRCA1 or BRCA2 for a woman who has an affected family member (one who has the genetic mutation) means that although she does not have the genetic mutation, her risk of developing breast cancer is the same as that of the general population. There are multiple genes as well as other factors associated with the risk of developing breast cancer that are not identified through genetic testing, such as the age at which a woman has her first child. In addition, most breast cancer is not believed to be hereditary, and most women diagnosed with breast cancer do not test positive for the BRCA1 or BRCA2 genetic mutation.

GENETIC COUNSELING

Generally, patients receive explicit counseling before undergoing genetic testing to ensure that they are able to make informed decisions about choosing to have the tests and the consequences of testing. Most genetic counseling is informative and nondirective—it is intended to offer enough information to allow families or individuals to determine the best courses of action for themselves but avoids making testing recommendations.

Patients undergoing tests to improve their care and treatment have different pretest counseling needs from those choosing susceptibility or predictive testing. In such instances genetic counselors do provide testing recommendations, particularly when a test offers an opportunity to prevent disease. As tests for genetic risk factors increasingly become routine in clinical medical practice, they are likely to be offered without formal pretest counseling. Genetic counselors urge physicians, nurses, and other health care professionals not to discount or rush through the process of obtaining informed consent to conduct a genetic test. They caution that potential psychological, social, and family implications should be acknowledged and addressed in advance of testing, such as the possibility that the predictive value of genetic information may be overestimated. The potentially life-changing consequences of genetic testing suggest that all health care professionals involved in the process should not only adhere to thoughtful informed consent procedures for genetic testing but also offer or make available genetic counseling when patients and families receive the test results.

CHAPTER 7
THE HUMAN GENOME PROJECT

When the full map of the human genome is known... we shall have passed through a phase of human civilisation as significant as, if not more significant than, that which distinguished the age of Galileo from that of Copernicus, or that of Einstein from that of Newton....We have crossed a boundary of unprecedented importance.... There is no going back....We are walking hopefully in the scientific foothills of a gigantic mountain range.

—Ian Lloyd, Official Record, British House of Commons, 1990

In 1953 James D. Watson (1928–) and Francis Crick (1916–2004) described the double helical structure of deoxyribonucleic acid (DNA). Their molecular DNA structure was published in "Molecular Structure of Nucleic Acids: A Structure for Deoxyribose Nucleic Acid" (*Nature*, vol. 171, no. 4356, April 25, 1953), an article that was only two pages long. Despite the brevity of this article, it ushered in a new age of discovery in genetics and laid the foundation for the sequencing of the human genome.

The word *genome* was derived from two words: *gene* and *chromosome*. In the 21st century the term *genome* is widely understood to be the entire complement of genetic material in the cell of an organism. A genome is composed of a series of four nitrogenous (containing nitrogen, a nonmetallic element that constitutes almost four-fifths of the air by volume) DNA bases: adenine (A), guanine (G), thymine (T), and cytosine (C). In each organism these bases are arranged in a specific order, or sequence, and this order constitutes the genetic code of the organism. In humans the genome is composed of approximately 3 billion bases. In 2001 a first-draft sequence of the entire human genome was completed and made available to the public for study and research. The Human Genome Project (HGP) of the National Human Genome Research Institute (NHGRI), which is part of the National Institutes of Health (NIH), completed the full human genome sequence in April 2003.

LAYING THE GROUNDWORK FOR THE SEQUENCING OF THE HUMAN GENOME

During the 1960s and 1970s the techniques that would enable the study of molecular genetics were developed. The American virologist Howard Temin (1934–1994) worked with ribonucleic acid (RNA) viruses and discovered in 1964 that Crick's central tenet—that DNA makes RNA and that RNA makes protein—did not always hold true. In 1965 Temin described the process of reverse transcriptase—that genetic information in the form of RNA can be copied into DNA. The enzyme called reverse transcriptase used RNA as a template for the synthesis of a complementary DNA strand. Throughout the 1960s the American biochemists Robert W. Holley (1922–1993), Har Gobind Khorana (1922–2011), and Marshall Warren Nirenberg (1927–2010), along with the American molecular geneticist Philip Leder (1934–), all contributed to deciphering the genetic code by determining the DNA sequence for each of the 20 most common amino acids. Holley, Khorana, and Nirenberg were awarded the 1968 Nobel Prize in Physiology or Medicine.

The American biochemist Paul Berg (1926–) created the first recombinant DNA in 1972, and his work paved the way for isolating and cloning genes. Recombinant DNA is formed by combining segments of DNA, frequently from different organisms. In 1975 the British molecular biologist Sir Edwin Southern (1938–) developed a method to isolate and analyze fragments of DNA that are still being used. Known as the Southern blot analysis, it is a procedure for separating DNA fragments by electrophoresis (a technique that separates molecules based on their size and charge) and identifying a specific fragment using a DNA probe. Figure 7.1 shows how the Southern blot analysis is performed. It is used in genetic research, forensic examinations of DNA evidence in legal proceedings, and clinical medical practice.

FIGURE 7.1

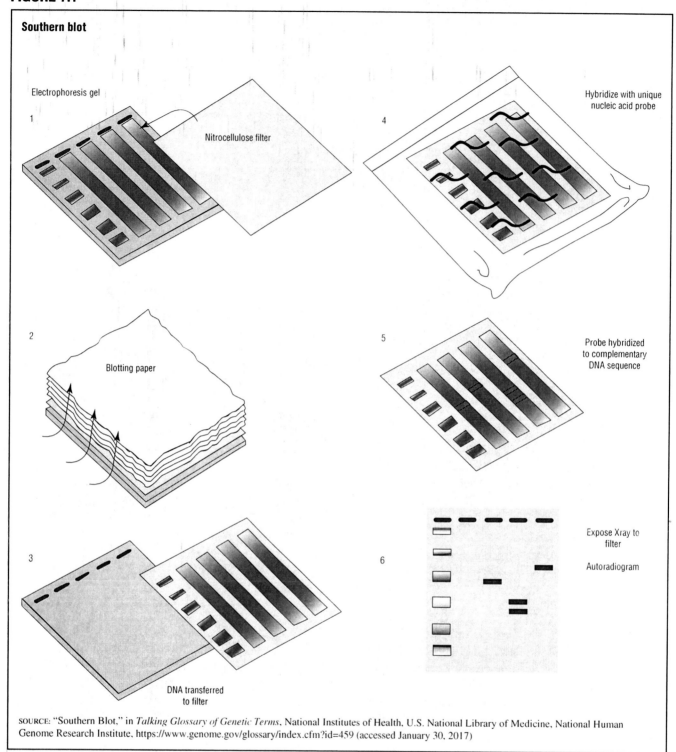

Southern blot

SOURCE: "Southern Blot," in *Talking Glossary of Genetic Terms*, National Institutes of Health, U.S. National Library of Medicine, National Human Genome Research Institute, https://www.genome.gov/glossary/index.cfm?id=459 (accessed January 30, 2017)

In 1977 the British biochemist Frederick Sanger (1918–2013), whose many accomplishments have been acknowledged by two Nobel Prizes in Chemistry in 1958 and 1980, and his colleagues developed techniques to determine the nucleic acid base sequence for long sections of DNA. In 1978 the American biologists Hamilton O. Smith (1931–) and Daniel Nathans (1928–1999) and the Swiss molecular geneticist Werner Arber (1929–) were awarded the Nobel Prize in Physiology or Medicine for an array of discoveries made during the 1960s, including the use of restriction enzymes, which ignited the biotechnology field. Restriction enzymes recognize and cut specific DNA sequences. That same year restriction fragment length polymorphisms (DNA sequence variants) were discovered. Figure 7.2 shows two single nucleotide polymorphisms—single base changes between homologous DNA fragments.

FIGURE 7.2

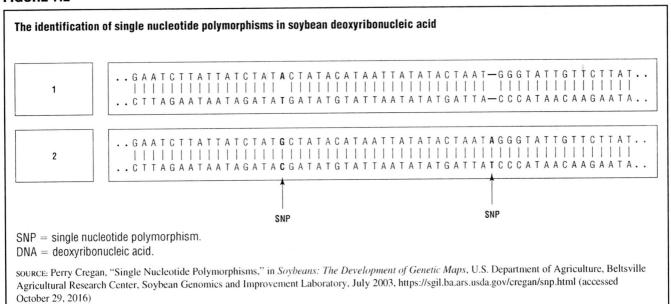

The identification of single nucleotide polymorphisms in soybean deoxyribonucleic acid

SNP = single nucleotide polymorphism.
DNA = deoxyribonucleic acid.

SOURCE: Perry Cregan, "Single Nucleotide Polymorphisms," in *Soybeans: The Development of Genetic Maps*, U.S. Department of Agriculture, Beltsville Agricultural Research Center, Soybean Genomics and Improvement Laboratory, July 2003, https://sgil.ba.ars.usda.gov/cregan/snp.html (accessed October 29, 2016)

Using these new techniques, researchers were able to identify several genes for serious human disorders during the 1980s. In 1982 the American molecular biologist James F. Gusella (1952–) and his colleagues at Harvard University studied patients with Huntington's disease and determined that the gene for this degenerative disorder was located on the short arm of chromosome 4. That same year a gene for neurofibromatosis type 1 was found on the long arm of chromosome 17. Neurofibromatoses are a group of genetic disorders that cause tumors (most of which are nonmalignant) to grow along various types of nerves and can affect the development of nonnervous tissues such as bones and skin. The disorder may also result in developmental abnormalities such as learning disabilities.

In 1985 the American biochemist Kary B. Mullis (1944–) and his colleagues at the Cetus Corporation in California pioneered the polymerase chain reaction, a fast, inexpensive technique that amplifies small fragments of DNA to make sufficient quantities available for DNA sequence analysis—that is, determining the exact order of the base pairs in a segment of DNA. Because it enabled researchers to make an unlimited number of copies of any piece of DNA, it was dubbed "molecular photocopying," and in 1993 Mullis was awarded the Nobel Prize in Chemistry for this tremendous breakthrough in gene analysis. By 1987 automated sequencers were developed, enabling even more rapid sequencing and analysis on large segments of DNA. Figure 7.3 shows the steps involved in a polymerase chain reaction.

In 1985 the Canadian molecular geneticist Lap-Chee Tsui (1950–) and his colleagues mapped the gene responsible for cystic fibrosis, the most common inherited fatal

FIGURE 7.3

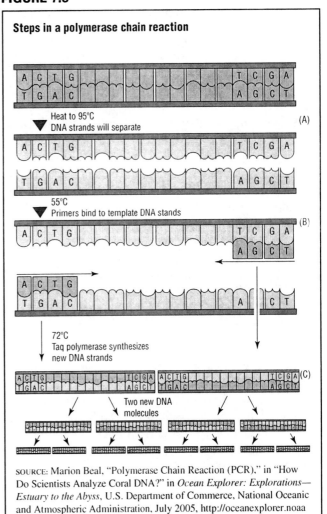

Steps in a polymerase chain reaction

SOURCE: Marion Beal, "Polymerase Chain Reaction (PCR)," in "How Do Scientists Analyze Coral DNA?" in *Ocean Explorer: Explorations—Estuary to the Abyss*, U.S. Department of Commerce, National Oceanic and Atmospheric Administration, July 2005, http://oceanexplorer.noaa.gov/explorations/04etta/background/dna/dna.html (accessed October 29, 2016)

disease of children and young adults in the United States, to the long arm of chromosome 7. The gene for cystic fibrosis was discovered in 1989, and it was determined that three missing nucleic acid bases occurred in the altered gene of 70% of patients with cystic fibrosis.

The mutations associated with Duchenne muscular dystrophy were identified in 1987. This gene is located close to the gene for chronic granulomatous disease (an X-linked autosomal recessive disorder that, if left untreated, is fatal in childhood) on the short arm of the X chromosome. In 1990 the American geneticist Mary-Claire King (1946–) found the first evidence that a gene on chromosome 17 (now known as BRCA1) could be associated with an inherited predisposition to breast and ovarian cancer.

The discoveries and technological advances made by researchers during the 1970s and 1980s gave rise to modern clinical molecular genetics. Cytogenetics, the study of chromosome structure and function, produced methods to view distinct bands on each chromosome. Figure 7.4 is a cytogenetic map of human chromosomes. Cytogenetic studies are applied in three broad areas of medicine: congenital (from birth) disorders, prenatal diagnosis, and neoplastic diseases (cancer).

THE BIRTH OF THE HUMAN GENOME PROJECT

The first meetings to discuss the feasibility of sequencing the human genome were organized by Robert Louis Sinsheimer (1920–), a molecular biologist and chancellor of the University of California, Santa Cruz, and were held on campus in 1985. The idea of sequencing the human genome generated excitement among the many well-known researchers in attendance—they considered the undertaking to be the "Holy Grail" of molecular biology. The following year Congress began considering the feasibility of human genome research. The initial funding from Congress was $17.2 million to the NIH and $10.7 million to the U.S. Department of Energy's (DOE) Office of Health and Environmental Science, with progressive increases over the next several years. (See Table 7.1.)

In 1988 Congress funded the NIH and the DOE to "coordinate research and technical activities related to the human genome." The NIH established the Office of Human Genome Research in September 1988. The following year the office was renamed the National Center for Human Genome Research (NCHGR). Watson served as its enthusiastic champion and director until April 1992. Larry Thompson of the NHGRI explains in "Background Paper on Internal Educational Activities at the National Human Genome Research Institute" (June 10, 2002, https://www.genome.gov/10005291/) that following Watson's appointment, the NIH and the DOE committed 5% of the project's budget to address ethical, legal, and social

FIGURE 7.4

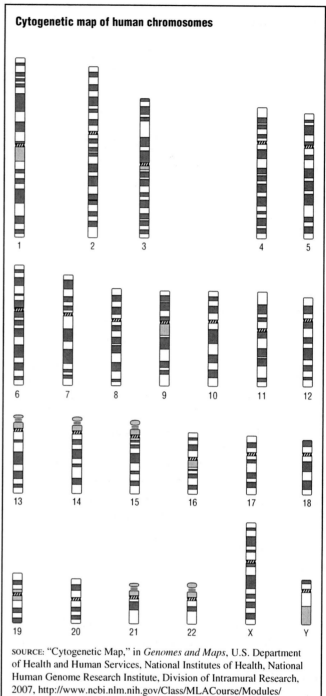

Cytogenetic map of human chromosomes

SOURCE: "Cytogenetic Map," in *Genomes and Maps*, U.S. Department of Health and Human Services, National Institutes of Health, National Human Genome Research Institute, Division of Intramural Research, 2007, http://www.ncbi.nlm.nih.gov/Class/MLACourse/Modules/Genomes/map_cytogenetic.html (accessed October 29, 2016)

issues that arose from the study of the human genome. This ambitious undertaking constituted the largest bioethics program, in terms of funding and human resources, in the world.

International genomic research was also under way. For example, in 1987 the Italian National Research Council launched a genome research project, and an international group of geneticists founded the Human Genome Organization in Switzerland in 1989. As individual scientists exchanged information in their quest for genetic

TABLE 7.1

U.S. Human Genome Project funding, fiscal years 1988–2003

[$ millions]

Fiscal year	Department of Energy	National Institutes of Health	U.S. total
1988	10.7	17.2	27.9
1989	18.5	28.2	46.7
1990	27.2	59.5	86.7
1991	47.4	87.4	134.8
1992	59.4	104.8	164.2
1993	63.0	106.1	169.1
1994	63.3	127.0	190.3
1995	68.7	153.8	222.5
1996	73.9	169.3	243.2
1997	77.9	188.9	266.8
1998	85.5	218.3	303.8
1999	89.9	225.7	315.6
2000	88.9	271.7	360.6
2001	86.4	308.4	394.8
2002	90.1	346.7	434.3
2003	64.2	372.8	437.0

Note: These numbers do not include construction funds, which are a very small part of the budget.

SOURCE: "Human Genome Project Budget," in *Human Genome Project Information Archive 1988–2003*, U.S. Department of Energy Office of Science, Office of Biological and Environmental Research, Human Genome Project, September 14, 2004, http://web.ornl.gov/sci/techresources/Human_Genome/project/budget.shtml (accessed October 29, 2016)

TABLE 7.2

Model organisms sequenced

Date sequenced[a]	Species	Total bases[b]
7/28/1995	*Haemophilus influenzae* (bacterium)	1,830,138
10/30/1995	*Mycoplasma genitalium* (bacterium)	580,073
5/29/1997	*Saccharomyces cerevisiae* (yeast)	12,069,247
9/5/1997	*Escherichia coli* (bacterium)	4,639,221
11/20/1997	*Bacillus subtillis* (bacterium)	4,214,814
12/31/1998	*Caenorhabditis elegans* (round worm)	97,283,371
		99,167,964[c]
3/24/2000	*Drosophila melanogaster* (fruit fly)	~137,000,000
12/14/2000	*Arabidopsis thaliana* (mustard plant)	~115,400,000
1/26/2001	*Oryza sativa* (rice)	~430,000,000
2/15/2001	*Homo sapiens* (human)	~3,200,000,000

[a]First publication date.
[b]Data excludes organelles or plasmids. These numbers should not be taken as absolute. Scientists are confirming the sequences; several laboratories were involved in the sequencing of a particular organism and have slightly different numbers; and there are some strain variations.
[c]The first number was originally published, and the second is a correction as of June 2000.

SOURCE: Richard Robinson, ed., "Model Organisms Sequenced," in *Genetics*, vol. 2, *E–I*, Macmillan Reference USA, 2000

This action ensured that gene sequences were in the public domain and could not be patented.

Worming Away

Although sequencing the human genome was the principal objective, the HGP also sought to sequence the genomes of other organisms. These other organisms served as models, enabling researchers to test and refine new methods and technologies that helped identify corresponding genes in the human genome. Table 7.2 is a list of some of the model organisms, including the roundworm, sequenced during the course of the HGP, along with the dates the sequences were published and the number of bases in each organism.

At Cambridge University, the British molecular biologist Sydney Brenner (1927–) was studying the nematode worm *Caenorhabditis elegans*. By 1989 Brenner and his colleagues had successfully produced a map of the entire *Caenorhabditis elegans* genome. The map consisted of multiple overlapping fragments of DNA, arranged in the correct order, and Brenner's research team printed the worm's genome on postcard-sized pieces of paper.

Watson believed the genomes of smaller organisms would not only help refine research methods and the technology but also provide valuable sources of comparison once the human genome project was under way. The worm map convinced Watson that *Caenorhabditis elegans* should be the first multicellular organism to have its complete genome accurately sequenced. When the worm-sequencing project began in 1990, the first automatic sequencing machines had just become available from Applied Biosystems, Inc. The sequencing machines enabled the worm pilot project to meet its objective of

links to disease, the Human Genome Organization developed an international framework to coordinate research projects and prevent wasted resources through duplication, creating a culture of sharing data. In 1990 the European Commission initiated a two-year human genome project. The Soviet Union funded its genome research project that same year.

In 1990 the initial planning stage of the HGP was completed with the publication of the joint research plan *Understanding Our Genetic Inheritance: The Human Genome Project—The First Five Years, Fiscal Years 1991–1995* (http://web.ornl.gov/sci/techresources/Human_Genome/project/5yrplan/index.shtml). Just two years into the five-year plan, Watson resigned from his leadership position with the NCHGR because he vehemently disagreed with NIH decisions about the commercialization, propriety, and legality of patenting human gene sequences. Watson maintained that data from the HGP should be in the public domain and freely available to all scientists as well as to the public. In 1993 Francis S. Collins (1950–) was named the director.

Many prominent researchers sided with Watson against the patenting and commercialization of HGP data. In 1996 scientists at leading research institutions throughout the world agreed to submit their findings and genome sequences to GenBank, a genome database maintained by the NIH. In a resounding and unanimous move, they required the publication of any submitted sequence data on the Internet within 24 hours of its receipt by GenBank.

sequencing 3 million bases in three years. Equally important, the worm project demonstrated that the technology could scale up—that is, more machines and more technologists could produce more sequences faster.

A DECADE OF ACCOMPLISHMENTS: 1993 TO 2003

When the HGP began in September 1990, its projected completion date was 2005, at a cost of $3 billion for just the U.S. portion of the research. Ever-improving research techniques—including the use of restriction fragment length polymorphisms, which is described in detail in Figure 7.5, as well as the polymerase chain reaction, bacterial and yeast artificial chromosomes, and pulsed-field gel electrophoresis—accelerated the progress of the project. The HGP researchers finished the mapping two years earlier than scheduled, and U.S. scientists spent just $2.7 billion.

Although the HGP was a collaborative, international effort, most of the sequencing was performed at the Whitehead Institute for Medical Research in Massachusetts, the Baylor College of Medicine in Texas, the University of Washington, the Joint Genome Institute in California, and the Sanger Centre in the United Kingdom. Along with U.S government funding, the HGP was supported by the Wellcome Trust, a charitable foundation in the United Kingdom.

Public and Private Initiatives Compete

The NIH (February 14, 2017, https://www.nih.gov/about-nih/what-we-do/nih-almanac/national-human-genome-research-institute-nhgri) notes that in 1993 the NCHGR established the Division of Intramural Research, which was charged with developing genome technology research of specific diseases. By 1996 eight NIH institutes and centers had also collaborated to create the Center for Inherited Disease Research to study the genetics of complex diseases. In 1997 the NCHGR gained full institute status at the NIH and was renamed the National Human Genome Research Institute. The following year a third five-year plan, "New Goals for the U.S. Human Genome Project: 1998–2003," was published in *Science* (vol. 282, no. 5389, October 23, 1998).

The mid-1990s also saw the birth of a privately funded genomics effort led by the American geneticist J. Craig Venter (1946–). Venter had been working in a laboratory at the NIH when he decided to concentrate his sequencing efforts not on the genome itself, but on the gene products—that is, messenger RNAs produced by each cell. Venter eventually left the NIH to establish

FIGURE 7.5

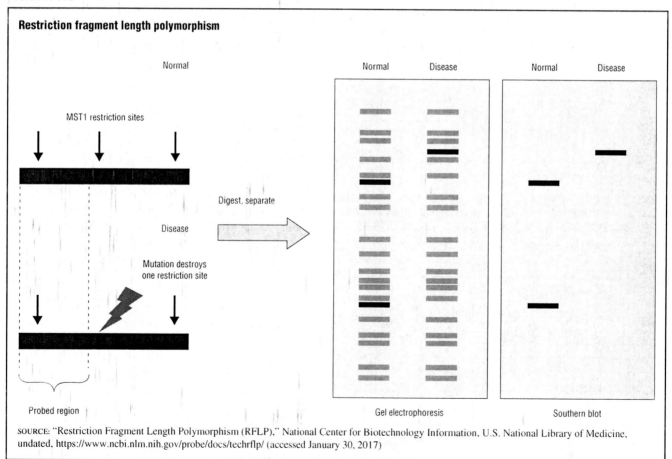

Restriction fragment length polymorphism

SOURCE: "Restriction Fragment Length Polymorphism (RFLP)," National Center for Biotechnology Information, U.S. National Library of Medicine, undated, https://www.ncbi.nlm.nih.gov/probe/docs/techrflp/ (accessed January 30, 2017)

the Institute for Genomic Research, a private, nonprofit organization aimed at collecting and interpreting vast numbers of expressed sequence tags. (Expressed sequence tags are random fragments of complementary DNAs derived from the information in the RNAs, which contain all the information that is actually expressed in a given cell type.) In 1995 the Institute for Genomic Research published the first sequenced genome of the bacteria *Haemophilus influenzae*. (See Table 7.2.)

The international partners in the genome project met in Bermuda in 1996 at a strategy meeting sponsored by the Wellcome Trust. There they created the "Bermuda Principles," conditions that governed access to data, including the standard that sequence information be released into public databases within 24 hours. To adhere to this agreement, participating scientists were to deposit base sequences into one of three databases within 24 hours of sequencing completion. The data contained in the three databases were exchanged daily. Because these were public databases, access to the stored sequences was free and unrestricted. The agreement was extended to data on other organisms at a meeting that same year.

In 1998 Venter announced that he was allying with the Perkin-Elmer Corporation to form a new company that would compete directly with the public effort to sequence the entire human genome by 2001. The new company, Celera Genomics, took a different approach from the one used by the HGP. Instead of the map-based, gene-by-gene approach taken by the HGP, Venter's firm broke the genome into random lengths, and then sequenced and reassembled it. This method saved time by eliminating the mapping phase, but it required robust computing capabilities to reassemble the human genome, which includes many repeated sequences. Venter's approach relied on the use of a supercomputer and 300 high-speed automatic sequencers manufactured by Perkin-Elmer; it was the precursor to the large-scale genomics studies that are now standard.

The competition between public and private initiatives began in earnest. Within one week of the launch of the private initiative, the Wellcome Trust increased its funding to the Sanger Centre to step up the production of raw sequence. In response to this increased support, the Sanger Centre revised its objective by aiming to sequence a full one-third of the entire genome rather than just one-sixth. The race between the public and private projects was on, and the milestones of the sequencing project came ever more rapidly.

Venter's firm energized the publicly funded project and inspired intensified efforts. HGP investigators feared that if their efforts were perceived as slow and inefficient, the HGP would lose congressional support and funding. The threat of genomic information ending up solely in the hands of a private firm was unacceptable to the researchers.

When Celera entered the race to sequence the human genome, it changed the landscape of the field. Celera resolved to make its data available only to paying customers and planned to patent some sequences before releasing them.

The publicly funded HGP released sequence information quickly both to provide the scientific community with timely data that were immediately usable and to place identified sequences beyond the reach of commercial companies wishing to patent them or charge for access to the data. The leadership of the HGP echoed the sentiments that had prompted Watson's resignation. It contended that patenting the human sequence was unethical and delayed the timely application of genomic information to medical disorders. Although the HGP staunchly opposed privileged access to the data, to speed completion of the project it had agreed to grant Celera specific rights to the data generated by the collaborative effort between the two groups.

THE FIRST DRAFT OF THE COMPLETED HUMAN GENOME

In April 2000 Celera announced that it was prepared to present the first draft of the human genome. Not unexpectedly, scientists and the public eagerly anticipated this "first look" at the human genome. Although geneticists and other scientists could better comprehend the future implications of this endeavor than the general public, the significance of this achievement was evident to professionals and laypeople alike. The professional literature and the mass media had successfully communicated the importance of this achievement, and it was understood that knowledge of the human genome held the key to the singularity of the human species. Furthermore, it was widely assumed that this information would be the basis for unprecedented advances in medicine and biomedical technology.

In "Rival Demands Sink Genome Alliance Plans" (*Nature*, vol. 404, no. 6774, March 9, 2000), Natasha Loder explains that all hopes for continuing cooperation and collaboration between the HGP and Celera were dashed when a letter from the Wellcome Trust was released to the public. The letter described the HGP's concerns about Celera's intent to have its commercial rights extended to research applications and to publish data from the collaborative efforts under its name alone. Celera said the release of the letter was unethical, and in a media interview Venter described it as "a low-life thing to do." The anger and bitterness between the two groups was not resolved and each vowed to publish the data separately.

In February 2001 the first working draft of the human genome was published in special issues of the journals *Nature* (vol. 409, no. 6822, February 15, 2001) and

Science (vol. 291, no. 5507, February 16, 2001). *Nature* detailed the initial analysis of the descriptions of the sequence generated by the HGP, and *Science* contained the draft sequence reported by Celera. *Nature* said its choice to publish the HGP data reflected its traditional preference for publicly funded research. *Science* editors Barbara R. Jasny and Donald Kennedy lauded Venter's group in the editorial "The Human Genome," writing "his colleagues made it possible to celebrate this accomplishment far sooner than was believed possible."

One of several surprises from the first draft was that previous estimates of gene number appeared to have been wildly inaccurate. Most pre–genome project estimates predicted that humans had as many as 60,000 to 150,000 genes. The first draft of the complete genome sequence indicated that the true number of genes required to make a human being was less than 40,000. (The current estimate is between 19,000 and 20,000 genes.) By comparison, yeast have 6,000 genes, fruit flies have 14,000, roundworms have 19,000, and the mustard weed plant has 26,000. Another surprise was the observation that humans share 99.9% of the nucleotide code in the human genome. Notably, human diversity at the genetic level is encoded by less than a 0.1% variation in DNA.

First Draft Is Headline News

In June 2000 the White House press release "President Clinton Announces the Completion of the First Survey of the Entire Human Genome" (http://web.ornl.gov/sci/techresources/Human_Genome/project/clinton1.shtml) predicted some of the anticipated medical outcomes of the project. These included the ability to:

- Alert patients that they are at risk for certain diseases. Once scientists discover which DNA sequence changes in a gene can cause disease, healthy people can be tested to see whether they risk developing conditions such as diabetes or prostate cancer later in life. In many cases, this advance warning can be a cue to start a vigilant screening program, to take preventive medicines, or to make diet or lifestyle changes that may prevent the disease.

- Reliably predict the course of disease. Diagnosing ailments more precisely will lead to more reliable predictions about the course of a disease. For example, a genetic fingerprint will allow doctors treating prostate cancer to predict how aggressive a tumor will be. New genetic information will help patients and doctors weigh the risks and benefits of different treatments.

- Precisely diagnose disease and ensure the most effective treatment is used. Genetic analysis allows us to classify diseases, such as colon cancer and skin cancer, into more defined categories. These improved classifications will eventually allow scientists to tailor drugs for patients whose individual response can be predicted by genetic fingerprinting. For example, cancer patients facing chemotherapy could receive a genetic fingerprint of their tumor that would predict which chemotherapy choices are most likely to be effective, leading to fewer side effects from the treatment and improved prognoses.

- Developing new treatments at the molecular level. Drug design guided by an understanding of how genes work and knowledge of exactly what happens at the molecular level to cause disease, will lead to more effective therapies. In many cases, rather than trying to replace a gene, it may be more effective and simpler to replace a defective gene's protein product. Alternatively, it may be possible to administer a small molecule that would interact with the protein to change its behavior.

The U.S. media celebrated the achievement with a flood of press releases and features. Efforts were also made to explain this monumental accomplishment to the public and to educate students. The DOE Human Genome Program provided a wealth of information about the HGP findings on the Internet. Figure 7.6 is an example of the information the DOE made available to the public. Figure 7.7 presents some of the potential environmental benefits of the HGP, including reducing the impact of climate change and developing sustainable energy sources that may be realized in the future.

PUFFER FISH AND MOUSE GENOMES ARE SEQUENCED

In 2002 the DOE Joint Genome Institute (JGI), operated by the Lawrence Berkeley National Laboratory, the Lawrence Livermore National Laboratory, and the Los Alamos National Laboratory, announced the draft sequencing, assembly, and analysis of the genome of the Japanese puffer fish *Fugu rubripes*. The Fugu Genome Project was initiated in 1989 in Cambridge, England, and in November 2000 the International Fugu Genome Consortium was formed, headed by the JGI. In 2001 the puffer fish genome was sequenced and assembled using the whole genome method pioneered by Celera. The puffer fish was the first vertebrate genome to be publicly sequenced and assembled in this manner and the first vertebrate genome published after the human genome. According to the JGI (2017, http://genome.jgi.doe.gov/Takru4/Takru4.home.html), the puffer fish has the smallest known genome among vertebrates (animals with bony backbones or cartilaginous spinal columns—fish, reptiles, birds, and mammals, including humans).

Comparison of the human and puffer fish genomes enabled investigators to predict the existence of nearly 1,000 previously unidentified human genes. Although the

FIGURE 7.6

From deoxyribonucleic acid to humans

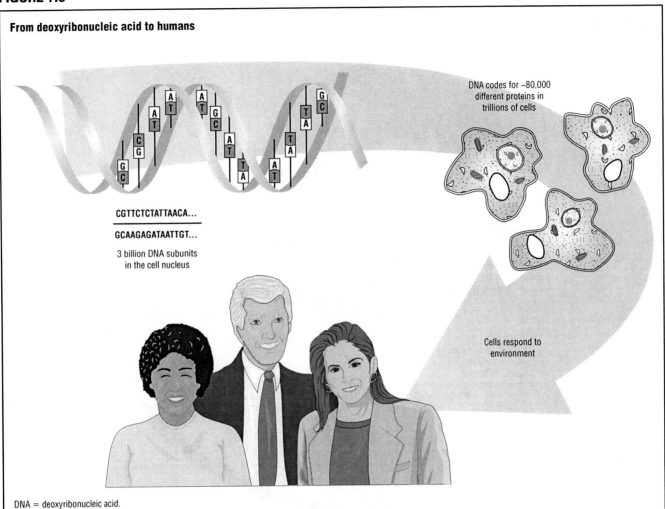

DNA codes for ~80,000 different proteins in trillions of cells

CGTTCTCTATTAACA...

GCAAGAGATAATTGT...

3 billion DNA subunits in the cell nucleus

Cells respond to environment

DNA = deoxyribonucleic acid.

SOURCE: "From DNA to Humans," in *Human Genome Project Information Basic Genomics Image Gallery*, U.S. Department of Energy Office of Science, Office of Biological and Environmental Research, Human Genome Project, undated, https://public.ornl.gov/site/gallery/detail.cfm?id=402&topic=& citation=&general=From%20DNA%20to%20Humans&restsection=all (accessed October 31, 2016)

function of these additional genes is as yet unknown, they contribute to the complete catalog of human genes. Ascertaining the existence and location of genes helped scientists begin to describe how they are regulated and function in the human body. With more than 30,000 puffer fish genes identified, the vast majority of human genes have counterparts in the puffer fish, with the most significant differences in genes of the immune system, metabolic regulation, and other physiological systems that are not alike in fish and mammals.

On December 5, 2002, the first draft of the sequence of the mouse genome was published in *Nature* (vol. 420, no. 6915). The mouse genome findings were deemed among the most important in terms of their comparability with humans. Mice and humans have about the same number of genes and about 90% of genes associated with medical disorders in humans have counterparts in mice. This finding means that mice are especially well suited for studying human diseases and for testing treatments.

THE HGP IS COMPLETED

After the publication of a first-draft human genome in 2001, researchers continued to fill in the blanks and produce a complete and accurate sequence. In January 2003 another milestone in the human genome sequencing effort was reported: the fourth human chromosome—chromosome 14, the largest one to date, with 107 million base pairs—had been sequenced. Roland Heilig et al. published their findings in "The DNA Sequence and Analysis of Human Chromosome 14" (*Nature*, vol. 421, no. 6923, February 6, 2003). They found two genes that are vital for immune responses on chromosome 14 and about 60 genes that, when defective, contribute to disorders such as spastic paraplegia and Alzheimer's disease.

In a remarkable coincidence that made the crowning achievement of the HGP even more poignant, the completion of the sequencing of the human genome occurred during the same year slated for celebrations of the 50th anniversary of the discovery of the DNA double helix. On April 14, 2003, the International Human Genome

FIGURE 7.7

Projected national benefits of genomics research

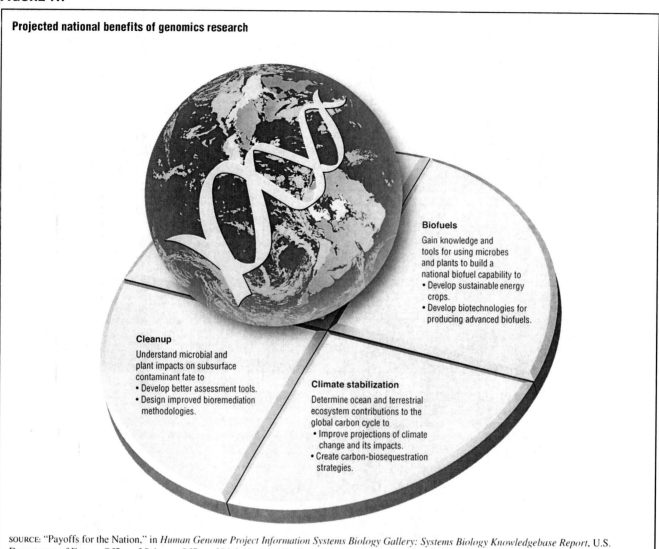

Biofuels

Gain knowledge and tools for using microbes and plants to build a national biofuel capability to
- Develop sustainable energy crops.
- Develop biotechnologies for producing advanced biofuels.

Cleanup

Understand microbial and plant impacts on subsurface contaminant fate to
- Develop better assessment tools.
- Design improved bioremediation methodologies.

Climate stabilization

Determine ocean and terrestrial ecosystem contributions to the global carbon cycle to
- Improve projections of climate change and its impacts.
- Create carbon-biosequestration strategies.

SOURCE: "Payoffs for the Nation," in *Human Genome Project Information Systems Biology Gallery: Systems Biology Knowledgebase Report*, U.S. Department of Energy Office of Science, Office of Biological and Environmental Research, Human Genome Project, March 2008, https://public.ornl.gov/ site/gallery/gallery.cfm?citation=10&restsection=public (accessed October 31, 2016)

Sequencing Consortium, which is directed by the DOE and the NHGRI, announced in the press release "All Goals Achieved; New Vision for Genome Research Unveiled" (https://www.genome.gov/11006929/) the successful completion of the HGP more than two years earlier than had been anticipated. The consortium included scientists at 20 sequencing centers in China, France, Germany, Japan, the United Kingdom, and the United States.

Nature, the same journal that had published the groundbreaking discoveries of Watson and Crick 50 years earlier, hailed the era of the genome in a special edition dated April 24, 2003 (vol. 422, no. 6934). As had been the practice since the inception of the HGP, the entirety of sequence data generated by the HGP was immediately entered into public databases and made freely available to the scientific community throughout the world, with no restrictions on its use or redistribution.

The data are used by researchers in academic settings and industry, as well as by commercial biotechnology firms. Figure 7.8 shows some of the landmarks in the sequenced human genome.

In "All Goals Achieved," the DOE and the NHGRI describe the international effort to sequence the 3 billion DNA base pairs in the human genome as "one of the most ambitious scientific undertakings of all time," comparing it to feats such as splitting the atom or traveling to the moon. Collins proudly declared that "the Human Genome Project has been an amazing adventure into ourselves, to understand our own DNA instruction book, the shared inheritance of all humankind. All of the project's goals have been completed successfully—well in advance of the original deadline and for a cost substantially less than the original estimates." In the same press release Eric Steven Lander (1957–), the director of the Whitehead Institute/Massachusetts Institute of Technology

FIGURE 7.8

Selected landmarks of the human genome

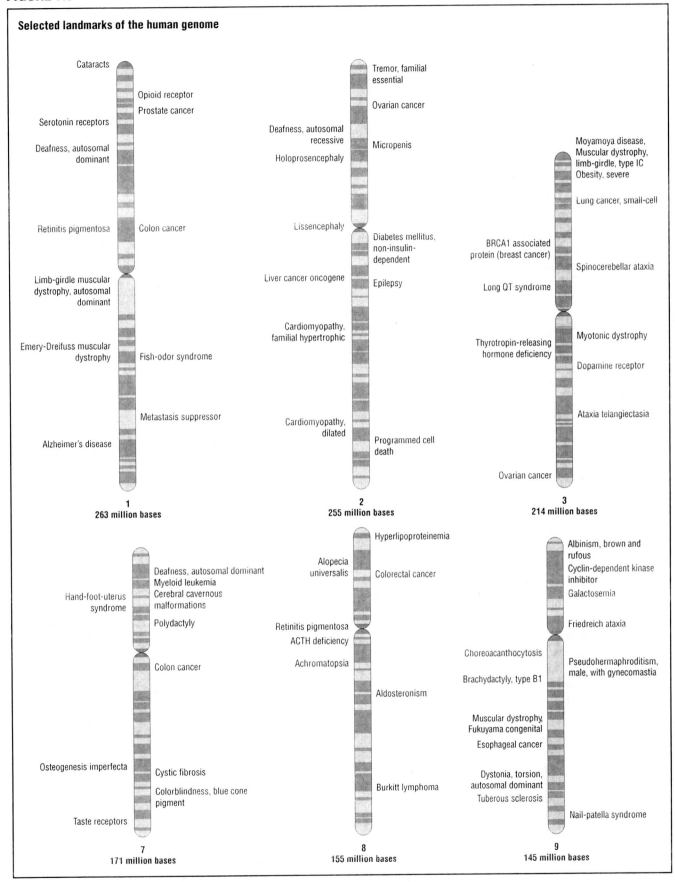

Center for Genome Research, predicted the postgenomic era when he asserted, "The Human Genome Project represents one of the remarkable achievements in the history of science. Its culmination this month signals the

FIGURE 7.8

Selected landmarks of the human genome [CONTINUED]

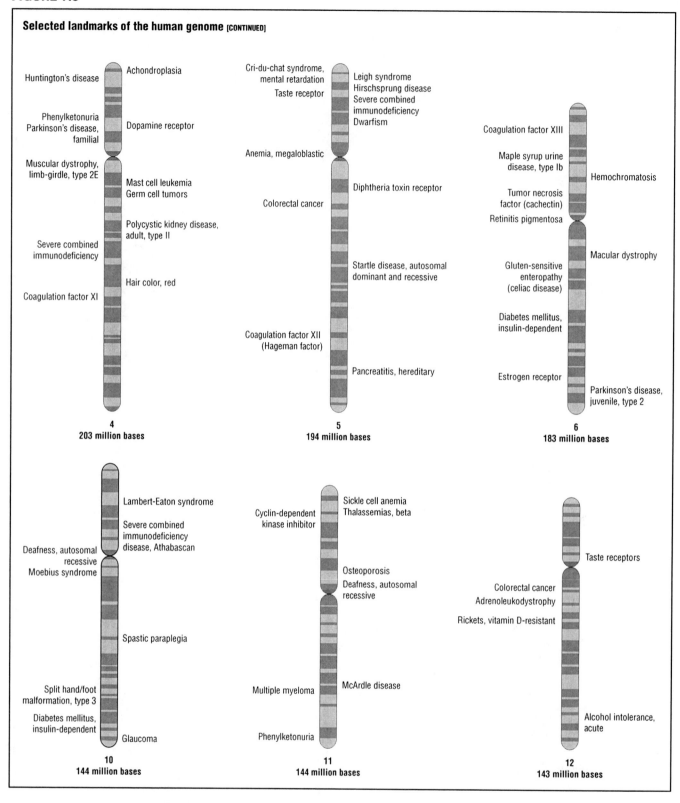

4
203 million bases

5
194 million bases

6
183 million bases

10
144 million bases

11
144 million bases

12
143 million bases

beginning of a new era in biomedical research. Biology is being transformed into an information science, able to take comprehensive global views of biological systems. With knowledge of all the components of the cells, we will be able to tackle biological problems at their most fundamental level."

Collins urged the scientific community not to rest on its laurels in the wake of this triumph, saying, "With this foundation of knowledge firmly in place, the medical advances promised from the project can now be significantly accelerated." The April 24, 2003, issue of *Nature* detailed the challenges researchers will face in the

FIGURE 7.8

Selected landmarks of the human genome [CONTINUED]

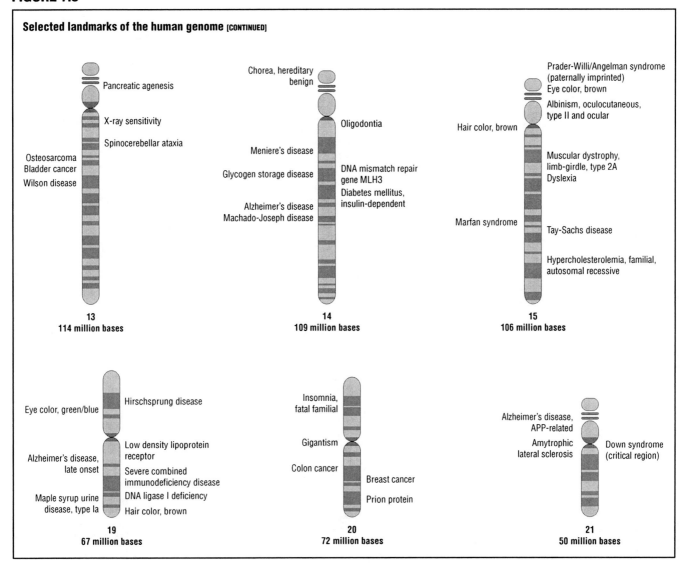

postgenomic era as they seek to employ the HGP data to treat disease and improve public health. Recommendations included collaborative efforts to produce:

- New tools to allow discovery in the not-too-distant future of the genetic contributions to frequently occurring diseases, including diabetes, heart disease, and mental illnesses such as schizophrenia.

- Improved methods for the early detection of disease and to enable timely treatment when it is likely to be effective.

- New technologies able to sequence the entire genome of any person affordably, ideally for less than $1,000. (Figure 7.9 shows the dramatic reduction in the cost of genome sequencing between 2001 and 2016; by October 2015 the cost per genome was just $1,245.)

- Wider access to tools and technologies of "chemical genomics" to enhance understanding of biological pathways and accelerate pharmaceutical and other treatment research.

Along with the special commemorative issue of *Nature*, the April 11, 2003, edition of *Science* (vol. 300, no. 5617) ran articles that described the multidisciplinary DOE plan dubbed "Genomes to Life," which aimed to use HGP data to understand the ways in which microbes can provide opportunities to develop clean energy, reduce climate change, and clean the environment.

THE HGP REVISES ITS ESTIMATE

In the October 2004 press release "International Human Genome Sequencing Consortium Describes Finished Human Genome Sequence" (https://www.genome.gov/12513430/), the NHGRI reduced its estimate of the number of human genes from between 30,000 and 35,000 to between 20,000 and 25,000. The refined human genome sequence, published in "Finishing the Euchromatic Sequence of the Human Genome" (*Nature*, vol. 431, no. 7011, October 21, 2004), was the most complete version to date. According to HGP scientists, it covered 99% of the gene-containing parts of

FIGURE 7.8

Selected landmarks of the human genome [CONTINUED]

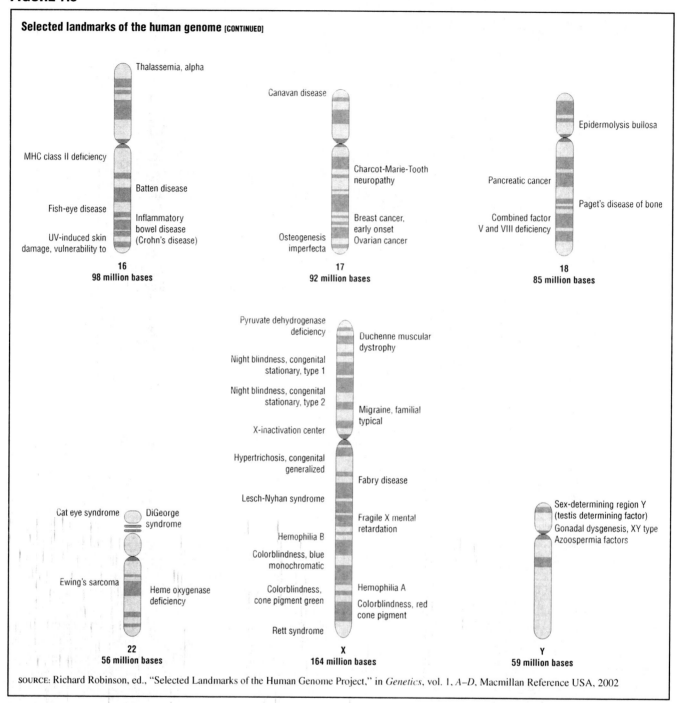

SOURCE: Richard Robinson, ed., "Selected Landmarks of the Human Genome Project," in *Genetics*, vol. 1, A–D, Macmillan Reference USA, 2002

the human genome, identified nearly all known genes, and was 99.9% accurate.

In November 2014 Iakes Ezkurdia et al. proposed in "Multiple Evidence Strands Suggest That There May Be as Few as 19,000 Human Protein-Coding Genes" (*Human Molecular Genetics*, vol. 23, no. 22) that the number of human genes should be revised downward to 19,000. Although the number of human genes may well decrease as scientists continue to study the human genome and refine their techniques, as of February 2017 the 19,000 to 20,000 estimate was most often cited in the scientific literature.

DAWN OF THE POSTGENOMIC ERA

Human Genome Project Information enumerates in "Potential Benefits of Human Genome Project Research" (July 23, 2013, http://web.ornl.gov/sci/techresources/Human _Genome/project/benefits.shtml) many of the potential benefits of HGP research. Besides its role in the practice of molecular medicine, other uses of HGP data and applications of human and other genomic research include:

- Microbial genomics—using bacteria to create new energy sources such as biofuels and safe, efficient toxic waste cleanup; enhancing understanding of how microbes cause disease; and protecting workers

FIGURE 7.9

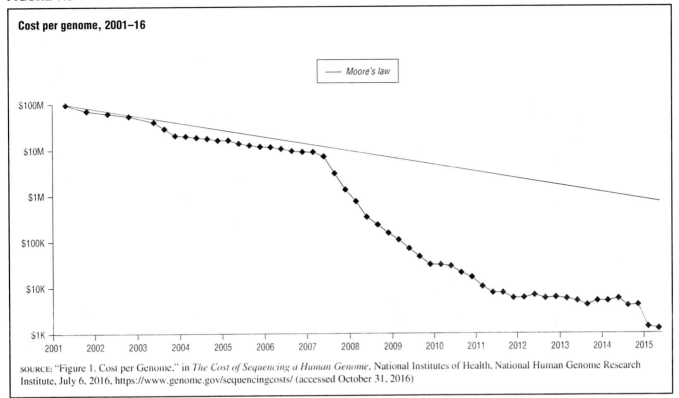

Cost per genome, 2001–16

— Moore's law

SOURCE: "Figure 1. Cost per Genome," in *The Cost of Sequencing a Human Genome*, National Institutes of Health, National Human Genome Research Institute, July 6, 2016, https://www.genome.gov/sequencingcosts/ (accessed October 31, 2016)

and the public from threats of biological and chemical terrorism and warfare. (See Figure 7.10.)

- Risk assessment—measuring the risks and health problems caused by exposure to radiation, carcinogens (cancer-causing agents), and mutagenic chemicals; and reducing the probability of heritable mutations.

- Archaeology, anthropology, evolution, and human migration—improving the understanding of germline (cells that give rise to eggs or sperm) mutations; studying migration based on female genetic inheritance; examining mutations on the Y chromosome to trace lineage and migration of males; and comparing the DNA sequences of entire genomes of different microbes to enhance the understanding of the relationships among archaebacteria (cells that do not contain nuclei), eukaryotes (cells that contain nuclei), and prokaryotes (single-celled organisms without nuclei).

- DNA forensics—identifying crime victims, potential suspects, and catastrophe victims through examination of DNA; confirming paternity and other family relationships; clearing people wrongly accused of crimes; identifying and protecting endangered species; detecting bacterial and other environmental pollutants; matching organ donors and recipients for transplant programs; and determining pedigrees for animals and plants.

- Agriculture and livestock breeding—developing healthier and stronger crops and farm animals that can resist insects, disease, and drought; creating safer

pesticides; growing more nutritious produce; incorporating vaccines into food products; and redeploying plants such as tobacco for use in environmental cleanup programs.

Genomic Medicine

The HGP and the technological advances it has generated, such as high-throughput technologies to detect genetic variation and gene expression, have moved the field of medicine forward with extraordinary speed. Charles Auffray et al. recount recent progress in "From Genomic Medicine to Precision Medicine: Highlights of 2015" (*Genome Medicine*, vol. 8, no. 1, January 29, 2016). They observe that whole-genome sequencing has not only improved understanding of disease but also its diagnosis and treatment. Auffray et al. cite advances in gene editing technologies that have the potential to cure rare single-gene diseases, increasing use of pharmacogenomics to personalize prescription drug treatment, and the rise of pathogen genomics, which may be used to rapidly identify disease-causing viruses and control the spread of infectious diseases.

In "Work on Telomeres Wins Nobel Prize in Physiology or Medicine for 3 U.S. Genetic Researchers [Update]" (*Scientific American*, October 13, 2009), Katherine Harmon reports that three American geneticists—Elizabeth H. Blackburn (1948–), Carol W. Greider (1961–), and Jack W. Szostak (1952–)—were awarded the Nobel Prize in Physiology or Medicine in 2009. The trio elaborated key

FIGURE 7.10

Projected solutions to energy challenges arising from genomics research

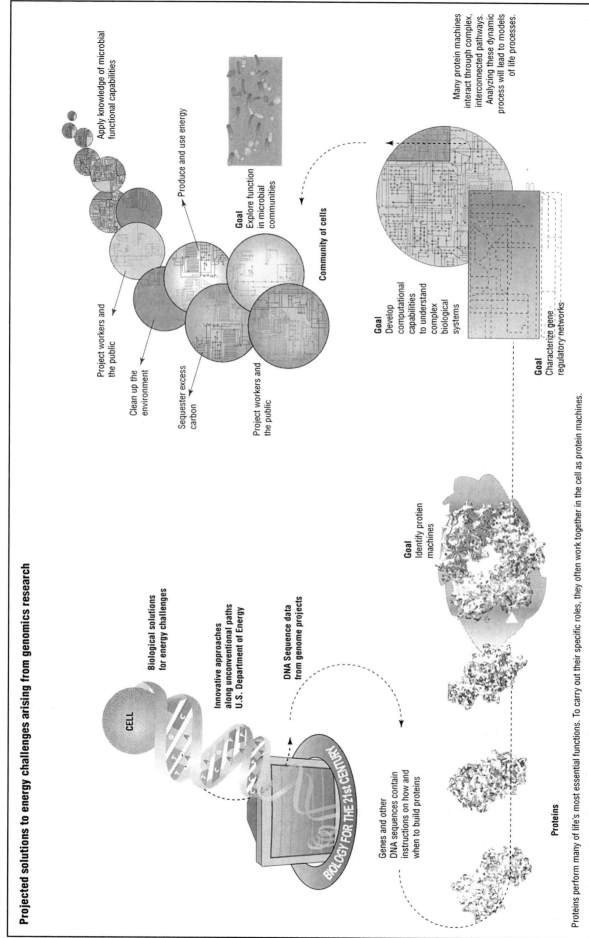

Proteins perform many of life's most essential functions. To carry out their specific roles, they often work together in the cell as protein machines.

SOURCE: "Genomics: GTL Program Pictorial," in *Human Genome Project Information Systems Biology Gallery*, U.S. Department of Energy Office of Science, Office of Biological and Environmental Research, Human Genome Project, https://public.oml.gov/site/gallery/gallery.cfm (accessed October 31, 2016)

aspects of the DNA replication process—how chromosomes protect themselves from degrading when cells divide. They discovered telomerase, the enzyme that creates telomeres, and identified the way that it adds DNA to the tips of chromosomes to replace genetic material that has eroded (telomeres may be compared with aglets, the plastic tips at the end of shoelaces that keep the laces from unraveling). Research is under way to determine whether drugs that block telomerase will be useful in combating disease. Scientists are also looking at the role of DNA erosion in some illnesses.

In 2010 three researchers who made strong contributions to the HGP—Collins, the director of the NIH; Lander, the president and director of the Broad Institute of the Massachusetts Institute of Technology and Harvard University; and David Botstein (1942–), the director of the Lewis-Sigler Institute for Integrative Genomics at Princeton University—were awarded the 10th annual Albany Medical Center Prize in Medicine and Biomedical Research for their insights that led to the mapping of the human genetic blueprint. The Albany Medical Center Prize, often called "America's Nobel," is the nation's top prize in medicine.

2016 AWARDS RECOGNIZE GENETICISTS' CONTRIBUTIONS. In 2016 the Albany Medical Center Prize in Medicine and Biomedical Research was awarded to F. Ulrich Hartl (1957–), Arthur L. Horwich (1951–), and Susan Lee Lindquist (1949–2016) for their elaboration of the mechanism of protein folding inside cells. Protein folding is the last step in transmitting genetic information from amino acids to proteins and the disruption of this process can cause disease.

The molecular biologists Michael Grunstein (1946–) and C. David Allis (1951–) were the 2016 Gruber Genetics Prize winners. They were recognized as being pioneers in epigenetics and lauded for their work that included identifying and describing how histone can alter chromatin structure and affect gene regulation.

In 2016 the Genetics Society of America bestowed the Thomas Hunt Morgan Medal for lifetime achievement in the field of genetics to Nancy E. Kleckner (1947–) of Harvard University for her contributions to present-day understanding of chromosomes and the mechanisms of inheritance. The society chose Maria Barna of Stanford University and Carolyn McBride of Princeton University to receive the Rosalind Franklin Young Investigator Award, which honors women who display original thinking and scientific creativity in the early years of their career.

The American Society of Human Genetics named James F. Gusella as recipient of the 2016 William Allan Award, which honors a scientist for scientific contributions to human genetics. Gusella pioneered the so-called

genetic research cycle model to translate basic science research into prevention, diagnostics, and treatments and with his colleagues identified genes that are associated with neurological conditions including amyotrophic lateral sclerosis, Alzheimer's disease, and autism.

Haplotype Mapping Project

In 2002 an international effort to develop a haplotype map of the human genome was launched. A haplotype is a set of alleles (particular forms of genes) or markers on one of a pair of homologous chromosomes, and a haplotype map shows human genetic variation. The premise of the International HapMap Project was that in the human genome different genetic variants within a chromosomal region (haplotypes) occur together far more frequently than others. Based on common haplotype patterns (combinations of DNA sequence variants that are usually found together), the haplotype map simplifies the search for medically important DNA sequence variations and offers new understanding of human population structure and history.

Given that any two people are 99.9% genetically identical, understanding the 0.1% difference is important because it helps explain why one person may be more susceptible to a certain disease than another. Researchers can use the HapMap to compare the genetic variation patterns of a group of people known to have a specific disease with a group of people without the disease. Finding a certain pattern more often in people with the disease identifies a genomic region that may contain genes that contribute to the condition. Researchers hope that identifying single nucleotide polymorphisms (SNPs), which are specific positions in the genome sequence that are occupied by one nucleotide in some copies and by a different nucleotide in others, will enable them to identify the alleles that are associated with increased or decreased susceptibility to common diseases, such as asthma, heart disease, or psychiatric illness. Figure 7.11 shows that most SNP variation (about 85%) occurs within all populations. Figure 7.12 shows an SNP that occurs when one base changes; this type of SNP is present in at least 1% of the population.

Researchers hypothesize that differences between haplotypes may be associated with varying susceptibility to disease, so mapping the haplotype structure of the human genome may be the key to identifying the genetic basis of many common disorders. The International HapMap Project serves as a resource for studying the genetic factors that contribute to variation in response to environmental factors, susceptibility to infection, and the identification of genetic variants that are associated with the effectiveness of, and adverse responses to, drugs and vaccines.

To create a haplotype map, researchers must have enough SNPs to be sure that regions containing disease

FIGURE 7.11

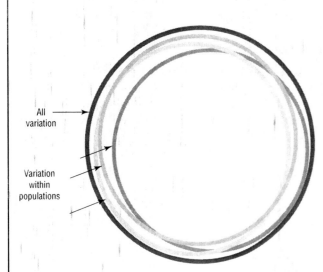

Most single nucleotide polymorphism variation occurs within all groups

All variation

Variation within populations

SOURCE: "Most SNP Variation Is within All Groups," in *Developing a Haplotype Map of the Human Genome for Finding Genes Related to Health and Disease*, U.S. Department of Health and Human Services, National Institutes of Health, National Human Genome Research Institute, Division of Extramural Research, 2001, http://www.genome.gov/10001665 (accessed October 31, 2016)

FIGURE 7.12

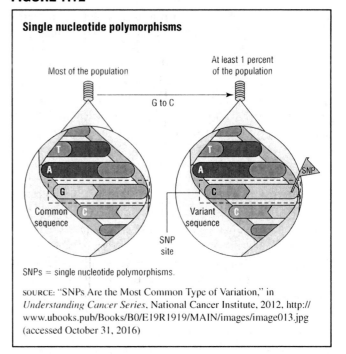

Single nucleotide polymorphisms

Most of the population

At least 1 percent of the population

G to C

T

A

G

Common sequence

T

A

SNP

C

Variant sequence

SNP site

SNPs = single nucleotide polymorphisms.

SOURCE: "SNPs Are the Most Common Type of Variation," in *Understanding Cancer Series*, National Cancer Institute, 2012, http://www.ubooks.pub/Books/B0/E19R1919/MAIN/images/image013.jpg (accessed October 31, 2016)

alleles have been found and that regions not containing disease alleles can be excluded from further consideration. The HapMap enables researchers to study the genetic risk factors underlying a wide range of disorders. For any given disease, researchers may perform an association study by using the HapMap tag SNPs to compare the haplotype patterns of a group of people known to have the disease to a group of people without the disease. If the association study finds a specific haplotype more frequently in those with the disease, researchers scrutinize the precise genomic region in their search for the specific genetic variant.

By mid-2005 a draft of the HapMap, consisting of 1 million markers of genetic variation, was released. The first draft enabled researchers to analyze the human genome in ways that were not possible with the human DNA sequence alone. Second-generation data from the project were released in 2006 and provided a denser map that enables scientists to narrow gene discovery more precisely to specific regions of the genome. The first- and second-generation HapMaps analyzed data from 270 people from four different populations: Yoruba in Ibadan, Nigeria; Japanese in Tokyo, Japan; Han Chinese in Beijing, China; and people with north and west European ancestry living in Utah.

In 2010 the International HapMap 3 Consortium published a third generation of the HapMap in "Integrating Common and Rare Genetic Variation in Diverse Human Populations" (*Nature*, vol. 467, no. 7311). The third-generation HapMap was the largest survey of human genetic variation performed. It contains data from 1,184 people from a total of 11 different populations, the four original populations and seven additional populations: people with African ancestry in the southwestern United States; people with Chinese ancestry in Denver, Colorado; Gujarati Indians in Houston, Texas; Luhya people in Webuye, Kenya; Maasai people in Kinyawa, Kenya; people with Mexican ancestry in Los Angeles, California; and Italians in Tuscany, Italy. Because the third-generation HapMap looked at 1.6 million SNPs in approximately 500 samples from the four original populations and more than 650 samples from the seven new populations, it was able to detect rarer variants than those identified in the two previous HapMaps.

In 2016 the National Center for Biotechnology Information took down the HapMap site in response to security concerns but noted that it had planned to decommission the site because the 1,000 Genomes Project (1KG; which operated between 2008 and 2015 and created the largest public catalog of human variation and genotype data) had become the research standard for population genetics and genomics. In "Human Genomics: The End of the Start for Population Sequencing" (*Nature*, vol. 526, no. 7571, October 1, 2015), Ewan Birney and Nicole Soranzo explain that the 1KG sought to find most genetic variants with frequencies of at least 1% in the populations studied. The 1KG leveraged advances in sequencing technology and was the first project to sequence the genomes of a large number of people and to provide a comprehensive resource on human genetic variation.

FIGURE 7.13

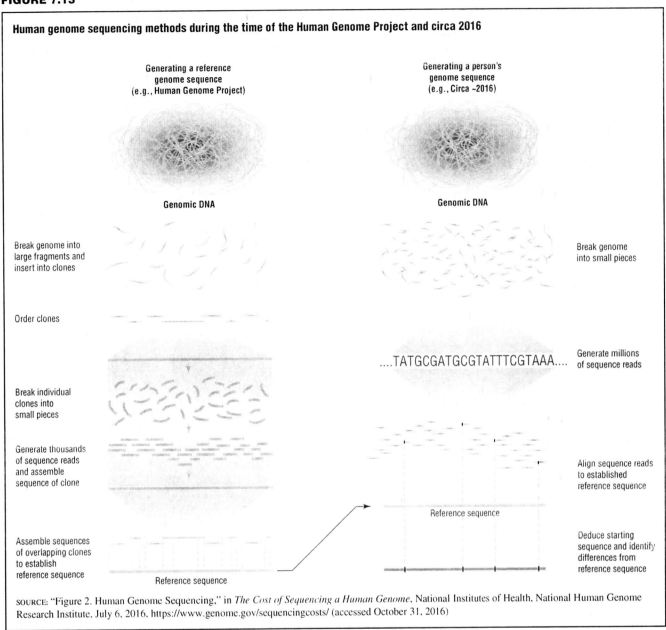

Human genome sequencing methods during the time of the Human Genome Project and circa 2016

SOURCE: "Figure 2. Human Genome Sequencing," in *The Cost of Sequencing a Human Genome*, National Institutes of Health, National Human Genome Research Institute, July 6, 2016, https://www.genome.gov/sequencingcosts/ (accessed October 31, 2016)

Figure 7.13 compares how human genome sequencing was performed by the HGP to the simpler, streamlined, and less costly method used in 2016. The data from the 1KG have been placed in public databases and are accessible to scientists worldwide.

The Future of Genomic Research

Established by the NHGRI in 2003, the Encyclopedia of DNA Elements (ENCODE) Project (February 17, 2017, https://www.genome.gov/10005107/) developed efficient ways to identify and locate all the protein-coding genes, nonprotein-coding genes, and other sequence-based functional elements contained in the human DNA sequence. This ambitious undertaking created an enormous resource for researchers seeking to use and apply

the human sequence to predict disease risk and to develop new approaches to prevent and treat disease.

In "The ENCODE (ENCyclopedia Of DNA Elements) Project" (*Science*, vol. 306, no. 5696, October 22, 2004), the ENCODE researchers described their plans to build a "parts list" of all sequence-based functional elements in the human DNA sequence. The researchers hoped to identify as-yet-unrecognized functional elements. During the pilot phase they developed and tested ways to identify functional elements efficiently. The conclusions from the pilot phase were published in "Identification and Analysis of Functional Elements in 1% of the Human Genome by the ENCODE Pilot Project" (*Nature*, vol. 447, no. 7146, June 14, 2007). The ENCODE researchers describe the human genome as an "elegant but cryptic store of information" and report that their analyses provide a "rich source of functional information...of the human genome" and that the pilot phase constituted "an impressive platform for future genome exploration efforts." During the next phase the researchers worked to develop new technologies to apply to the ENCODE Project.

Susan E. Celniker et al. observe in "Unlocking the Secrets of the Genome" (*Nature*, vol. 459, no. 7249, June 18, 2009) that free access to not only the human genome but also to the genomes of five model organisms—*Escherichia coli*, yeast (*Saccharomyces cerevisiae*), worm (*Caenorhabditis elegans*), fly (*Drosophila melanogaster*), and mouse (*Mus musculus*)—has provided new insights into how the information encoded in the genome produces complex multicultural organisms. The pilot and first phases of the ENCODE analysis produced new insights, pointed out the complexity of the biology, and raised new questions. In 2007 the NHGRI expanded the ENCODE Project to include the entire genome and initiated the model organism ENCODE project to catalog the genomic elements of the worm and fly, which share similarities with other organisms, even humans. The advantage of cataloging these genomes is that they are small enough to be thoroughly examined using currently available technologies.

In 2013 the NIH released the first human genomic data for Alzheimer's disease (AD). In the press release "NIH Deposits First Batch of Genomic Data for Alzheimer's Disease" (December 2, 2013, https://www.genome.gov/27555537/), Collins explains, "Providing raw DNA sequence data to a wide range of researchers proves a powerful crowd-sourced way to find genomic changes that put us at increased risk for this devastating disease." In 2014 a second batch of data was released and made available at the Alzheimer's Disease Sequencing Project website (https://www.niagads.org/adsp). The project aims to:

- Identify new genomic variants contributing to increased risk of developing late-onset AD

- Identify new genomic variants contributing to protection against developing AD

- Provide insight as to why individuals with known risk factor variants escape from developing AD

- Examine these factors in multiethnic populations as applicable to identify new pathways for disease prevention

Another NIH initiative is the creation of publicly available libraries, such as the National Chemical Genomics Center (NCGC), which collects and catalogs data about chemical compounds for scientists who are developing chemical probes and charting biological pathways. These chemical compounds have many promising applications in genomic research. For example, their ability to enter cells readily makes them natural vehicles for drug development and drug delivery system design. The NCGC uses technologies such as robotic-enabled, high-output screening to create large libraries that contain up to a million chemical compounds.

The United Kingdom's Wellcome Trust, Canadian-funded organizations, and the global pharmaceutical company GlaxoSmithKline established the charitable organization Structural Genomics Consortium in 2003 to round out international efforts in structural genomics. Structural genomics is the systematic, high-volume generation of the three-dimensional structure of proteins. The goal of examining the structural genomics of any organism is the complete structural description of all proteins encoded by the genome of that organism. These descriptions are important for drug design, diagnosis, and treatment of disease. Like the HGP and the NCGC, the Structural Genomics Consortium is placing all the protein structures in public databases that are accessible to scientists worldwide.

The NHGRI cites in "Important Events in NHGRI History" (March 17, 2016, https://www.nih.gov/about-nih/what-we-do/nih-almanac/national-human-genome-research-institute-nhgri) some of the institute's milestones in the postgenomic era. In 2015 the NHGRI celebrated the 25th anniversary of the HGP and hosted the seminar series "A Quarter Century after the Human Genome Project: Lessons beyond Base Pairs," during which HGP participants shared their views about the project and its influence on their career. That same year 1KG scientists completed the world's largest catalog of genomic differences among humans, offering researchers with information to help them determine differences in susceptibility to diseases.

The NHGRI also reports that in 2015 Shawn Burgess and his colleagues developed a transgenic zebrafish as a live animal model of metastasis (the spread of cancer), which provides cancer researchers with a new way to screen drugs and identify new therapeutic targets. Another achievement in 2015 was NIH researchers' finding that genomic switches of a blood cell are important regulators of immune system function. This finding may inform drug development and treatment for people with autoimmune disorders (when the body's immune system mistakenly attacks and destroys healthy tissue).

ENCODE Finds There Is No Junk DNA

Besides the genes themselves, the genome is composed of a much larger amount of DNA whose function remained unknown for many years. Until 2012 this portion of the genome was often referred to as "junk DNA," because its true value and function had not yet been determined. It was widely held that repeated sequences that do not code for proteins (junk DNA) made up at least 50% of the human genome, that just 2% to 10% of human DNA was actually used, and that the balance was essentially inactive.

In 2012 four studies—"Presenting ENCODE" (*Nature*, vol. 489, no. 7414, September 6, 2012) by Magdalena Skipper, Ritu Dhand, and Philip Campbell; "Massive Study Compilation Illuminates Regulatory Role of Non-gene-Encoding DNA" (*Journal of the American Medical Association*, vol. 308, no. 14, October 10, 2012) by Bridget M. Kuehn; "Cracking ENCODE" (*Lancet*, vol. 380, no. 9846, September 2012); and "Genomics. ENCODE Project Writes Eulogy for Junk DNA" (*Science*, vol. 337, no. 6099, September 7, 2012) by Elizabeth Pennisi—that were overseen by Ewan Birney (1972–) of the European Bioinformatics Institute described a huge portfolio of at least 4 million genetic switches, signals, and markers that control how cells, organs, and other tissues behave and are implanted like hieroglyphics throughout human DNA.

The portion of the DNA previously called "junk" has been redubbed the "dark matter" of DNA. Researchers are enthused about the potential to use the information contained in this dark matter to find cures for previously incurable diseases. If these previously undetected genetic switches can keep the body functioning and healthy, then they must also play a role in the development of disease. In "Don't Trash These Genes" (Time.com, October 22, 2012), Alice Park quotes Eric Green, the director of the NHGRI, who asserts, "This is a powerful resource for exploring the fundamental question of how life is

encoded. It will help us understand the genomic basis of human disease."

Because only about 2% of the human genome encodes proteins, researchers are scrutinizing noncoding regions. Susan Jenks reports in "Navigating the Genome's 'Dark Matter'" (*Journal of the National Cancer Institute*, vol. 105, no. 10, May 15, 2013) that ENCODE "found that at least three-quarters of the genome's DNA is transcribed into RNA molecules of many types, not just messenger RNAs, which turn DNA into proteins." Genetic differences in the genome's noncoding region that were previously considered unimportant may prove to be significant. This noncoding RNA is now believed to be an as yet uncharted source of genetic information with the potential to inform diagnostic and therapeutic medical genetics. Jenks spoke to John Mattick of the Garvin Institute of Medical Research in Sydney, Australia, who surmised that "the complexity and diversity of human beings and other eukaryotes lies in their RNA, not in the number of protein-coding genes."

By 2016 the importance of noncoding regions of DNA was established. Whether regulating gene expression and transcription or providing protein attachment sites, this once-dismissed portion of the genome is considered to be vitally important in development and disease. In "Global Mapping of Human RNA-RNA Interactions" (*Molecular Cell*, vol. 62, no. 4, May 19, 2016), Eesha Sharma et al. describe their development of a method that enables researchers to detect and understand the actions of noncoding RNAs in human cells.

CHAPTER 8

REPRODUCTIVE CLONING

Human reproductive cloning should not now be practiced. It is dangerous and likely to fail. The panel therefore unanimously supports the proposal that there should be a legally enforceable ban on the practice of human reproductive cloning. For this purpose, we define human reproductive cloning as the placement in a uterus of a human blastocyst derived by the technique that we call nuclear transplantation.

—Committee on Science, Engineering, and Public Policy Board on Life Sciences, *Scientific and Medical Aspects of Human Reproductive Cloning* (2002)

The Human Genome Project defines three distinct types of cloning. The first is the use of highly specialized deoxyribonucleic acid (DNA) technology to produce multiple, exact copies of a single gene or other segment of DNA to obtain sufficient material to examine for research purposes. This process produces cloned collections of DNA known as clone libraries. The second kind of cloning involves the natural process of cell division to create identical copies of the entire cell. These copies are called a cell line. The third type of cloning, reproductive cloning, is the one that has received the most attention in the mass media. This is the process that generates complete, genetically identical organisms such as Dolly, the famous Scottish sheep cloned in 1996 and named after the entertainer Dolly Parton (1946–).

Cloning may also be described by the technology that is used to perform it. For example, the term *recombinant DNA technology* describes the technology and mechanism of DNA cloning. Also known as molecular cloning or gene cloning, it involves the transfer of a specific DNA fragment of interest to researchers from one organism to a self-replicating genetic element of another species such as a bacterial plasmid. The DNA under study may then be reproduced in a host cell. This technology has been in use since the 1970s and is a standard practice in molecular biology laboratories.

Figure 8.1 displays and describes the dissimilarities between different types of cloning. It also presents some of the current and potential applications of therapeutic cloning, which produces stem cells that may be used to improve human health. This chapter focuses on cloning genes and reproductive cloning and the following chapter describes therapeutic cloning and stem cell research.

Just as GenBank is a free online public repository of the human genome sequence, the Clone Registry database is a sort of "public library." It is used by genome sequencing centers to record which clones have been selected for sequencing, which sequencing efforts are currently under way, and which are finished and represented by sequence entries in GenBank. To coordinate all of this information, a standardized system of naming clones is essential. The nomenclature used is shown in Figure 8.2.

CLONING GENES

Molecular cloning is performed to enable researchers to have many copies of genetic material available in the laboratory for the purpose of experimentation. Cloned genes allow researchers to examine encoded proteins and are used to sequence DNA. Gene cloning also allows researchers to isolate and experiment on the genes of an organism. This is particularly important in terms of human research; in instances where direct experimentation on humans might be dangerous or unethical, experimentation on cloned genes is often practical and feasible.

Figure 8.3 shows how a gene is cloned. First, a DNA fragment containing the gene being studied is isolated from chromosomal DNA using restriction enzymes. It is joined with a plasmid (a small ring of DNA found in many bacteria that can carry foreign DNA) that has been cut with the same restriction enzymes. When the fragment of chromosomal DNA is joined with its cloning vector (cloning vectors, such as plasmids and artificial

FIGURE 8.1

Types of cloning and use of stem cells

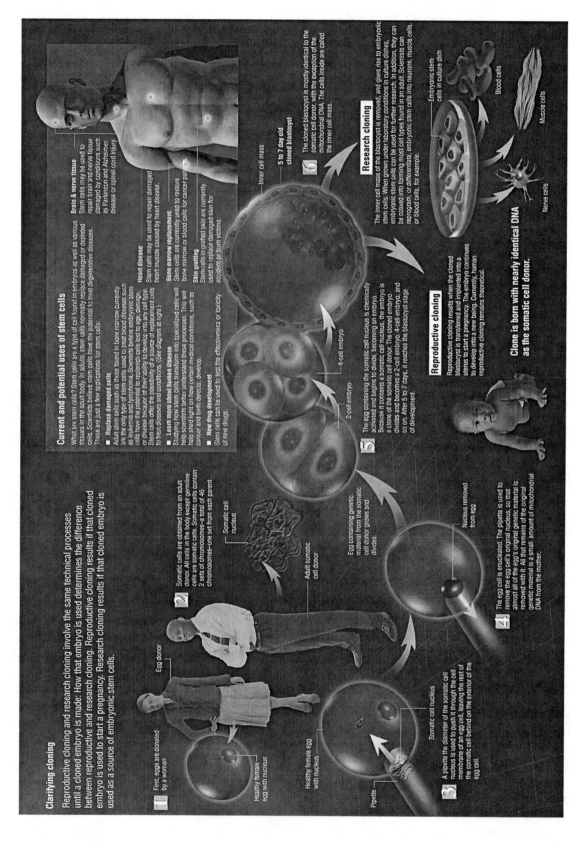

SOURCE: Christina Ullman, "Genes & Society: Cloning," in *Genes & Society Graphics*, Genetics & Public Policy Center at Johns Hopkins University with support from the Pew Charitable Trusts, 2010. Courtesy of the Genetics & Public Policy Center with support from The Pew Charitable Trusts.

FIGURE 8.2

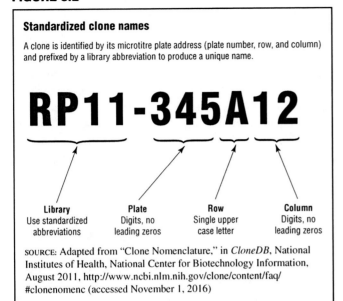

Standardized clone names

A clone is identified by its microtitre plate address (plate number, row, and column) and prefixed by a library abbreviation to produce a unique name.

RP11-345A12

Library
Use standardized abbreviations

Plate
Digits, no leading zeros

Row
Single upper case letter

Column
Digits, no leading zeros

SOURCE: Adapted from "Clone Nomenclature," in *CloneDB*, National Institutes of Health, National Center for Biotechnology Information, August 2011, http://www.ncbi.nlm.nih.gov/clone/content/faq/#clonenomenc (accessed November 1, 2016)

FIGURE 8.3

Cloning deoxyribonucleic acid in plasmids

By fragmenting DNA of any origin (human, animal, or plant) and inserting it in the DNA of rapidly reproducing foreign cells, billions of copies of a single gene or DNA segment can be produced in a very short time. DNA to be cloned is inserted into a plasmid (a small, self-replicating circular molecule of DNA) that is separate from chromosomal DNA. When the recombinant plasmid is introduced into bacteria, the newly inserted segment will be replicated along with the rest of the plasmid.

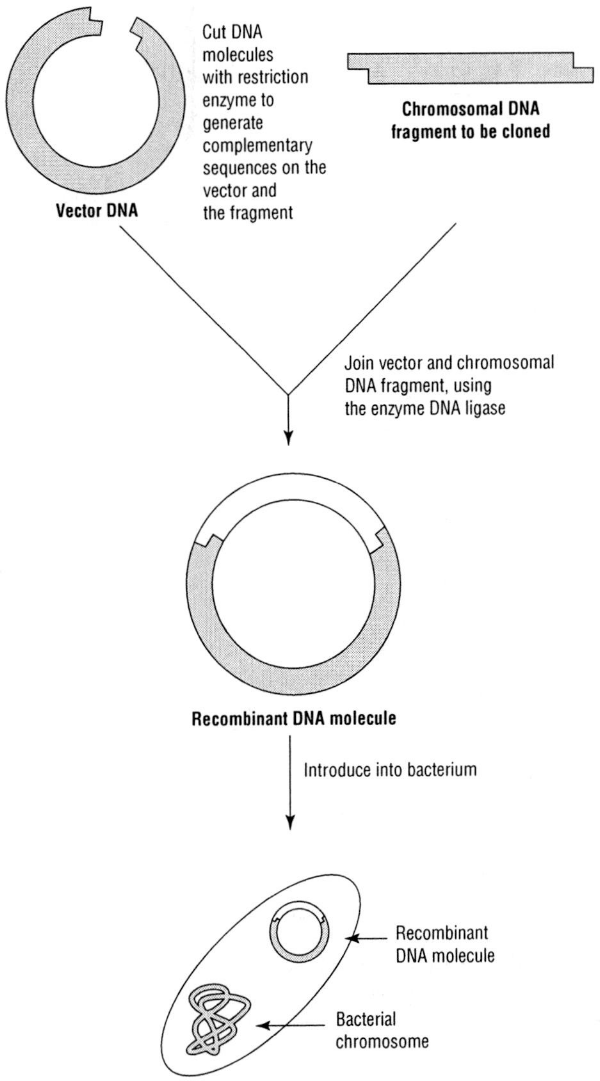

Cut DNA molecules with restriction enzyme to generate complementary sequences on the vector and the fragment

Vector DNA

Chromosomal DNA fragment to be cloned

Join vector and chromosomal DNA fragment, using the enzyme DNA ligase

Recombinant DNA molecule

Introduce into bacterium

Recombinant DNA molecule

Bacterial chromosome

DNA = deoxyribonucleic acid.

SOURCE: Denise Casey, "Fig. 11a. Cloning DNA in Plasmids," in *Primer on Molecular Genetics*, U.S. Department of Energy Office of Science, Office of Biological and Environmental Research, Human Genome Project, 1992, http://web.ornl.gov/sci/techresources/Human_Genome/publicat/primer2001/primer.pdf (accessed November 1, 2016)

yeast chromosomes, introduce foreign DNA into host cells), it is called a recombinant DNA molecule. Once it has entered the host cell, the recombinant DNA can be reproduced along with the host cell DNA.

Cloned genes are used to produce pharmaceutical drugs, insulin (a pancreatic hormone that regulates blood glucose levels), clotting factors, human growth hormone, and industrial enzymes. Before the widespread use of molecular cloning, these proteins were difficult and expensive to manufacture. For example, before the development of recombinant DNA technology, insulin was extracted and purified from cow and pig pancreases. Because the amino acid sequences of insulin from cows and pigs are slightly different from those in human insulin, some patients with diabetes experienced adverse immune reactions to the nonhuman "foreign insulin." The recombinant human version of insulin is identical to human insulin so it does not produce an immune reaction. Figure 8.4 shows how recombinant DNA is used to produce insulin.

Another molecular cloning technique that is simpler and less expensive than the recombinant cloning method is the polymerase chain reaction (PCR). PCR has also been dubbed "molecular photocopying" because it amplifies DNA without the use of a plasmid. Figure 6.3 in Chapter 6 shows how PCR is used to generate a virtually unlimited number of copies of a piece of DNA.

A collection of clones of chromosomal and vector DNA (a small piece of DNA that contains regulatory and coding sequences of interest) is called a library. These libraries of clones containing partly overlapping regions are constructed to show that two particular clones are next to one another in the genome. Figure 8.5 shows how, by dividing the inserts into smaller fragments and determining which clones share the same DNA sequences, clone libraries are constructed.

ORGANISMAL OR REPRODUCTIVE CLONING

Another way to describe and classify cloning is by its purpose. Organismal or reproductive cloning is a technology used to produce a genetically identical

FIGURE 8.4

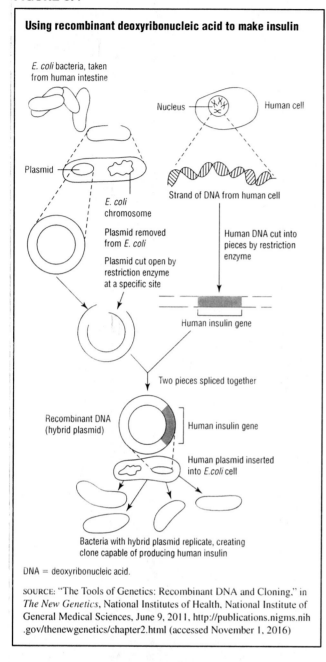

Using recombinant deoxyribonucleic acid to make insulin

E. coli bacteria, taken from human intestine

Human cell

Nucleus

Plasmid

E. coli chromosome

Strand of DNA from human cell

Plasmid removed from E. coli

Human DNA cut into pieces by restriction enzyme

Plasmid cut open by restriction enzyme at a specific site

Human insulin gene

Two pieces spliced together

Recombinant DNA (hybrid plasmid)

Human insulin gene

Human plasmid inserted into E.coli cell

Bacteria with hybrid plasmid replicate, creating clone capable of producing human insulin

DNA = deoxyribonucleic acid.

SOURCE: "The Tools of Genetics: Recombinant DNA and Cloning," in *The New Genetics*, National Institutes of Health, National Institute of General Medical Sciences, June 9, 2011, http://publications.nigms.nih.gov/thenewgenetics/chapter2.html (accessed November 1, 2016)

organism—an animal with the same nuclear DNA as an existing animal.

The reproductive cloning technology used to create animals is called somatic cell nuclear transfer (SCNT). In SCNT scientists transfer genetic material from the nucleus of a donor adult cell to an enucleated egg cell (a cell without a nucleus). This eliminates the need for fertilization of an egg by a sperm. The reconstructed egg containing the DNA from a donor cell is treated with chemicals or electric current to stimulate cell division. Once the cloned embryo reaches a suitable stage, it is transferred to the uterus of a surrogate (female host), where it continues to grow and develop until birth. Figure 8.6 shows the SCNT process that culminates in the transfer of the embryo into the surrogate mother and ultimately the birth of a cloned animal.

Organisms or animals generated using SCNT are not perfect or identical clones of the donor organism or "parent" animal. The clone's nuclear DNA is identical to the donor's, but some of the clone's genetic materials come from the mitochondria in the cytoplasm of the enucleated egg. Mitochondria, the organelles that serve as energy sources for the cell, contain their own short segments of DNA called mtDNA. Acquired mutations in the mtDNA contribute to differences between clones and their donors and are believed to influence the aging process.

As of 2017 there was no U.S. federal regulation of therapeutic and reproductive cloning; however, some states had enacted legislation that prohibited therapeutic and/or reproductive cloning. According to the Center for Genetics and Society, in "Human Cloning Policies" (2017, http://www.geneticsandsociety.org/article.php?id=325), human cloning has been completely banned in 46 countries, including Canada, France, Germany, Russia, and the United Kingdom.

Dolly the Sheep Paves the Way for Other Cloned Animals

In 1952 scientists transferred a cell from a frog embryo into an unfertilized egg, which then developed into a tadpole. This process became the prototype for cloning. Ever since, scientists have been cloning animals. During the 1980s the first mammals were also cloned from embryonic cells. In 1996 cloning became headline news when, after more than 250 failed attempts, Ian Wilmut (1944–) and his colleagues at the Roslin Institute in Edinburgh, Scotland, successfully cloned a sheep, which they named Dolly. Dolly was the first mammal cloned from the cell of an adult animal, and since then researchers have used cells from adult animals and modifications of nuclear transfer technology to clone a range of animals, including a gaur, sheep, goats, cows, horses, mules, oxen, deer, mice, rats, pigs, cats, dogs, and rabbits.

To create Dolly, the researchers transplanted a nucleus from a mammary gland cell of a Finn Dorsett sheep into the enucleated egg of a Scottish blackface ewe and used electricity to stimulate cell division. The newly formed cell divided and was placed in the uterus of a blackface ewe to gestate. Born several months later, Dolly was a true clone—genetically identical to the Finn Dorsett mammary cells and not to the blackface ewe, which served as her surrogate mother. Her birth revolutionized the world's understanding of molecular biology, ignited worldwide discussion about the morality of generating new life through cloning, resulted in legislation in dozens of countries, and prompted an ongoing political debate in Congress.

FIGURE 8.5

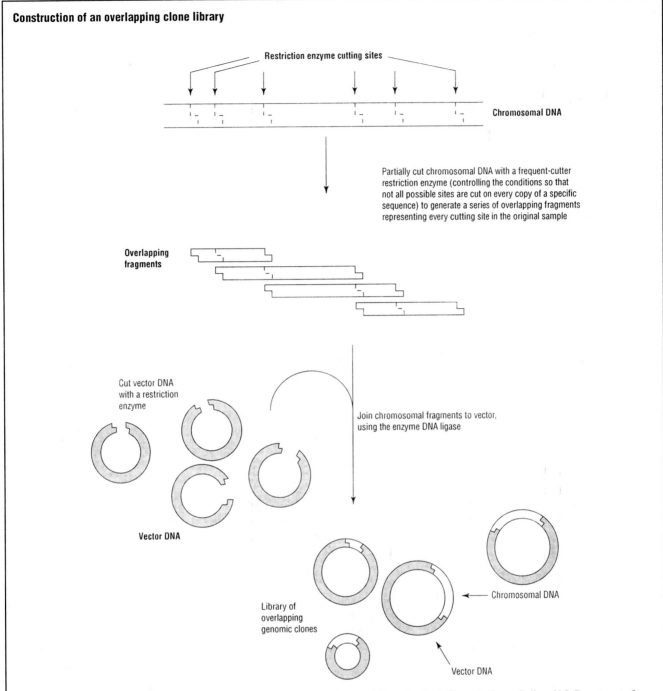

Construction of an overlapping clone library

Restriction enzyme cutting sites

Chromosomal DNA

Partially cut chromosomal DNA with a frequent-cutter restriction enzyme (controlling the conditions so that not all possible sites are cut on every copy of a specific sequence) to generate a series of overlapping fragments representing every cutting site in the original sample

Overlapping fragments

Cut vector DNA with a restriction enzyme

Join chromosomal fragments to vector, using the enzyme DNA ligase

Vector DNA

Library of overlapping genomic clones

Chromosomal DNA

Vector DNA

SOURCE: "Construction of an Overlapping Clone Library," in *Human Genome Project Information Basic Genomics Image Gallery*, U.S. Department of Energy Office of Science, Office of Biological and Environmental Research, Human Genome Project, 2008, https://public.ornl.gov/site/gallery/gallery.cfm (accessed November 1, 2016)

Dolly was the object of intense media and public fascination. She proved to be a healthy clone and produced six lambs of her own through normal sexual means. Before her death by lethal injection in 2003, Dolly suffered from lung cancer and arthritis. An autopsy (postmortem examination) revealed that, other than cancer and arthritis, which are common diseases in sheep, Dolly was anatomically like other sheep.

In 1997 Don Paul Wolf (1939–) and his colleagues at the Oregon Regional Primate Center cloned two rhesus monkeys using laboratory techniques that had previously produced frogs, cows, and mice. It was the first time that researchers used a nuclear transplant to generate monkeys. The monkeys were created using different donor blastocysts (early-stage embryos), so they were not clones of one another—each monkey was a clone of the original blastocyst that developed from a

FIGURE 8.6

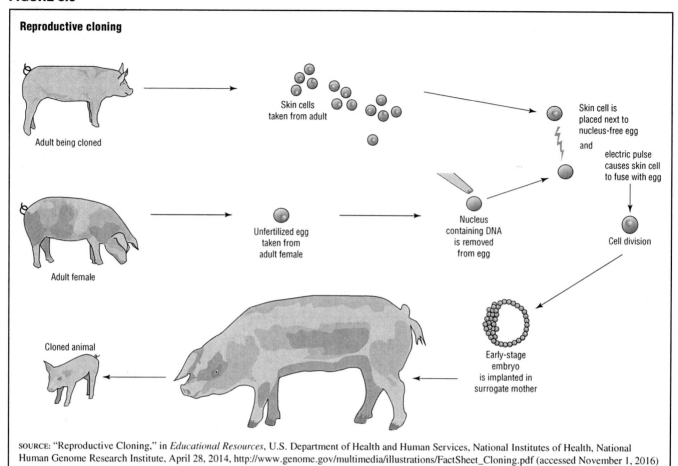

Reproductive cloning

SOURCE: "Reproductive Cloning," in *Educational Resources*, U.S. Department of Health and Human Services, National Institutes of Health, National Human Genome Research Institute, April 28, 2014, http://www.genome.gov/multimedia/illustrations/FactSheet_Cloning.pdf (accessed November 1, 2016)

fertilized egg. Neither of the cloned monkeys survived past the embryonic stage.

An important distinction between the process that created Dolly and the one that produced the monkeys was that a specialized adult cell was used to create Dolly, whereas unspecialized embryonic cells were used to create the monkeys. The Oregon experiment was followed closely in the scientific and lay communities because, in terms of evolutionary biology and genetics, primates are closely related to humans.

In 2000 the Oregon researchers succeeded when one of four embryos, which were created by splitting a blastocyst four ways and implanting the pieces into surrogate mothers, survived. The survivor was named Tetra, from the Greek prefix for the number four. Researchers and the public speculated that if monkeys could be cloned, it might become feasible to clone humans.

In 2001 BresaGen Limited, an Australian biotechnology firm, announced the birth of that country's first cloned pig. The pig was cloned from cells that had been frozen in liquid nitrogen for more than two years, and the company employed technology that was different from the process used to clone Dolly the sheep. The most immediate benefit of this new technology was to improve livestock—cloning enables breeders to take some animals with superior genetics and rapidly produce more. Biomedical scientists were especially attentive to this research because of its potential for xenotransplantation (the use of animal organs for transplantation into humans). Pig organs that are genetically modified so that they will not be rejected by the human immune system could prove to be a boon to medical transplantation.

That same year the first cat was cloned, and the following year rabbits were successfully cloned. In January 2003 researchers at Texas A&M University reported that cloned pigs behaved normally—as expected for a litter of pigs—but were not identical to the animals from which they were cloned in terms of food preferences, temperament, and how they spent their time. The researchers explained the variation as arising from the environment and epigenetic (not involving DNA sequence change) factors, causing the DNA to line up differently in the clones. Epigenetic activity is any gene-regulating action that does not involve changes to the DNA code and that persists through one or more generations, and it may explain why abnormalities such as fetal death occur more frequently in cloned species.

In May 2003 a cloned mule (the first successful clone of any member of the horse family) was born in Idaho. The clone was not just any mule, but the brother of the world's second-fastest racing mule. Named Idaho Gem, the cloned mule was created by researchers at the University of Idaho and Utah State University.

In August 2003 scientists at the Laboratory of Reproductive Technology in Cremona, Italy, were the first to clone a horse. Cesare Galli et al. describe their cloning technique in "Pregnancy: A Cloned Horse Born to Its Dam Twin" (*Nature*, vol. 424, no. 6949, August 7, 2003).

The mule was cloned from cells that were extracted from a mule fetus, whereas the cloned horse's DNA came from adult horse skin cells. There were other differences as well. The researchers harvested fertile eggs from mares, removed the nucleus of each egg, and inserted DNA from cells of a mule fetus. The reconstructed eggs were then surgically implanted into the wombs of female horses. In contrast, Galli et al. harvested hundreds of eggs from mare carcasses, cultured the eggs, removed their DNA, and replaced it with DNA that was taken from either adult male or female horse skin cells. The eggs were then placed in the wombs of female horses to gestate.

In 2004 the first bull was cloned from a previously cloned bull in a process known as serial somatic cell cloning or recloning. Before the bull, the only other successful recloning efforts involved mice. Chikara Kubota, X. Cindy Tian, and Xiangzhong Yang describe their cloning technique in "Serial Bull Cloning by Somatic Cell Nuclear Transfer" (*Nature Biotechnology*, vol. 22, no. 6, June 2004). Their effort was also cited in the *Guinness Book of World Records* as the largest clone in the world.

At the close of 2004 researchers in South Korea reported cloning macaque monkey embryos, which would be used as a source of stem cells. In 2005 Texas A&M University announced the first successfully cloned foal in the United States. That same year scientists at Seoul National University cloned a dog they named Snuppy. In 2006 ferrets were cloned using SCNT.

In 2009 the first camel was cloned in Dubai. In 2010 researchers in Spain reported cloning the first fighting bull. In 2013 researchers in Japan cloned a mouse using a single drop of blood from a living donor mouse. In "Mouse Cloning Using a Drop of Peripheral Blood" (*Biology of Reproduction*, vol. 89, no. 2, August 1, 2013), Satoshi Kamimura et al. describe using blood from live mice for SCNT to produce genetic copies of mice. In 2016 scientists in India reported cloning a buffalo.

World's Largest Animal Cloning Center in China

Charles Clover and Clive Cookson report in "Chinese-Korean Joint Venture to Mass-Produce Cloned Beef Cattle" (FT.com, November 23, 2015) that in 2015 the BoyaLife Group, a Chinese company, partnered with Sooam Biotech, a South Korean firm, in a joint venture to produce 100,000 cloned cattle per year to supply beef. In 2016 the joint venture oversaw the construction of a 151,000-square-foot (14,000-sq-m) cloning facility in Tianjin, China. When completed, it will be the largest animal cloning center in the world. In "World's Biggest Animal Cloning Center Set for '16 in a Skeptical China" (NYTimes.com, November 26, 2015), Owen Guo reports that the center will produce "a million beef cattle embryos a year, as well as sniffer dogs, racehorses and other animals."

Cloning Endangered Species

Reproductive cloning technology may also be used to repopulate endangered species such as the African bongo antelope, the Sumatran tiger, and the giant panda, or animals that reproduce poorly in zoos or are difficult to breed. Cloning an endangered animal, however, is different from cloning a more common animal because cloned animals need surrogate mothers to be carried to term. Furthermore, embryo transfer is risky, and researchers are reluctant to put an endangered animal through the rigors of surrogate motherhood, so they opt to use non-endangered domesticated animals whenever possible.

In 2001 scientists at Advanced Cell Technology, Inc. (later renamed Ocata Therapeutics, Inc.), a biotechnology company in Marlborough, Massachusetts, announced the birth of the first clone of an endangered animal: a bull gaur (a large wild ox from India and Southeast Asia). The gaur was cloned using the nuclei of frozen skin cells taken from an adult male gaur that had died eight years earlier. The skin cell nuclei were joined with enucleated cow eggs, one of which was implanted into a surrogate cow. The cloned gaur died from an infection within days of its birth. That same year scientists in Italy successfully cloned an endangered wild sheep.

In 2003 Advanced Cell Technology announced the birth of a healthy Javan banteng (an endangered cattlelike animal native to the bamboo forests of Asia) clone. The clone was created from a single skin cell that was taken from another banteng before it died in 1980. The skin cell was kept frozen until it was used to create the clone. The banteng embryo gestated in a standard beef cow in Iowa.

Born in 2003, the cloned banteng developed normally, growing its characteristic horns and reaching an adult weight of about 1,800 pounds (820 kg). The banteng lived at the San Diego Zoo until it died in 2010 at age seven, less than half of the anticipated life span of an uncloned banteng. Hunting and habitat destruction reduced the number of banteng by more than 75% between 1983 and 2003. In "Ecological-Economic Models of Sustainable Harvest for an Endangered but Exotic Megaherbivore in Northern Australia" (*Natural Resource*

Modeling, vol. 20, no. 1, March 2007), Corey J. A. Bradshaw and Barry W. Brook of Charles Darwin University report that in 2007 a population of between 8,000 and 10,000 banteng lived on an isolated peninsula in northern Australia.

In 2005 the Audubon Nature Institute in New Orleans, Louisiana, reported that two unrelated endangered African wildcat clones had given birth to eight kittens. These births confirmed that clones of wild animals can breed naturally, which is vitally important for protecting endangered animals on the brink of extinction.

Rob Waters describes in "Animal Cloning: The Next Phase" (Businessweek.com, June 10, 2010) the Frozen Zoo, a laboratory repository of frozen skin cells and DNA from about 800 species that are housed at the San Diego Zoo's Institute for Conservation Research. In 2010 tissue from the Frozen Zoo was used to create stem cells of an endangered African monkey and the stem cells successfully developed into brain cells, giving rise to the hope that in the not-too-distant future it will be possible to clone endangered animals and save them from extinction.

The article "World's First Handmade Cloned Transgenic Sheep Born in China" (ScienceDaily.com, April 19, 2012) notes that in 2011 scientists in China cloned transgenic minipigs and micropigs. In 2012 Chinese scientists announced the birth of the world's first transgenic sheep, named Peng Peng. The scientists used a technique called handmade cloning, which requires less sophisticated and less costly equipment and is more efficient than previous methods, to take cells from a Chinese Merino sheep and genetically manipulate them.

In 2013 researchers in South Korea cloned endangered Jeju black cattle (JBC). In "Post-death Cloning of Endangered Jeju Black Cattle (Korean Native Cattle): Fertility and Serum Chemistry in a Cloned Bull and Cow and Their Offspring" (*Journal of Reproductive Development*, vol. 59, no. 6, December 17, 2013), Eun Young Kim et al. describe using SCNT to produce a JBC bull and cow, which were found to have normal fertility.

The article "Save Endangered Species by Cloning?" (SMDP.com, September 12, 2016) explains that cloning endangered species has a very low success rate (less than 1%), so it is not yet a viable way to save endangered species. However, as the technology improves, it may become possible to resurrect endangered or even extinct animals using frozen genetic material.

Cloning Extinct Species

Cloning extinct animals is even more challenging than cloning living animals because the egg and the surrogate mother used to create and harbor the cloned embryo are not the same species as the clone. Furthermore, for most already extinct animal species such as the woolly mammoth (*Mammuthus primigenius*) or the smilodon (*Smilodon populator*), there is insufficient intact cellular and genetic material from which to generate clones. In the future, carefully preserving intact cellular material of imperiled species may allow for their preservation and propagation.

Sariah Cottrell, Jamie L. Jensen, and Steven L. Peck observe in "Resuscitation and Resurrection: The Ethics of Cloning Cheetahs, Mammoths, and Neanderthals" (*Life Sciences, Society and Policy*, vol. 10, no. 3, January 10, 2014) that cloning for conservation purposes is already occurring and cloning extinct species may soon follow. They discuss the ethics of cloning endangered and extinct species, citing the examples of the endangered cheetah (*Acinonyx jubatus*) and the extinct woolly mammoth and Neanderthal (*Homo neanderthalensis*). Cottrell, Jensen, and Peck opine that cloning a cheetah is "an ethically admirable practice" because it potentially would preserve a large number of species. By contrast, they deem cloning a mammoth "problematical" because it would be very expensive and would create another endangered species. They also oppose cloning Neanderthals because they are closely related to humans and argue, "If it is unethical to clone humans, and if cloned Neanderthals would be morally equivalent to humans and receive human rights, then it is likewise unethical to clone Neanderthals."

Reproductive Human Cloning

In 2002 a group known as the Raelians made news when their private biotechnology firm, Clonaid, announced that it had successfully delivered the world's first cloned baby. The announcement, which was not independently verified or substantiated, generated unprecedented media coverage and was condemned in the scientific and lay communities.

Clonaid's announcement, which ultimately was viewed as a hoax, brought attention to the fact that several laboratories around the world had embarked on clandestine efforts to clone a human embryo. For example, in 2002 Panayiotis Zavos (1944–), a U.S. fertility specialist, claimed to be collaborating with about two dozen international researchers to produce human clones. Another doctor, Severino Antinori (1945–), attracted media attention when he maintained that hundreds of infertile couples in Italy and thousands in the United States had already enrolled in his human cloning initiative. As of February 2017, neither these researchers nor anyone else had offered proof of successful reproductive human cloning.

Moral and Ethical Objections to Human Cloning

The difficulty and low success rate of much animal reproductive cloning (an average of just one or two viable offspring result from every 100 attempts) and the as-yet-incomplete understanding of reproductive cloning have prompted many scientists to deem it unethical to attempt to clone humans. Many attempts to clone mammals fail, and about one-third of clones suffer from anatomical, physiological, or developmental abnormalities that are often debilitating. Some cloned animals die prematurely from infections and other complications at rates higher than conventionally bred animals, and researchers anticipate comparable outcomes from human cloning. Furthermore, scientists cannot yet characterize how cloning influences intellectual and emotional development. Although the attributes of intelligence, temperament, and personality may not be as important for cattle or other primates, they are vital for humans. Without considering the myriad religious, social, and other ethical concerns, the presence of so many unanswered questions about the science of reproductive cloning has prompted many scientists to consider any attempts to clone humans as scientifically irresponsible, unacceptably risky, and morally unallowable.

People who oppose human cloning are as varied as the interests and institutions they support. Religious leaders, scientists, politicians, philosophers, and ethicists argue against the morality and acceptability of human cloning. Nearly all objections hinge, to various degrees, on the definition of human life, beliefs about its sanctity, and the potentially adverse consequences for families and society as a whole.

In an effort to stimulate consideration of and debate about this critical issue, the President's Council on Bioethics examined the moral and ethical objections to human cloning in *Human Cloning and Human Dignity: An Ethical Inquiry* (July 2002, https://repository.library .georgetown.edu/bitstream/handle/10822/559368/pcbe _cloning_report.pdf?sequence=1&isAllowed=y). The council's report distinguished between therapeutic and reproductive cloning and outlined key concerns by trying to respond to many as yet unresolved questions about the ethics, morality, and societal consequences of human cloning.

The council determined that the key moral and ethical objections to therapeutic cloning (cloning for biological research) center on the moral status of developing human life. When therapeutic cloning involves the deliberate production, use, and destruction of cloned human embryos, many contend that cloned embryos produced for research are no different from those that could be used in attempts to create cloned children. Even when therapeutic cloning is performed using cells from adults, as was reported by several research teams in 2014, concerns remain that acceptance of therapeutic cloning will lead society down a slippery slope to reproductive cloning, a prospect that is almost universally viewed as unethical

and morally unacceptable. (For a detailed examination of the risks and benefits of therapeutic cloning, see Chapter 9.)

The unacceptability of human reproductive cloning stems from the fact that it challenges the basic nature of human procreation by redefining having children as a form of manufacturing. Human embryos may then be viewed as products and commodities rather than as sacred and unique human beings. Furthermore, reproductive cloning might substantially change fundamental issues of human identity and individuality, and allowing parents unprecedented genetic control of their offspring may significantly alter family relationships across generations.

The council concluded that "the right to decide '*whether* to bear or beget a child' does not include a right to have a child *by whatever means*. Nor can this right be said to imply a corollary—the right to decide what *kind* of child one is going to have ....Our society's commitment to freedom and parental authority by no means implies that all innovative procedures and practices should be allowed or accepted, no matter how bizarre or dangerous."

Nearly universal opposition to human cloning persists well into the second decade of the 21st century. For example, in "Guest Post by Arthur Caplan on Human Reproductive Cloning" (January 8, 2013, http://www.ge neticsandsociety.org/article.php?id=6626), the American bioethicist Arthur L. Caplan (1950–), the 2014 recipient of the National Science Board Public Service Award, suggests that human reproductive cloning will never become popular because it does not offer immortality. Instead, Caplan believes that interest will center on creating children outside the womb in safer environments and perhaps genetically enhancing such children. He also posits that reproductive cloning will be widely used in agricultural applications—to create plants and animals. Caplan concludes that "human cloning for reproduction will, in my view, be of ethical interest only as a matter of historical folly."

Objections to Editing the Human Genome

In 2015 a group of biologists called for a global ban on the use of CRISPR-Cas9, a new genome editing technique that was invented by Jennifer Doudna (1964–) of the University of California, Berkeley, and Emmanuelle Charpentier (1968–) of Umea University. CRISPR-Cas9 changes human DNA in a way that can be inherited. David Baltimore et al. explain in "A Prudent Path forward for Genomic Engineering and Germline Gene Modification" (*Science*, vol. 348, no. 6230, April 3, 2015) that there is concern that the technique, which can repair or enhance any gene, might be used before its safety can be assured. Although the technique may be used to cure genetic diseases, scientists caution that replacing a defective gene with a normal one may not be entirely harmless and could have unanticipated consequences. They also note that the technique could be used to alter characteristics such as beauty, intelligence, and athletic ability. Baltimore et al. support continuing research using the technique and express confidence that its use will be strictly regulated in the United States and Europe, but they are concerned about its use in countries where genomic research is less regulated.

Public Opinions about Cloning

According to Art Swift of Gallup, Inc., in *Birth Control, Divorce Top List of Morally Acceptable Issues* (June 8, 2016, http://www.gallup.com/poll/192404/birth -control-divorce-top-list-morally-acceptable-issues.aspx), a 2016 poll found that 13% of Americans supported cloning humans, whereas 81% opposed it. In contrast, Americans were more favorable to cloning animals— 34% supported it and 60% opposed it.

CHAPTER 9
THERAPEUTIC CLONING: STEM CELL RESEARCH

Therapeutic cloning is the creation of stem cells for use in biomedical research. Stem cells are "master cells" that are capable of differentiating into multiple other cell types. This potential is important to biomedical researchers because stem cells may be used to generate any type of specialized cell, such as nerve, muscle, blood, or brain cells. Stem cells not only provide a ready supply of replacement tissue but also may hold the key to developing more effective treatments for common disorders such as heart disease and cancer as well as for degenerative diseases such as Alzheimer's and Parkinson's.

Advocates of therapeutic cloning point to other treatment benefits such as using stem cells to generate bone marrow for transplants. They observe that therapeutic cloning can produce perfectly matched bone marrow using the patient's own skin or other cells. This effectively eliminates the problem of rejection of foreign tissue that is associated with bone marrow transplant and other organ transplantation.

The National Institutes of Health (NIH) explains in "Stem Cell Basics V" (2016, https://stemcells.nih.gov/info/basics/5.htm) that because adult stem cells are derived from a patient's own tissue, they are less likely to be rejected following transplantation.

EMBRYONIC STEM CELLS

Embryonic stem cells used in research are harvested from the blastocyst after it has divided for five days, during the earliest stage of embryonic development. Embryonic stem cells are derived from embryos that have been created in vitro and then donated for research. They are not from embryos conceived in a woman's body, nor are they embryos destined to gestate. Nonetheless, because many people regard human embryos as potential human beings, they consider their destruction, or even using techniques to obtain stem cells that might imperil their future viability, as immoral or unethical.

Historically, embryonic stem cells were thought to have advantages over adult stem cells. Because embryonic stem cells are pluripotent, they can be readily differentiated to become any type of cell in the body, whereas until 2012 adult stem cells were thought to be limited to developing into cell types from their tissue of origin. In 2012 María Berdasco et al. reported the transformation of adult stem cells from body fat into muscle and bone cells in "DNA Methylation Plasticity of Human Adipose-Derived Stem Cells in Lineage Commitment" (*American Journal of Pathology*, vol. 181, no. 6, December 2012).

Therapeutic Cloning to Produce Stem Cells

This section considers selected milestones in therapeutic cloning (also known as somatic cell nuclear transfer [SCNT]) and stem cell research from 2001, when the first cloned human embryo was reported in the medical literature, to 2016, when stem cells were created from adults using cloning techniques similar to the techniques used to create Dolly, the cloned sheep.

In 2001 Jose B. Cibelli et al. of Advanced Cell Technology, Inc. (later renamed Ocata Therapeutics, Inc.), reported in "Somatic Cell Nuclear Transfer in Humans: Pronuclear and Early Embryonic Development" (*E-biomed: The Journal of Regenerative Medicine*, vol. 2, November 26, 2001) the creation of a cloned human embryo. To clone the human embryos, Cibelli et al. collected women's eggs and removed the genetic material from the eggs with a thin needle. A skin cell was inserted inside each of eight enucleated eggs, which were then chemically stimulated to divide.

In 2003 a Chinese research team reported that it had made human embryonic stem cells by combining human skin cells with rabbit eggs. In "Embryonic Stem Cells Generated by Nuclear Transfer of Human Somatic Nuclei into Rabbit Oocytes" (*Cell Research*, vol. 13, no. 4, August 2003), Ying Chen et al. describe how they removed the rabbit eggs' deoxyribonucleic acid (DNA) and injected human skin cells into them. The eggs then grew to form embryos containing human genetic material. After several days the embryos were dissected to extract their stem cells.

In 2005 Ian Wilmut (1944–) was granted a license by the British government to clone human embryos to generate stem cell lines to study motor neuron disease (MND). By observing how stem cells grow into neurons, the researchers hoped to discover what causes the cells to degenerate. Their research compared stem cells with healthy and diseased cells from MND patients to improve understanding of the illness.

In 2006 Deepa M. Deshpande et al. restored movement to paralyzed rats using a method that demonstrates the potential of embryonic stem cells to restore function to humans with neurological disorders. They published their results in "Recovery from Paralysis in Adult Rats Using Embryonic Stem Cells" (*Annals of Neurology*, vol. 60, no. 1, July 2006). In 2008 other researchers reported successfully treating mice for Parkinson's disease using therapeutic cloning. In "Therapeutic Cloning in Individual Parkinsonian Mice" (*Nature Medicine*, vol. 14, no. 4, April 2008), Viviane Tabar et al. explain that they used skin cells from the tails of the mice to generate the missing nerve cells in Parkinson's disease.

In 2006 Kevin A. D'Amour et al. of Novocell Inc. in San Diego, California, reported in "Production of Pancreatic Hormone-Expressing Endocrine Cells from Human Embryonic Stem Cells" (*Nature Biotechnology*, vol. 24, no. 11, October 2006) the development of a process to turn human embryonic stem cells into pancreatic cells capable of producing insulin.

Three studies—Volker Schächinger et al.'s "Intracoronary Bone Marrow–Derived Progenitor Cells in Acute Myocardial Infarction," Ketil Lunde et al.'s "Intracoronary Injection of Mononuclear Bone Marrow Cells in Acute Myocardial Infarction," and Birgit Assmus et al.'s "Transcoronary Transplantation of Progenitor Cells after Myocardial Infarction"—describing the use of stem cells in the treatment of heart disease were published in the *New England Journal of Medicine* (vol. 355, no. 12, September 21, 2006). The studies produced conflicting results: Schächinger et al. reported benefits for patients who had suffered from heart attacks. Lunde et al. found no benefit from stem cell treatment of such patients. Assmus et al. studied patients with chronic heart failure, who did show improvement after treatment. In the editorial "Cardiac Cell Therapy—Mixed Results from Mixed Cells" in the same issue of the journal, Anthony Rosenzweig writes that the three studies "provide a realistic perspective on this approach while leaving room for cautious optimism and underscoring the need for further study."

Lina Dahl et al. report in "Lhx2 Expression Promotes Self-Renewal of a Distinct Multipotential Hematopoietic Progenitor Cell in Embryonic Stem Cell-Derived Embryoid Bodies" (*PLoS One*, vol. 3, no. 4, April 23, 2008) that they used genetically modified embryonic stem cells to isolate the stem cell that produces blood cells. This finding not only promises enhanced ability to develop treatment for diseases of the blood such as anemia and leukemia but is also key to developing techniques to replace damaged or diseased tissue with new, healthy tissue.

For years scientists questioned whether cloned fetuses would have more mutations than naturally conceived fetuses. In 2009, after five years of research comparing mutation rates in mouse fetuses, Patricia Murphey et al. reported in "Epigenetic Regulation of Genetic Integrity Is Reprogrammed during Cloning" (*Proceedings of the National Academy of Sciences*, vol. 106, no. 12, March 24, 2009) that the rate of spontaneous mutations was similar in fetuses produced by either cloning or natural conception, which means that cloned fetuses do not acquire mutations more rapidly than naturally conceived fetuses. The researchers even speculate that there "may be a relationship between pluripotency and enhanced maintenance of genetic integrity" for cloned fetuses.

One of the challenges of culturing human embryonic stem cells has been to grow them without adding other animal proteins to the culture, which then makes them unusable for treating humans. In "Long-Term Self-Renewal of Human Pluripotent Stem Cells on Human Recombinant Laminin-511" (*Nature Biotechnology*, vol. 28, no. 6, June 2010), Sergey Rodin et al. report the first successful culture of human embryonic stem cells using a chemical environment that was free of animal proteins or other contaminants. The researchers cultured the human embryonic stem cells in a matrix of a single human protein called laminin-511.

Other research sheds light on how embryonic stem cells behave and change. Maurice A. Canham et al. observe in "Functional Heterogeneity of Embryonic Stem Cells Revealed through Translational Amplification of an Early Endodermal Transcript" (*PLoS Biology*, vol. 8, no. 5, May 25, 2010) that embryonic stem cells change in response to certain environmental cues, which prompt them to become a specific tissue cell type. They not only can morph into precursors for adult cells but also are able to change into more primitive cells, such as those associated with the placenta (the lining of the uterus that surrounds the developing fetus). Understanding when embryonic stem cells commit to becoming a specific cell type is key to producing specific types of cells in the laboratory.

In 2012 two pioneering scientists were recognized for their contributions to reproductive and therapeutic cloning. John B. Gurdon (1933–), the first researcher to clone an animal, and Shinya Yamanaka (1962–), who identified four transcription factors (proteins that can reprogram an adult cell back to its stem cell form), were awarded the Nobel Prize in Physiology or Medicine.

In 2013 researchers reported the successful use of SCNT techniques (described in Chapter 8) to create a source of embryonic stem cells from the skin cells of a patient. In "Human Embryonic Stem Cells Derived by Somatic Cell Nuclear Transfer" (*Cell*, vol. 153, no. 6, June 6, 2013), Masahito Tachibana et al. detail how they reprogrammed human skin cells to become embryonic stem cells capable of transforming into any other cell type in the body.

In 2014 Young Chung et al. reported using SCNT to create stem cells using older adult donors aged 35 and 75 in "Human Somatic Cell Nuclear Transfer Using Adult Cells" (*Cell Stem Cell*, vol. 14, no. 6, June 2014). Although the intent of this research is not to clone a human, the technique could, theoretically, be used to create a baby with the same genetic makeup as the donor. As such, it renewed controversies about human cloning and techniques such as SCNT, which has a legitimate scientific purpose, but may also be misused.

In 2016 researchers reported using resident fibrocartilage stem cells to repair cartilage and the temporomandibular joint, which articulates the jaw bone to the skull, in animals suffering from temporomandibular joint degeneration. In "Exploiting Endogenous Fibrocartilage Stem Cells to Regenerate Cartilage and Repair Joint Injury" (*Nature Communications*, October 10, 2016), Mildred C. Embree et al. report that stimulating the body's own stem cells to repair tissue overcomes many of the barriers to stem cell transplantation such as immune rejection or transmission of pathogens (disease-causing agents).

Other animal studies have successfully used stem cell therapy to treat many different disorders, including replacing proteins that the body cannot produce, speeding up muscle repair, creating skin grafts to treat burn wounds, and treating heart disease. Many researchers believe similar types of therapies will eventually be used on humans.

Therapeutic Benefits of Stem Cells without Cloning

The ability to make precise genetic changes in human stem cells may be used to boost their therapeutic potential or make them more compatible with patients' immune systems. Some researchers assert that the success of this bioengineering feat will eliminate the need to pursue the hotly debated practice of therapeutic cloning, but others caution that such research could heighten concerns among those who fear that stem cell technology will lead to the creation of "designer babies," which are bred for specific characteristics such as appearance, intelligence, or athletic prowess.

Karin Hübner et al. report in "Derivation of Oocytes from Mouse Embryonic Stem Cells" (*Science*, vol. 300, no. 5623, May 23, 2003) the transformation of mouse embryo cells into egg cells in laboratory dishes. The researchers selected stem cells that bore certain genetic traits suggesting the potential to become eggs and isolated those cells in laboratory dishes. Eventually, the cells morphed into two kinds of cells, including young egg cells. The eggs matured normally and were healthy in terms of their appearance, size, and gene expression. When cultured for a few days, the eggs underwent spontaneous division and formed structures resembling embryos, a process called parthenogenesis. This finding implies that the eggs were fully functional and likely could be fertilized with sperm.

This technology enables scientists to engineer traits into animals and helps conservationists rebuild populations of endangered species. It offers researchers the chance to observe mammalian egg cells as they mature, a process that occurs unseen within the ovary. The technology also offers an unparalleled opportunity to learn about meiosis, the process of cell division during which an egg or sperm disgorges half of its genes so that it can join with a gamete of the opposite sex. There are many potential medical benefits as well. For example, women who cannot make healthy eggs could use this technology to ensure healthy offspring.

Like many new technologies, transforming cells into eggs simultaneously resolves existing ethical issues and creates new ones. For example, because the embryonic stem cells spontaneously transformed themselves into eggs, this procedure overcomes many of the ethical objections to cloning, which involves creating offspring from a single parent. However, it also paves the way for the creation of "designer eggs" from scratch and, if performed with human cells, could redefine the biological definitions of mothers and fathers.

In 2012 researchers found that certain stem cells in the ovaries may be able to generate egg cells. It was previously thought that women were born with all the eggs they will ever have; however, this research suggests that more eggs may be made by using stem cells taken from the ovaries. Evelyn E. Telfer and David F. Albertini report in "The Quest for Human Ovarian Stem Cells" (*Nature Medicine*, vol. 18, no. 3, March 2012) that these special cells, called oogonial stem cells, spontaneously generated apparently normal oocytes when cultured in the lab. The researchers opine that in the future, "the ability to harvest such cells from human ovaries could change the options available for fertility preservation and the treatment of infertility."

Katsuhiko Hayashi et al. report in "Offspring from Oocytes Derived from In Vitro Primordial Germ Cell–Like Cells in Mice" (*Science*, vol. 338, no. 6109, October

4, 2012) that they successfully induced sperm and oocytes from mouse embryonic stem cells and induced pluripotent stem cells (iPSCs). A potential application of this technology would be to assist infertile couples. In "Ethical and Legal Issues Arising in Research on Inducing Human Germ Cells from Pluripotent Stem Cells" (*Cell Stem Cell*, vol. 13, no. 2, August 1, 2014), Tetsuya Ishii, Renee A. Reijo Pera, and Henry T. Greely pose some of the ethical issues that may arise from use of this technology. For example, using germ cells generated from someone who has died might produce unique social concerns. The technology could also be used to produce many embryos, allowing prospective parents to choose their embryo from scores of options, or to create a sibling primarily to provide genetically matched transplantation therapy for a relative. Many ethicists advise consideration of such issues before permitting human experimentation.

Baby with DNA from Three People Is Born

In 2016 physicians used a technique known as mitochondrial manipulation technology (MMT) to isolate the nucleus from a mother's oocyte, transfer it to a donor's oocyte that was fertilized, and then implanted it in the mother. In "First Live Birth Using Human Oocytes Reconstituted by Spindle Nuclear Transfer for Mitochondrial DNA Mutation Causing Leigh Syndrome" (*Fertility and Sterility*, vol. 106, no. 3, September 2016), John Zhang et al. explain that MMT was performed because the mother carried a mitochondrial mutation for Leigh syndrome (a rare inherited disorder that affects the central nervous system and is often fatal) and her two previous children were born with the disorder and died before their first birthday.

Although the popular media described the baby as having three parents, Richard J. Paulson, the president-elect of the American Society for Reproductive Medicine, told Gina Kolata in "Birth of Baby with Three Parents' DNA Marks Success for Banned Technique" (NYTimes.com, September 27, 2016) that MMT does not produce a baby with three parents. He explained, "Mitochondria do not define who you are. The genes for traits that make up a person's appearance and other characteristics are carried in the nuclear DNA. If a white woman got mitochondria from an Asian woman, for example, her babies would be white, with no traces of the Asian mitochondrial donor."

In "First Spindle Nuclear Transfer Baby Has Low Mutant DNA Load" (Medscape.com, October 20, 2016), Kate Johnson reports that although MMT is banned in the United States, U.S. reproductive scientists hope that the success of this technique, along with the United Kingdom's recent move to permit it, will inspire the United States to change its stance. However, MMT and its recent success was met with mixed reviews at the American

Society for Reproductive Medicine's 2016 Scientific Congress. According to Johnson, reactions "ranged from dismay to admiration," with some condemning it as "changing the heredity of humans forever" and others lauding it as a "great [example] of preventive medicine."

Methods to Obtain Stem Cells without Destroying Embryos

In the press release "Advanced Cell Technology Develops First Human Embryonic Stem Cell Line without Destroying an Embryo" (June 21, 2007, http://www.businesswire.com/news/home/20070621005536/en/Advanced-Cell-Technology-Develops-Human-Embryonic-Stem), Advanced Cell Technology announced that it had successfully produced a human embryonic stem cell line without destroying an embryo. The technique was similar to that used for preimplantation genetic diagnosis. The stem cells were genetically normal and differentiated into various cell types, including blood cells, neurons, heart cells, cartilage, and other cell types with significant therapeutic potential. In 2008 Young Chung et al. reported in "Human Embryonic Stem Cell Lines Generated without Embryo Destruction" (*Cell Stem Cell*, vol. 2, no. 2, February 2008) the creation of fully functional liver cells from embryonic stem cells. That same year Shi-Jiang Lu et al. described in "Biological Properties and Enucleation of Red Blood Cells from Human Embryonic Stem Cells" (*Blood*, vol. 112, no. 12, December 1, 2008) success in transforming human embryonic stem cells into red blood cells. This finding is very promising because it may be possible to use this technique to produce blood for transfusion, which is often in short supply.

In "Breakthrough of the Year: Reprogramming Cells" (*Science*, vol. 322, no. 5909, December 19, 2008), Gretchen Vogel reports that scientists have successfully transformed human skin cells into stem cells. The method that is used to transform the skin cells into stem cells is called induced pluripotent stem cells. iPSCs are created by genetically reprogramming adult cells, which involves disrupting their DNA, so that they have the same characteristics as embryonic stem cells. Because their DNA is disrupted, administering iPSCs to people is likely to increase their risk of developing cancer. Luigi Warren et al. report in "Highly Efficient Reprogramming to Pluripotency and Directed Differentiation of Human Cells with Synthetic Modified mRNA" (*Cell Stem Cell*, vol. 7, no. 5, November 5, 2010) a significant advance in the technique of creating iPSCs that does not involve genetic modification. When produced this way, iPSCs can be used in patients for transplants and for other therapeutic purposes without the increased risk of cancer.

In "Patient-Specific Induced Pluripotent Stem-Cell Models for Long-QT Syndrome" (*New England Journal of Medicine*, vol. 363, no. 15, October 7, 2010), Alessandra Moretti et al. report that they were able to generate heart muscle cells from human skin cells using iPSCs. Anthony Rosenzweig observes in the editorial "Illuminating the Potential of Pluripotent Stem Cells" (*New England Journal of Medicine*, vol. 363, no. 15, October 7, 2010) that "patient-specific cellular models derived from induced pluripotent stem cells provide an important additional tool for the study of human disease." However, Rosenzweig also cautions that there are many challenges to overcome and much additional work to do before these findings can be expected to benefit patients: "Working with induced pluripotent stem cells remains technically demanding, and the process of differentiation into specific cell types is incompletely understood, often inefficient, and somewhat variable."

Robin McKie reports in "Cloning Scientists Create Human Brain Cells" (Guardian.co.uk, January 28, 2012) that scientists at the Centre for Regenerative Medicine in Edinburgh, Scotland, where the technology to make Dolly, the first cloned mammal, was developed, announced another breakthrough: creating brain tissue in the laboratory from skin cells that were taken from people suffering from mental and neurological disorders. The scientists coaxed the skin cells to become stem cells and then directed the stem cells to become brain cells. The newly created brain cells enable researchers to test novel treatments for disorders ranging from schizophrenia and bipolar depression to multiple sclerosis and Parkinson's disease.

In 2014 researchers at Brown University reported a new method for extracting a variety of potential bone-producing cells from human fat. The researchers developed a fluorescent tag that finds and identifies cells expressing a gene called ALPL, which indicates bone-making potential. In "Gene Expression-Based Enrichment of Live Cells from Adipose Tissue Produces Subpopulations with Improved Osteogenic Potential" (*Stem Cell Research & Therapy*, vol. 5 , no. 12, October 6, 2014), Hetal D. Marble et al. report that their method produced more than twice the yield of potential bone makers compared with a method that involves sorting cells based on surface proteins thought to indicate that a cell is a stem cell.

INTERNATIONAL POLICIES AND REGULATIONS

Many countries, including Australia, Belgium, China, Finland, India, Israel, Japan, Singapore, South Africa, South Korea, Spain, Sweden, and the United Kingdom, and some U.S. states permit the creation of embryos for research. All limit the culture period of the created embryo to about the 14th day of embryo development. In some countries researchers must justify their need to create embryos for research. In countries that allow the creation of embryos for research, therapeutic cloning using SCNT is generally permitted.

Jonathan Kimmelman et al. note in "Policy: Global Standards for Stem-Cell Research" (*Nature*, vol. 533, no. 7603, May 19, 2016) that in 2016 the International Society for Stem Cell Research, an independent nonprofit organization, updated its guidelines for research involving stem cells and the translation of that research into medical therapies. The guidelines were developed by 25 scientists and ethicists from Asia, Europe, North America, and Australia, with contributions from more than 100 individuals and groups, including regulators, funding bodies, journal editors, patient advocates, researchers, and members of the public. According to Kimmelman et al., mitochondrial transfer techniques (MRTs) that may replace impaired mitochondria in eggs or embryos with those obtained from healthy donors may be performed in the United Kingdom, and in 2016 the National Academy of Medicine recommended clinical testing of MRT in the United States.

Kimmelman et al. report that the International Society for Stem Cell Research's updated guidelines prohibit attempts to "modify the nuclear genome of human embryos for the purpose of human reproduction" but advocate continued laboratory research on human embryos and the stem cell lines derived from them. The guidelines also call for intensified efforts to ensure safety and transparency, advising that the results of all preclinical (laboratory and animal) studies be reported in peer-reviewed journals and that investigators "conduct a systematic review to capture all relevant information before initiating a clinical trial."

U.S. Public Policy

Although it is widely believed that stem cell research is the key to developing effective treatment for a range of serious and chronic diseases, public policy issues, especially funding for stem cell research, have been shaped by a heated controversy that pits scientific advancement against the sanctity of life. Opposition to stem cell research, and federal funding for it, arises from the concern that it may involve destroying human life contained in embryonic stem cells and the fear that embryos might be created solely for use in research, which could result in their destruction.

In the press release "President Discusses Stem Cell Research" (August 9, 2001, https://georgewbush-whitehouse.archives.gov/news/releases/2001/08/20010809-2.html), President George W. Bush (1946–) announced his decision to allow federal funds to be used for research on existing human embryonic stem cell lines as long as the derivation process (which begins with the removal of the inner cell mass from the blastocyst) had already been initiated and the embryo from which the stem cell line was derived no longer had the possibility of developing into a human being. The president established criteria that limited research to only 21 cell lines that had been created before August 9, 2001, and banned federal funding for hundreds of embryonic stem cell lines, including some that carried genetic mutations for specific conditions and diseases that scientists wanted to study in the hope of finding cures. Many people argued that because embryonic stem cells are not, in fact, embryos, research on such cells should be allowed. Others, such as President Bush, believed that any research that creates or destroys human embryos at any stage, such as creating new embryos through cloning or destroying an embryo to create a stem cell line, should not be allowed. Despite the ban on federal funding, such research remained legal as long as private funding was used.

In 2002 the Panel on Scientific and Medical Aspects of Human Cloning was convened by the National Academy of Sciences; the National Academy of Engineering; the Institute of Medicine Committee on Science, Engineering, and Public Policy; and the National Research Council, Division on Earth and Life Studies Board on Life Sciences. Following the panel, the report *Scientific and Medical Aspects of Human Reproductive Cloning* (2002, https://www.nap.edu/read/10285/chapter/1) was issued; it called for a ban on human reproductive cloning.

The panel recommended a legally enforceable ban with substantial penalties as the best way to discourage human reproductive cloning experiments in both the public and private sectors. It cautioned that a voluntary measure might be ineffective because many of the technologies needed to accomplish human reproductive cloning are widely accessible in private clinics and in other organizations that are not subject to federal regulations.

The panel did not, however, conclude that the ban on human reproductive cloning was applicable to SCNT to produce stem cells. In view of the potential to generate new treatments for life-threatening diseases and advance biomedical knowledge, the panel recommended that biomedical research using SCNT to produce stem cells be permitted.

According to the press release "President Bush Calls on Senate to Back Human Cloning Ban" (https://georgewbush-whitehouse.archives.gov/news/releases/2002/04/200 20410-4.html), in April 2002 President Bush called on the U.S. Senate to back legislation that banned all types of human cloning.

On September 25, 2002, Elias A. Zerhouni (1951–; https://olpa.od.nih.gov/hearings/107/session2/testimonies/stemcelltest.asp), the director of the NIH, testified before the Senate Appropriations Subcommittee on Labor, Health and Human Services, and Education in favor of advancing the field of stem cell research. Zerhouni

exhorted Congress to continue to fund both human embryonic stem cell research and adult stem cell research simultaneously to learn as much as possible about the potential of both types of cells to treat human disease. Despite Zerhouni's impassioned plea and subsequent efforts by Congress to reverse limitations on embryonic stem cell research, President Bush vetoed such attempts, and, at the close of 2008, U.S. law continued to ban federal funding of any research that might harm human embryos.

The Obama Administration's Stance on Stem Cell Research

On March 9, 2009, shortly after entering office in January, Barack Obama (1961–) issued an executive order that lifted the ban on federal funding for embryonic stem cell research. The NIH then established policies and procedures that enabled federally funded researchers to work with selected batches of human embryonic stem cells (largely stem cell lines that had been created with private, as opposed to federal, funds).

In June 2010 James Sherley of Boston Biomedical Research Institute and Theresa Deisher of AVM Biotechnology filed a suit seeking to stop federal funding of human embryonic stem cell research because they claimed such research destroys human embryos and diverts funds from researchers working with adult stem cells. On August 23, 2010, Judge Royce C. Lamberth (1943–) of the U.S. District Court for the District of Columbia issued a preliminary injunction that halted federal funding of embryonic stem cell research. The injunction also banned research on the stem cell lines that had been approved for research by President Bush in 2001.

The following day the Obama administration vowed to appeal the ruling, which would suspend $54 million for 22 projects by the end of September 2010 if it was upheld. The U.S. Department of Justice requested that the injunction be lifted, pending its appeal. Sheryl Gay Stolberg and Gardiner Harris note in "Stem Cell Ruling Will Be Appealed" (NYTimes.com, August 24, 2010) that Francis S. Collins (1950–), the director of the NIH, asserted, "This decision has the potential to do serious damage to one of the most promising areas of biomedical research, just at the time when we were really gaining momentum."

On September 9, 2010, the U.S. Circuit Court of Appeals for the District of Columbia suspended the district court's preliminary injunction and stated that federal funding of stem cell research could proceed while the government's appeal was being considered.

In 2012 the U.S. Circuit Court of Appeals for the District of Columbia dismissed the lawsuit and ruled that federal funding of embryonic stem cell research was permissible. In "Obama Administration Can Fund Embryonic Stem Cell Research: Federal Court" (HuffingtonPost.com, August 24, 2012), Jesse J. Holland quotes Collins, who stated shortly after the ruling, "NIH will continue to move forward, conducting and funding research in this very promising area of science. The ruling affirms our commitment to the patients afflicted by diseases that may one day be treatable using the results of this research."

In 2013 the U.S. Supreme Court refused to hear a lawsuit that would have blocked federal funding of all research on human embryonic stem cells, putting an end to more than three years of uncertainty about their use and federal funding to support research. During this period many researchers chose to work with iPSCs to circumvent the controversy.

Future of Stem Cell Research

The administration of Donald Trump (1946–) may jeopardize the future of stem cell research. Although Trump did not state his position on embryonic stem cell research during his campaign, Mike Pence (1959–), the vice president, has consistently opposed embryonic stem cell research, asserting that iPSCs are a viable substitute for embryonic stem cells. Kelly Servick reports in "Embryonic Stem Cells and Fetal Tissue Research—Will Trump Intervene?" (Sciencemag.org, November 11, 2016) that many researchers who work with embryonic stem cells lament their loss. For example, Timothy Kamp, a cardiologist at the University of Wisconsin, Madison, opined that embryonic stem cells are "part of the tool set that we like to use to show that our finding is robust, reproducible, and not an artifact of reprogramming or keeping cells in culture. I think it would be quite devastating to take those lines out of federally funded protocols." Servick suggests that if a federal ban on embryonic stem cell research was enacted, it might stimulate support for funding at the state level. In "How Trump Could Slow Medical Progress" (TheAtlantic.com, December 28, 2016), Olga Khazan notes that the U.S. secretary of health and human services Tom Price (1954–), who has cosponsored bills that describe "a 'human being' from 'the moment of fertilization,'" could halt funding for stem cell research. As of February 2017, the Trump administration had not acted to curtail stem cell research.

Legislation Aims to Completely Ban Human Cloning

In 2003 the U.S. House of Representatives voted to outlaw all forms of human cloning. The measure did not pass in the Senate, which was closely divided about whether therapeutic cloning should be prohibited along with reproductive cloning. In the Senate two bills were introduced: S. 245 was a complete ban, intended to amend the Public Health Service Act to prohibit all human cloning, and S. 303 was a less sweeping measure that also prohibited cloning but protected stem cell research. S. 245 was referred to the Senate Committee on Health, Education, Labor, and Pensions and S. 303 was referred to the Senate Committee on the Judiciary. In 2007 the Human Cloning Ban and Stem Cell Research Protection Act (S. 812) was introduced and referred to the Senate Committee on the Judiciary. As of February 2017, none of these bills, nor any comparable proposed legislation, had emerged from the Senate committees.

Although nearly all lawmakers concur that Congress should ban reproductive cloning, many disagree about whether legislation should also ban the creation of cloned human embryos that serve as sources of embryonic stem cells. Many legislators agree with scientists that stem cells derived from cloned human embryos have medical and therapeutic advantages over those derived from conventional embryos or adults. Those who oppose the legislation calling for a total ban assert that the aim of allowing research is to relieve the suffering of people with degenerative diseases. They say that the legislation's sponsors are effectively thwarting advances in medical treatment and biomedical innovation.

Supporters of the total ban contend that Congress must send a clear message that cloning research and experimentation will not be tolerated. They consider cloning immoral and unethical, fear unintended consequences of cloning, and feel they speak for the public when they assert that it is not justifiable to create human embryos simply for the purpose of experimenting on them and then destroying them.

Although President Obama lifted the ban on federal funding for all embryonic stem cell research on lines other than those President Bush approved (those created before August 9, 2001), his executive order did not address reproductive cloning. When he issued the executive order, President Obama expressed his strong support of embryonic stem cell research but stated that under no circumstances should the funding for embryonic stem cell research be used to explore human cloning.

State Human Cloning Laws

In the United States, federal laws only prohibit federal funding of research that results in the destruction or risk of damage to human embryos. Researchers with nonfederal funding are permitted to create human embryos for research. Several states have, however, enacted statutes that address human embryo research. For example, California, Connecticut, Illinois, Iowa, Maryland, Massachusetts, New Jersey, and New York encourage embryonic stem cell research, with various restrictions. Other states, such as Louisiana and Michigan, discourage or ban human embryo research including embryonic stem cell research.

According to the article "State Laws on Human Cloning" (TheNewAtlantis.com, Summer 2015), as of 2015 seven states—Arizona, Arkansas, Michigan, North Dakota, Oklahoma, South Dakota, and Virginia—prohibit cloning for any purpose. Ten states—California, Connecticut, Illinois, Iowa, Maryland, Massachusetts, Missouri, Montana, New Jersey, and Rhode Island—prohibit cloning to produce children but permit cloning for biomedical research. One state, Minnesota, has a statute that appears to prohibit cloning for biomedical research but does not cover cloning to produce children.

California Leads the Way in Stem Cell Research

State funding of stem cell research is supported by many individual states even when federal funds are not available. As one of the states that support stem cell research, California has been at the forefront of this technology. In 2002 the California state legislature passed a law encouraging therapeutic cloning. In November 2004 Californians approved Proposition 71, a ballot measure to make public funding available to support stem cell research and therapeutic cloning. Proposition 71 enabled the state to create the California Institute for Regenerative Medicine (CIRM; https://www.cirm.ca.gov/). The proposition prohibits reproductive cloning but funds human cloning projects that are designed to create stem cells and allocated $3 billion over 10 years in research funds.

The CIRM was officially established in 2005 and an independent oversight committee of public officials was created to govern the agency. The institute funds biomedical research based in California that focuses on developing diagnostics and therapies and on other vital research that will lead to lifesaving medical treatments. For example, the CIRM notes in the press release "CIRM Creates First of Its Kind Center to Accelerate Stem Cell Development" (https://www.cirm.ca.gov/about-cirm/news room/press-releases/06152016/cirm-creates-first-its-kind -center-accelerate-stem-cell) that in June 2016 it invested $15 million in a partnership with Quintiles Transnational Inc. to create a stem cell accelerating center, which will help promising stem cell therapies advance from the laboratory to human clinical trials.

Public Opinions about Stem Cell Research

In *Stem Cell Research* (http://www.gallup.com/poll/21676/stem-cell-research.aspx), Gallup, Inc., reports that there was considerable (60%) support for stem cell research in 2016, although the percentage of Americans that considered stem cell research morally acceptable fell from a peak of 65% in 2014. More than half of Americans have deemed stem cell research morally acceptable since 2002.

CHAPTER 10

GENETIC ENGINEERING AND BIOTECHNOLOGY

The first decades of the new millennium saw explosive advances in biotechnology. Technological breakthroughs offered scientists and physicians unprecedented opportunities to develop previously inconceivable solutions to pressing problems in agriculture, environmental science, and medicine. Simultaneously, researchers, politicians, ethicists, theologians, and the public were challenged to analyze and determine the feasibility of using new biotechnology in view of the opportunities, possibilities, risks, benefits, and diverse viewpoints about the safe, effective, and ethical applications of genetic research.

This chapter describes several examples of applications of genetic research and biotechnology, including uses that address pressing problems such as environmental issues and world hunger as well as the role of genetic engineering in developing lifesaving medical therapeutics. It also considers industry and consumer viewpoints as well as recommendations scientists, policy makers, and ethicists have made regarding the wise, judicious, and equitable use of these technologies.

AGRICULTURAL APPLICATIONS OF GENETIC ENGINEERING

Genetically modified (GM) or transgenic crops (sometimes also called genetically engineered [GE] crops) contain one or more genes that have been artificially inserted instead of received through pollination (fertilization by the transfer of pollen from an anther to a stigma of a plant). The inserted gene sequence, called the transgene, may be introduced to produce different results—either to overexpress or silence (direct a gene not to synthesize a specific protein) an existing plant gene, and it may come from another unrelated plant or from a completely different species. Transgenics is the science of inserting a foreign gene into an organism's genome to produce a transgenic organism.

For example, the transgenic corn that produces its own insecticide contains a gene from a bacterium, and Macintosh apples with a gene from a moth that encodes an antimicrobial protein are resistant to a bacterial infection called fire blight. Although all crops have been genetically modified from their original wild state by domestication, selection, and controlled breeding over long periods, the terms *transgenic crops*, *GE crops*, *GM crops*, and *biotech crops* refer to plants with transgenes (inserted gene sequences).

Genes are introduced into a crop plant to make it as useful and productive as possible by acting to protect the plant, improve the harvest, or enable the plant to perform a new function or acquire a new trait. Specific objectives of genetically modifying a plant include increasing its yield, improving its quality, or enhancing its resistance to pests or disease and its tolerance for heat, cold, or drought. Some of the GM traits that have been introduced into food crops are enhanced flavor, slowed ripening, reduced reliance on fertilizer, self-generating insecticide, and added nutrients. Examples of transgenic food crops include frost-resistant strawberries and tomatoes; slow-ripening bananas, melons, and pineapples; and insect-resistant and herbicide-tolerant corn, cotton, and soybeans. Figure 10.1 shows the increasing adoption of insect-resistant corn and cotton and herbicide-tolerant soybeans, cotton, and corn between 1996 and 2016.

Transgenic technology enables plant breeders to bring together in one plant useful genes from a wide range of living sources, not just from within the crop species or from closely related plants. It provides a reliable means for identifying and isolating genes that control specific characteristics in one kind of organism and enables researchers to move copies of these genes into another organism that will then develop the chosen characteristics. This technology gives plant breeders the ability to generate more useful and productive crop varieties

FIGURE 10.1

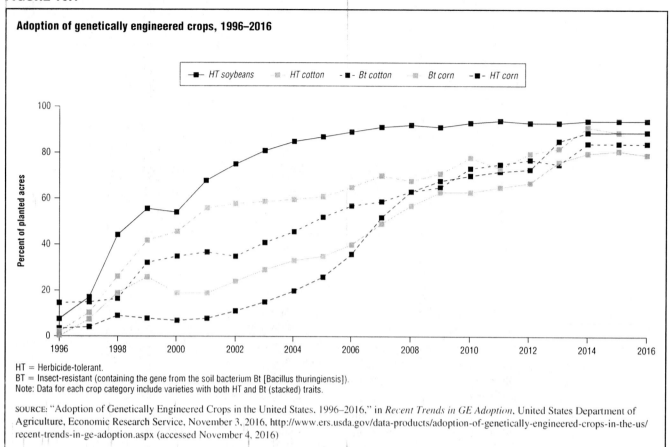

Adoption of genetically engineered crops, 1996–2016

HT = Herbicide-tolerant.
BT = Insect-resistant (containing the gene from the soil bacterium Bt [Bacillus thuringiensis]).
Note: Data for each crop category include varieties with both HT and Bt (stacked) traits.

SOURCE: "Adoption of Genetically Engineered Crops in the United States, 1996–2016," in *Recent Trends in GE Adoption*, United States Department of Agriculture, Economic Research Service, November 3, 2016, http://www.ers.usda.gov/data-products/adoption-of-genetically-engineered-crops-in-the-us/recent-trends-in-ge-adoption.aspx (accessed November 4, 2016)

containing new combinations of genes, and it significantly expands the range of trait manipulation and enhancements well beyond the limitations of traditional cross-pollination and selection techniques.

Although genetic modification of plants generates the same types of changes that are produced by conventional agricultural techniques, the results are often more rapid and more complete because it precisely alters a single gene. Traditional breeding techniques may require an entire generation or more to introduce or remove a single gene, and using conventional methods for breeding a polygenic trait into crops with multiyear generations could take several decades.

Creating Transgenic Crops

The first step in creating a transgenic plant is locating genes with the traits that growers, marketers, and consumers consider important. These are usually genes that increase productivity and yield and improve resistance to environmental stresses such as frost, heat, drought, salt, and insects. Identifying the gene associated with a specific trait is necessary but not sufficient; researchers must determine how the gene is regulated, its other influences on the plant, and its interactions with other genes to express or silence various traits. Researchers must then isolate and clone the gene to have sufficient quantities to

modify. Establishing the genomic sequence, called plant genomics, and the functions of genes of the most important crops is a priority of plant genomic research projects.

Most genes that are introduced into plants come from bacteria; however, increasing understanding of plant genomics is anticipated to permit greater use of plant-derived genes to genetically engineer crops. In 2000 the first entire plant genome *Arabidopsis thaliana* (commonly called thale cress or mouse-ear cress; a small flowering plant often considered to be a weed and found along roadsides) was sequenced, which provided researchers with new insight into the genes that control specific traits in many other agricultural plants. There are several approaches to introducing genes into plant cells: vector- or carrier-mediated transformation, particle-mediated transformation, and direct deoxyribonucleic acid (DNA) insertion.

Vector-mediated transformation involves infecting plant cells with a virus or bacterium that during the process of infection inserts foreign DNA into the plant cell. The most convenient vector is the soil bacterium *Agrobacterium tumefaciens*, which infects tomatoes, potatoes, cotton, and soybeans. This bacterium attacks cells by inserting its own DNA. When genes are added to the bacterium, they are transferred to the plant cell along with the other DNA.

Particle-mediated transformation involves gene transfer using a special particle tool known as a gene gun, which shoots tiny metal particles that contain DNA into the cell.

To perform direct DNA insertion or electroporation, cells are immersed in the DNA and electrically shocked to stimulate DNA uptake. The cell wall then opens for less than a second, allowing DNA to seep into the cell. Following gene insertion, the cell incorporates the foreign DNA into its own chromosomes and undergoes normal cell division. The new cells ultimately form the organs and tissues of the "regenerated" plant. To ensure the systematic sequence of these steps, other genes, called promoters, may be added along with the gene that is associated with the desired trait. They encourage the growth of cells that have integrated the inserted DNA, provide resistance to stresses (such as toxins present in the medium that is used to grow the cells), and may help regulate the functions of the gene that is linked to the desired trait.

To be certain that the new gene is in the organism, marker genes are sometimes inserted along with the gene for the desired trait. One common marker gene confers resistance to the antibiotic kanamycin. When this gene is used as a marker, investigators can confirm that the transfer was successful when the organism resists the antibiotic. The ultimate success of gene insertion is measured by whether the inserted gene functions properly by expressing, amplifying, or silencing the desired trait.

Are GM Crops Helpful or Harmful?

In "ISAAA Brief 51-2015: Top Ten Facts about Biotech/GM Crops in Their First 20 Years, 1996 to 2015" (2017, http://www.isaaa.org/resources/publications/briefs/51/toptenfacts/default.asp), the International Service for the Acquisition of Agri-biotech Applications (ISAAA), a nonprofit organization that delivers new agricultural biotechnologies to developing countries, observes that the genetic modification of crops continues to be a rapidly adopted technology. (See Figure 10.1.) The ISAAA contends that "from 1996 to 2014, biotech crops contributed to Food Security, Sustainability and the Environment/Climate Change." During this period U.S. crop production increased $150 billion, and the environment benefited from the use of 644,000 tons (584 million kg) less of pesticides. In 2014 alone, biotech crops reduced carbon dioxide emissions by 29.8 million tons (27 billion kg), which is equivalent to taking 12 million cars off the road for one year, and helped combat poverty for nearly 16.5 million small farmers and their families.

However, in "Doubts about the Promised Bounty of Genetically Modified Crops" (NYTimes.com, October 29, 2016), Danny Hakim reports that GM crops have failed to deliver on their promise to increase crop yields or reduce pesticide use. Using United Nations data and a National Academy of Science report, the *New York Times* performed an analysis and found the GM crops have not produced greater yields than conventional crops. Furthermore, use of herbicide has increased 21% in the United States since the introduction of GM crops. Hakim notes that although western Europe rejected GM crops, it still outpaces Canada, which uses GM crops, in canola oil production. Similarly, growth in yields of GM sugar beets in the United States have not kept pace with yields of non-GM sugar beets in western Europe.

There is other opposition to GM crops. Bioethicists contend that freedom of choice is a central tenet of ethical science and oppose what they deem to be interference with other forms of life. Environmentalists argue that transgenic technology poses the risk of altering delicately balanced ecosystems (biological communities and their environments) and causing unintended harm to other organisms. They are concerned that transgenic crops will replace traditional crop varieties, especially in developing countries, resulting in the loss of biological diversity.

Among the environmental concerns is the risk that pests may develop resistance to transgenics in much the same way that certain pests have become resistant to some types of pesticides that once effectively eradicated them. Critics also fear the infiltration of transgenic crops beyond their intended areas and inadvertent gene transfer to species not targeted for transgenics.

Although a review of 147 studies performed by Wilhelm Klümper and Matin Qaim, in "A Meta-analysis of the Impacts of Genetically Modified Crops" (*PLoS One*, vol. 9, no. 11, 2014), found that use of GM technology increased farm profits 68%, reduced pesticide use 37%, and increased crop yields 22%, Qaim told Hakim that the benefits were greatest in the developing world, particularly in India. Qaim explained that GM crops would not yield major gains in Europe, and said of herbicide-resistant crops, "I don't consider this to be the miracle type of technology that we couldn't live without."

Health risks also concern those who object to transgenic crops. They call for the labeling of GM food to alert consumers that they are purchasing foods that contain GM organisms. In "Myths & Realities of GE Crops" (2017, http://www.centerforfoodsafety.org/issues/311/ge-foods/myths-and-realities-of-ge-crops#), the Center for Food Safety (CFS), a nonprofit public interest and environmental advocacy membership organization that was established by the International Center for Technology Assessment for the purpose of challenging harmful food production technologies and promoting sustainable alternatives, asserts that there is no long-term research demonstrating that GE foods or crops are safe.

Terminator Technology

One of the most controversial developments in agricultural bioengineering is terminator technology (also known as genetic use restricted technology) because it is designed to genetically switch off a plant's ability to germinate a second time. Traditionally, farmers saved seeds for the next harvest; however, the use of terminator technology effectively prevents this practice, forcing farmers to purchase a fresh supply of seeds each year.

Advocates of terminator technology are generally corporations and the organizations that represent them. They contend that the practice protects corporations from corrupt farmers. Controlling seed germination helps prevent growers from pirating the corporations' licensed or patented technology. If crops remained fertile, there is a chance that farmers could use any saved transgenic seed from a previous season. This would result in reduced profits for the companies that own the patents.

Opponents of terminator technology believe it threatens the livelihood of farmers in developing countries such as India, where many poorer farmers have been unable to compete and some have been forced out of business. Opponents considered it a victory when the Monsanto Company, a major investor in this technology, decided not to market terminator technology in 1999. However, even without terminator technology, patent laws in the United States, Canada, and many other industrialized nations require farmers to sign a licensing agreement to reuse patented seed or grow Monsanto's GM seed. This agreement has the same effect on poor farmers as terminator technology; it renders them unable to compete.

Andrew Porterfield of the Genetic Literacy Project, a nonprofit scientific research and public policy group, questions in "Why Activists, but Few Farmers, Complain They Can't Save Patented Seeds" (August 17, 2016, https://www.geneticliteracyproject.org/2016/08/17/why-activists-but-few-farmers-complain-they-cant-save-patented-seeds/) opponents' contention that terminator technology and patents imperil farmers by preventing them from using saved seeds. Porterfield explains that using saved seeds is less reliable. He asserts that now that the patents for some seeds have expired, not all farmers will save the seeds because the saved seeds are likely to produce crops with reduced yields and increased susceptibility to disease.

Legal Actions Taken by and against Monsanto

In "Saved Seed and Farmer Lawsuits" (2017, http://www.monsanto.com/newsviews/pages/saved-seed-farmer-lawsuits.aspx), Monsanto notes that between 1997 and 2016 it filed 147 lawsuits against U.S. farmers. The largest judgment awarded to Monsanto was more than $3 million.

Darren Hauck describes in "Supreme Court Hands Monsanto Victory over Farmers on GMO [genetically modified organism] Seed Patents, Ability to Sue" (RT.com, January 15, 2014) Monsanto's more than 140 lawsuits against U.S. farmers and reports that a January 2014 U.S. Supreme Court decision in the federal lawsuit *Organic Seed Growers and Trade Association v. Monsanto* (No. 13-303) upheld the company's continued ability to sue farmers whose fields inadvertently become contaminated with Monsanto seeds.

In 2016 three farmers and an agronomist (scientist who studies soil, crops, and plant genetics) who were diagnosed with cancer sued Monsanto, claiming the company misled the public about the dangers of glyphosate, the herbicide that Monsanto markets as Roundup. In "Farmers Sue Monsanto over Alleged Roundup Cancer Link" (JournalStar.com, May 15, 2016), Nicholas Bergin notes that in 2015 the World Health Organization identified glyphosate as a probable cause of cancer and indicated that "farmers, farm workers and others with workplace exposure to Roundup" were at risk. According to Bergin, the lawsuit contends that "Monsanto championed falsified data and has attacked legitimate studies that revealed Roundup's dangers. Monsanto led a campaign of misinformation to convince government agencies, farmers and the general population that Roundup is safe." Bergin also reports that Monsanto has sued the state of California to keep glyphosate off its list of known carcinogens.

U.S. BIOTECHNOLOGY REGULATORY SYSTEM

The U.S. government operates a rigorous regulatory process for determining the safety of agricultural products of biotechnology. The process ensures that all biotechnology products that are commercially grown, processed, sold, and consumed are as safe as their conventional counterparts. Within the U.S. Department of Agriculture (USDA) the agencies responsible for regulation are the Animal and Plant Health Inspection Service (APHIS) and the U.S. Food and Drug Administration (FDA). The U.S. Environmental Protection Agency (EPA), which is not a part of the USDA, also plays a role in regulating biotechnology. Policies, processes, and regulations are continuously reviewed, evaluated, and, when necessary, revised to meet the challenges of this evolving technology.

APHIS oversees U.S. agriculture, protecting against pests and diseases. It is the lead agency that regulates the field-testing of new biotechnology-derived plant varieties and certain microorganisms. As such, APHIS grants approval and licenses for veterinary biological substances including animal vaccines that may be the products of biotechnology.

The FDA ensures that foods derived from new bioengineered plant varieties are safe and nutritious—they must meet or exceed the same high standards of safety that are applied to any food product. The FDA is also responsible for issuing and enforcing regulations to guarantee that all food and feed labels, including those related to biotechnology, are truthful and do not mislead consumers. Table 10.1 lists the genes, gene fragments, and GM products (with their intended effects) that were submitted to and approved by the FDA between 2015 and 2016.

The EPA approves new herbicidal and pesticidal substances. It issues permits for testing herbicides and biotechnology-derived plants that contain new pesticides. When the EPA makes a determination about whether to register a new pesticide, it considers human safety, environmental impact, its effectiveness on the target pest, and any consequences for other, nontarget species. The EPA establishes and enforces the guidelines that ensure the safe use of GE products that are classified as pesticides.

The Institute of Medicine Issues Reports

In 2004 the Institute of Medicine (IOM) of the National Academy of Sciences (NAS), an organization created by Congress in 1863 to advise the government about scientific and technical matters, published *Safety of Genetically Engineered Foods: Approaches to Assessing Unintended Health Effects* (https://www.nap.edu/read/

10977/chapter/1#ii). The USDA, FDA, and EPA commissioned the NAS to assess the potential for adverse health effects from GE foods compared with foods altered in other ways and to provide guidance on how to identify and evaluate the likelihood of these effects. Although adverse health effects from GE foods have not been detected in the human population, the technology is relatively new and concerns about its safety persist.

The IOM urges federal agencies to assess the safety of genetically altered foods on a case-by-case basis to determine whether unintended changes in their composition have the potential to adversely affect human health. It also calls for greater scrutiny of foods containing new compounds or unusual amounts of naturally occurring substances.

The IOM presents a framework to guide federal agencies in selecting the course and intensity of safety assessment. A new GM food whose composition is similar to a commonly used conventional version may warrant little or no additional safety evaluation. If, however, an unknown substance has been detected in a food, more detailed analyses should be conducted to determine whether an allergen or toxin may be present. Similarly,

TABLE 10.1

Bioengineered foods approved by the U.S. Food and Drug Administration, 2015–16

BNF No.	Traits	Food	Event designation unique identifier	FDA letter date (sorted Z-A)
152	Change in composition (other)	Potato	V11 SPS-ØØV11-6	Aug 19, 2016
151	Insect resistance and herbicide tolerance	Corn	MZIR098 SYN-ØØØ98-3	Apr 29, 2016
148	Herbicide tolerance	Corn	MON 87419 MON-87419-8	Mar 11, 2016
150	Herbicide tolerance	Corn	MZHGOJG SYN-ØØØJG-2	Feb 23, 2016
146	Altered composition (other) and blight resistance	Potato	W8 SPS-ØØØW8-4	Jan 12, 2016
147	Altered growth properties	Corn	MON 87403 MON-87403-1	Jun 19, 2015
144	Insect resistance	Soybean	MON 87751 MON-87751-7	May 27, 2015
132	Change in composition (other)	Apple	GD743 OKA-NBØØ1-8	Mar 20, 2015
132	Change in composition (other)	Apple	GS784 OKA-NBØØ2-9	Mar 20, 2015
141	Change in composition (other)	Potato	F10 SPS-ØØF10-7	Mar 20, 2015
141	Change in composition (other)	Potato	E12 SPS-ØØE12-8	Mar 20, 2015
141	Change in composition (other)	Potato	J3 SPS-ØØØJ3-4	Mar 20, 2015
141	Change in composition (other)	Potato	J55 SPS-ØØJ55-2	Mar 20, 2015
141	Change in composition (other)	Potato	G11 SPS-ØØG11-9	Mar 20, 2015
141	Change in composition (other)	Potato	H50 SPS-ØØH50-4	Mar 20, 2015

BNF = Biotechnology notification file.

SOURCE: Adapted from "The FDA List of Completed Consultations on Bioengineered Foods," U.S. Food and Drug Administration, August 31, 2016, http://www.accessdata.fda.gov/scripts/fdcc/?set=Biocon (accessed November 4, 2016)

foods with nutrient levels that fall outside the normal range should be assessed for their potential impact on consumers' health.

The IOM is also charged with examining the safety of foods from cloned animals. It recommends that the safety evaluation of these foods should focus on the product itself rather than on the process that is used to create it and advises that the evaluations compare foods from cloned animals with comparable food products from noncloned animals. Although there is no evidence that foods from cloned animals pose an increased risk to consumers, the IOM cautions that cloned animals engineered to produce pharmaceuticals should not be permitted to enter the food chain.

In 2010 the IOM issued the consensus report *Enhancing Food Safety: The Role of the Food and Drug Administration* (https://www.nap.edu/read/12892/chapter/1) in response to a request from Congress to assess current U.S. food safety practices and recommend strategies to improve food safety. The IOM concludes that "the FDA lacks a comprehensive vision for food safety and... should change its approach in order to properly protect the nation's food." Although the report focuses on efforts to prevent foodborne diseases and does not directly address GM foods, the IOM's recommendations include the development of a risk-based food safety system that can also be applied to GM food products. The IOM recommends the adoption of a risk-based approach that uses data and expertise to identify areas with the greatest potential for contamination along the food production, distribution, and handling chains. It calls for strengthening the FDA's capacity to gather, manage, and use data and reassessing its policies for sharing data with other agencies and organizations. The IOM also suggests increasing the FDA's authority by amending the Federal Food, Drug, and Cosmetic Act to strengthen the agency's role in preventive controls, risk-based inspection, mandatory recalls, and food import bans.

Criticisms of the U.S. Regulatory Approach

Opponents of GM foods do not believe there are sufficient government regulations in place to control U.S. production and distribution of these foods. They argue there has not been enough research or long-term experience with these foods, and as a result the health consequences of growing and eating such foods as well as the environmental impact are not yet known. Proponents claim the benefits of transgenic foods—improved flavor, increased nutritional value, longer shelf life, and greater yields—surpass any potential risks. They also discount the health risks, observing that 94% of the soybean crop and 92% of all corn grown in the United States consists of transgenic varieties, meaning that Americans have been consuming transgenic food products for years and, as of February 2017, there were no reports of adverse health effects as a result.

Since the late 1990s there have been protests staged to oppose the widespread use and consumption of GM foods as well as attacks on facilities conducting research on transgenic crops and companies marketing GM products. Protesters have dubbed the transgenic crops "Frankenfoods," likening them to Frankenstein's monster, a fictional character in the novel *Frankenstein; or, The Modern Prometheus* (1818) by the British author Mary Shelley (1797–1851). Greenpeace, an organization that opposes the creation of GM foods, and other activists staged a historic protest at the World Trade Organization (WTO) meeting in Seattle, Washington, in November 1999. Greenpeace also hung an anti-GM banner on the Kellogg Company's Cereal City USA museum in Battle Creek, Michigan, in 2000. It hopes that such actions will move U.S. lawmakers to require the labeling of GM foods. Greenpeace is inspired by the example of European consumers, who demanded product labeling that enables them to choose whether to purchase and consume GM foods. On the website "GM Contamination Register" (http://www.gmcontaminationregister.org/), Greenpeace documents "incidents of contamination arising from the intentional or accidental release of GMOs (which are also known as genetically engineered [GE] organisms)" and "illegal plantings of GM crops and the negative agricultural side-effects that have been reported." For example, the register indicates that in March 2016 Germany received insect-resistant rice from Asia.

In August 2006 a federal court issued in *Center for Food Safety et al. v. Johanns et al.* (Civil No. 03-00621 JMS/LEK) the first ruling ever on molecular farming, the controversial practice of genetically altering food crops to produce experimental drugs and industrial compounds. The court ruled that the USDA violated the Endangered Species Act when it permitted the cultivation of drug-producing GE crops in Hawaii. That same month the CFS reported in the press release "Unapproved, Genetically Engineered Rice Found in Food Supply" (August 18, 2006, http://www.centerforfoodsafety.org/press-releases/884/unapproved-genetically-engineered-rice-found-in-food-supply) that the USDA announced that an unapproved GE rice was found contaminating commercial long-grain rice supplies. The presence of this strain of GE rice, which was genetically altered to survive application of the powerful herbicide glufosinate, in the food supply was illegal because it had not undergone USDA review for potential environmental impacts and FDA review for possible harm to human health. In 2007 the USDA halted distribution and planting of the affected rice to minimize the spread of the unapproved GE strain.

In the press release "FDA Opens 'Pandora's Box' by Approving Food from Clones for Sale" (January 15, 2008, http://www.centerforfoodsafety.org/press-releases/827/fda-opens-pandoras-box-by-approving-food-from-clones-for-sale), the CFS reports that it and other food industry, consumer, and animal welfare groups condemned the FDA's determination that milk and meat from cloned animals are safe for sale to the public and the administration's decision to allow the sale of such products without labeling that identified them as produced by cloning. In September 2008 the CFS announced in the press release "20 Leading Food Companies and Retailers Reject Ingredients from Cloned Animals in Their Products" (http://www.centerforfoodsafety.org/press-releases/850/20-leading-food-companies-and-retailers-reject-ingredients-from-cloned-animals-in-their-products) that 20 leading U.S. food producers and retailers vowed not to use cloned animals in their foods. Lisa Bunin of the CFS asserted, "This rejection of food from clones sends a strong message to biotech firms that their products may not find a market. American consumers don't want to eat food from clones or their offspring, and these companies have realistically anticipated low market acceptance for this new and untested technology."

Consumers also are leery about consuming genetically engineered salmon, and there is concern about its environmental impact. The article "FDA: Genetically Engineered Fish Would Not Harm Nature" (Associated Press, December 21, 2012) notes that in December 2012 the FDA declared that genetically engineered salmon (GM salmon) is not only safe for consumption but also "unlikely to harm the environment," despite the fact that GM salmon grows twice as fast as natural salmon. Opponents fear that the GM salmon, which contains an added growth hormone from the Pacific Chinook salmon, might compromise human health and eventually overtake the native salmon population and that approval of the salmon may open a floodgate of approvals for other GM livestock such as pigs.

In November 2015 the FDA approved the introduction of AquAdvantage salmon, which has a growth hormone gene that prompts the salmon to grow big enough for consumption in just 18 months rather than the usual three years. In "FDA Bans Imports of Genetically Engineered Salmon—for Now" (WashingtonPost.com, January 29, 2016), Brady Dennis reports that just months after its approval, the FDA banned the importation and sale of AquAdvantage salmon until rules governing how it should be labeled are finalized.

U.S. Consumers' Opinions about Biotech Foods

Every year the International Food Information Council Foundation conducts a survey that tracks trends on public awareness and perceptions of various aspects of plant and animal biotechnology, confidence in the safety of the U.S. food supply, and U.S. consumer attitudes toward food labeling. The foundation indicates in *Food Decision 2016* (http://www.foodinsight.org/sites/default/files/2016-Food-and-Health-Survey-Report_%20FINAL_0.pdf) that

in 2016 consumers were quite familiar with the term *biotechnology*. Twenty-two percent said they feel favorably toward agricultural biotechnology, and 26% said they were neither favorable nor unfavorable toward it. More than a quarter viewed GMOs as helping provide food globally (28%) and developing foods that are more nutritious (26%). A quarter (25%) said GMOs help reduce the environmental impacts of farming.

According to the foundation, U.S. consumers appear to be satisfied with the information currently available on food labels. Just 16% said they would like to see additional information on labels. Of this group, 20% said they would like food labels to include information about GMO ingredients.

AN INTERNATIONAL FOOD FIGHT
Cartagena Protocol on Biosafety

In 2000, after participating in the negotiation of the International BioSafety Protocol, a treaty developed by the United Nations Convention on Biological Diversity (CBD) in which all signatories concurred that GM crops are significantly different from traditional crops, the United States refused to join other countries in signing. Known as the Cartagena Protocol on Biosafety (http://bch.cbd.int/protocol), the treaty aims to ensure the safe transfer, handling, and use of living modified organisms—such as GE plants, animals, and microbes—across international borders. The protocol is also intended to prevent adverse effects on the conservation and sustainable use of biodiversity without unnecessarily disrupting world food trade. By participating in the treaty negotiations, the U.S. government formally acknowledged that GM crops are not, as many U.S. federal agencies had previously maintained, substantially equivalent to traditional crops.

The Cartagena Protocol on Biosafety provides countries the opportunity to obtain information before new bioengineered organisms are imported. It acknowledges each country's right to regulate bioengineered organisms, subject to existing international obligations. It also aims to improve the capacity of developing countries to protect biodiversity. The protocol does not, however, cover processed food products or address the safety of GM food for consumption. Instead, the protocol is intended to protect the environment from the potential effects of introducing bioengineered products (referred to in the treaty as living modified organisms).

In 2003 the Cartagena Protocol on Biosafety entered into force, empowering the more than 100 countries that had ratified the protocol to bar imports of live GM organisms that they believe carry environmental or health risks. If the government of the importing country has concerns about safety, it can ask the exporting country to provide a risk assessment. The protocol also established a central online database of information on the potential risks of several types of GM organisms.

The CBD (http://bch.cbd.int/protocol/parties/) indicates that as of September 2016, 170 countries and the European Union (EU) had ratified the protocol. The article "Progress in Implementing the Strategic Plan for the Cartagena Protocol on Biosafety (2011–2020)" (April 2016, https://bch.cbd.int/protocol/cpb_newsletter.shtml) contains regional responses to achieving the objectives of the protocol. For example, eastern Europe, Asia, and Africa recounted efforts to improve communication, public awareness, education, and participation. Most regions cited challenges in implementing biosafety systems such as lack of financial, technical, and institutional capacities.

Europe Bans Biotech Foods and Halts U.S. Trade

U.S. supermarket shelves are stocked with GM foods, but historically European consumers and farmers have nearly driven GM foods and crops out of the EU market, which is the largest market in the world. Europeans were so wary of GM foods that, beginning in 1998, the EU's European Commission and member states unofficially began a suspension on imports of biotech foods.

In 1999 EU members called for a moratorium on new approvals of agricultural biotech products. The EU Environmental Council contended that the administration of new approvals should be linked to new labeling rules for biotech foods. Ministers from Denmark, France, Greece, Italy, and Luxembourg vowed to suspend approvals until new rules were established. The following year the EU environmental ministers moved to continue the moratorium at least until the European Commission prepared proposals for labeling and for tracing minute amounts of biotech products in foods such as vegetable and corn oils. The European Commission assured the United States that it would develop its proposals by the end of 2000 and promptly resume the approval process.

The labeling and traceability requirements were not presented until July 2001, when the European Commission promised to lift the moratorium within weeks. Frustrated by the many delays and effects of these sanctions, the USDA determined that the EU moratorium on agricultural biotech products violated international law. In the fact sheet "Agricultural Biotechnology: WTO Case on Biotechnology" (September 2006, https://ustr.gov/archive/assets/Trade_Sectors/Agriculture/Biotechnology/asset_upload_file690_8901.pdf), the USDA asked the EU to apply a scientific, rules-based review and approval process to agricultural biotech product applications. Furthermore, the USDA contended that it was not attempting to force European consumers to accept biotech foods or products. Instead, it denounced Europe's actions, which it considered whimsical rather than based on scientific, health, or environmental evidence, as depriving consumers of choice.

Europe's moratorium on GM foods came to an end in 2004, when the European Commission agreed to import worm-resistant GM corn known as BT-11 that was developed by the Swiss firm Syngenta. In 2006 the WTO condemned the long approval process for GE products entering the European market as unscientific and determined that the process amounted to a trade embargo against agriculture producers in the United States, Canada, and Argentina. EU officials declined to revise the approval process, and European consumers continued to view GE products with suspicion. However, the EU did somewhat soften its stance on GM crops.

According to GeneWatch, a nonprofit policy research and public interest group, the EU imports GM crops, food, animal feed, and biofuels. In "GM Crops and Foods in Britain and Europe" (2017, http://www.genewatch.org/sub-568547), GeneWatch observes that GM food and feed must be labeled, but food products produced from animals fed GM feed do not have to be labeled. In 2011 the EU permitted some use of unapproved GM crops in animal feed. Even so, policies and practices vary by country. For example, the article "15 EU Nations Opt to Stay GMO-Free" (October 1, 2015, https://phys.org/news/2015-10-eu-nations-opt-gmo-free.html) indicates that 15 of the 28 EU members vowed in 2015 to prevent GMOs from entering all or parts of their countries. Germany and France had a total ban on GMO cultivation, while Belgium chose to prevent GMOs from being cultivated in Wallonia, the country's French-speaking region.

In the press release "Joint PR: Warning That Transgenic Maize in Spain Could Spread Uncontrollably" (http://www.genewatch.org/article.shtml?als[cid]=568547&als[itemid]=576453), GeneWatch reports that in April 2016 British, German, and Spanish organizations sent a letter to the European Commission calling for measures to prevent GE maize, which produces an insecticidal protein, from spreading in the environment. The groups also asked that the cultivation of the GE maize, which is produced by Monsanto and cultivated in Spain, be halted.

Molecular Farming Harvests New Drugs

Molecular farming (also known as phytomanufacturing or biopharming) is the production of pharmaceutical proteins or other materials in GE plants and animals. It runs the gamut from tobacco plants harboring drugs to treat acquired immunodeficiency syndrome (AIDS) to plants intended to yield fruit-based hepatitis vaccines. In 2012 the FDA approved Elelyso, the first drug produced in a GE plant cell. (Elelyso is used to treat Gaucher disease, a genetic disease that results from the lack of a specific enzyme in which a lipid accumulates in cells and certain organs.) In 2014 an experimental plant-derived biotech drug was given to two U.S. medical workers who were infected with Ebola (an infectious viral disease that is often fatal). By 2017 a range of plant-derived (e.g., Enbrel [an anti-inflammatory agent that is used to treat arthritis]) and animal-derived (e.g., Kanuma [a drug produced by genetically modifying a chicken to produce the drug in its eggs to treat lysosomal acid lipase deficiency, a progressive, inherited disease that leads to organ failure]) biotech drugs were being prescribed in the United States.

Opponents call molecular farming "Pharmageddon," and environmentalists fear that the artificially combined genes will have unintended, untoward consequences for the environment. Consumer advocates, wary about the proliferation of GM foods, dread the possibility that plant-grown drugs and industrial chemicals will end up in food crops. To prevent such a scenario, the FDA issued new regulations to safeguard the food supply in 2003. Suzanne White Junod notes in "Celebrating a Milestone: FDA's Approval of First Genetically-Engineered Product" (April 10, 2009, https://www.fda.gov/AboutFDA/WhatWeDo/History/ProductRegulation/SelectionsFromFDLIUpdateSeriesonFDAHistory/ucm081964.htm) that in 2007 the FDA celebrated the 25th anniversary of the agency's first approval of a recombinant DNA drug product: human insulin. This celebration was intended not only to highlight the beneficial uses of GM products but also to help counter the fears expressed by consumer advocates.

In "Recent Advances on Host Plants and Expression Cassettes' Structure and Function in Plant Molecular Pharming" (*BioDrugs*, vol. 28, no. 2, April 2014), Abdullah Makhzoum et al. describe biopharming as "a promising system to produce important recombinant proteins such as therapeutic antibodies, pharmaceuticals, enzymes, growth factors, and vaccines." The researchers predict that this technology "will flourish in the near future when it is able to provide a high quality and quantity of pharmaceutical products at low costs and with basic infrastructure."

THE PROGRESS AND PROMISE OF GENOMIC MEDICINE

Progress in understanding the genetic basis of disease has arrived at a rapid-fire pace. Genetic and genomic information gained from the Human Genome Project promises to revolutionize prevention and treatment of disease in the 21st century. Physicians are increasingly able to predict patients' risks of acquiring specific diseases and advise them of actions they may take to reduce their risks, prevent disease, and protect their health. There are promising therapeutic applications of genetic research including custom-tailored treatment that relies on knowledge of the patient's genetic profile and development of highly specific and effective medications to combat diseases.

According to Joanne E. Wade, David H. Ledbetter, and Marc S. Williams, in "Implementation of Genomic Medicine in a Health Care Delivery System: A Value Proposition?" (*American Journal of Medical Genetics Part C: Seminars in Medical Genetics*, vol. 166, no. 1, March 2014), genomic medicine has the potential to advance the practice of medicine from diagnostics and treatment that rely on population data to the provision of personalized "precision medicine"—highly accurate diagnoses of diseases, whose causes are understood, and that may be treated with evidence-based therapies that are predictably effective.

Although Wade, Ledbetter, and Williams hope that genomic research will catalyze a shift from reactive patient care to predictive, preventive, and personalized care, this goal has not yet been fully realized. They point out that there are still scientific obstacles to overcome before these genomic data can be effectively translated into improved patient care. They also indicate that effective implementation of genomic medicine requires "a broad range of expertise including quality improvement, systems re-engineering, informatics, health economics, bioethics, and healthcare policy."

Preclinical Detection of Disease

Just as pharmacogenomics uses genetic information to predict how a patient may respond to a particular drug or whether a drug will help make a patient well, genomic susceptibility testing predicts a patient's likelihood of defending against or developing a disease in response to an environmental or other trigger. Called preclinical detection, this ability to predict and as a result intervene to prevent or avert serious disease requires an understanding of three levels of disease detection involving genomes, transcriptomes (the transcribed messenger ribonucleic acid [RNA] complement), and proteomes (the full range of translated proteins).

In "Biomarkers: Refining Diagnosis and Expediting Drug Development—Reality, Aspiration and the Role of Open Innovation" (*Journal of Internal Medicine*, vol. 276, no. 3, September 2014), Hugh Salter and Robert Holland assert that diagnostic biomarkers have exerted a tremendous impact on medical practice and drug development.

Genes Explain Disease, Drug Resistance, and Treatment Failures and Successes

The discovery of predictive genes and proteins that reflect sensitivity and/or resistance to treatment and genes that predict the risk of relapse has personalized the treatment of certain cancers, such as colorectal cancer. Salter and Holland observe that biomarkers predict how patients will respond to cancer chemotherapy and radiotherapy. Other markers can successfully predict toxic effects and other responses to standard therapies. For example, the drug Kalydeco only benefits the 5% of patients with cystic fibrosis who have the G551D mutation.

In 2016 Vienna Ludovini et al. reported in "Gene Identification for Risk of Relapse in Stage I Lung Adenocarcinoma Patients: A Combined Methodology of Gene Expression Profiling and Computational Gene Network Analysis" (*Oncotarget*, vol. 7, no. 21, May 24, 2016) that genome-wide sequencing and computational analyses identified characteristic gene profiles and predictive genes to identify which lung cancer patients are at risk for early relapse of their disease after surgery.

Responses to treatment for depression, especially with antidepressant medication, vary widely. In "Genome-wide Microarray Analysis of Gene Expression Profiling in Major Depression and Antidepressant Therapy" (*Progress in Neuro-Psychopharmacology and Biological Psychiatry*, no. 64, January 4, 2016), Eugene Lin and Shih-Jen Tsai report that genome-wide expression studies have been used to investigate the determinants of depression as well as response to antidepressant therapy. The researchers observe that because depression is strongly influenced by environment and genetics, epigenetic mechanisms, which are influenced by environmental factors, may be associated with response to antidepressant drugs.

Gene Therapy

Gene therapy aims to correct defective or faulty genes with one of several techniques. Most gene therapy involves the insertion of functioning genes into the genome to replace nonfunctioning genes. Other techniques entail swapping an abnormal gene for a normal one in a process known as homologous recombination, restoring an abnormal gene to normal function through selective reverse mutation, or changing the regulation of a gene by influencing the extent to which the gene is "turned on" or "turned off."

Early work in gene therapy focused on replacing a gene that was defective in a specific well-defined genetic disease such as cystic fibrosis. According to Clinicaltrials .gov (https://clinicaltrials.gov/ct2/results?term=gene+ therapy&Search=Search), a service of the National Institutes of Health (NIH), as of February 2017 nearly 5,000 gene therapy clinical trials were under way for a variety of cancers (including pancreatic, liver, ovarian, and prostate cancer), for thyroid and brain tumors, as well as for Alzheimer's disease, severe combined immunodeficiency syndrome (which results in reduced or faulty T- and B-lymphocytes, the white blood cells that combat infection), muscular dystrophies, cystic fibrosis, and many other diseases.

To treat each of these disorders, the appropriate therapeutic genes are inserted by using in vivo (in a living organism, rather than in the laboratory) gene therapy with adenoviral vectors. (Adenoviruses have double-stranded DNA genomes and cause respiratory, intestinal, and eye infections in humans.) A vector is a carrier molecule that is used to deliver the therapeutic gene to the target cells. The most frequently used vectors are viruses that have been genetically altered to contain normal human DNA. Researchers exploit viruses' natural abilities to encapsulate and deliver their genes to human cells. An unanticipated benefit of this type of gene therapy is that highly specific immunity can be induced by injecting into skeletal muscle DNA that is manipulated to carry a gene encoding a specific antigen. Because this form of therapy does not lead to integration of the donor gene into the host DNA, it eliminates potential problems that result from the disruption of the host's DNA, a phenomenon that was observed by investigators using ex vivo (outside a living organism, usually in the laboratory) gene therapy. Figure 10.2 shows how ex vivo cells are used to create a gene therapy product.

Figure 10.3 shows the steps that are involved in gene therapy using a retrovirus (a virus that has RNA rather than DNA as its genetic material and is able to become part of the host cell's DNA using an enzyme called reverse transcriptase) as the vector. In this process the vector discharges its DNA into the affected cells, which then begin to produce the missing or absent protein and are restored to their normal state. In this example the patient's own bone marrow cells are used as the vector to deliver severe combined immune deficiency–repaired genes to restore the function of the immune system.

There also are nonviral techniques for gene delivery. Nonviral carriers were developed to overcome concerns related to viral carriers' potential to trigger immune

FIGURE 10.2

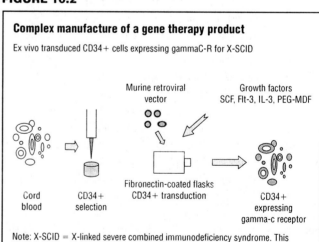

Complex manufacture of a gene therapy product

Ex vivo transduced CD34+ cells expressing gammaC-R for X-SCID

Murine retroviral vector

Growth factors SCF, Flt-3, IL-3, PEG-MDF

Cord blood

CD34+ selection

Fibronectin-coated flasks CD34+ transduction

CD34+ expressing gamma-c receptor

Note: X-SCID = X-linked severe combined immunodeficiency syndrome. This syndrome can cause death within the first year of life from severe recurrent infections.

SOURCE: Adapted from Philip D. Noguchi, "Slide 7. Complexity of (a) Gene Therapy Product," in *Simple Complexity in an Evolving World: Rising to the Challenge*, U.S. Food and Drug Administration Division of Dockets Management, Office of Cellular, Tissue, and Gene Therapies, October 2002, http://www.fda.gov/ohrms/dockets/ac/02/ slides/3902s1-07-noguchi/sld007.htm (accessed November 7, 2016)

FIGURE 10.3

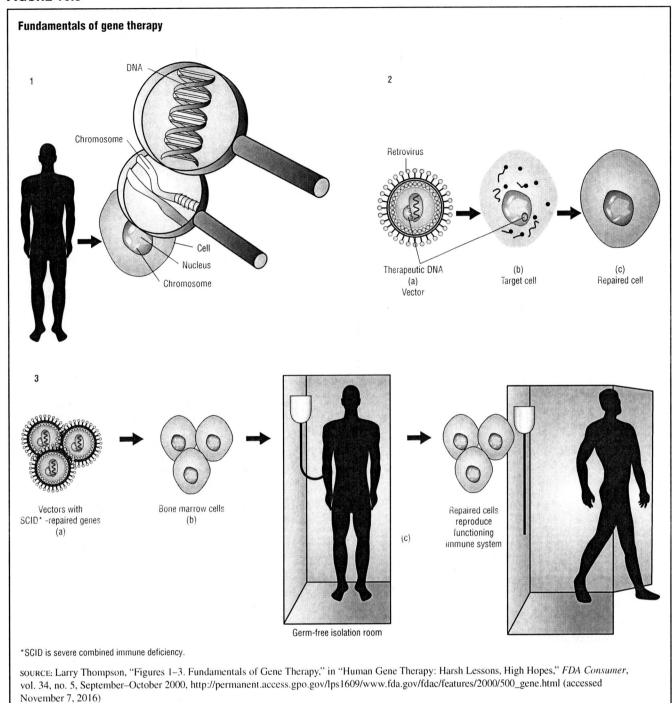

Fundamentals of gene therapy

1

DNA

Chromosome

Cell

Nucleus

Chromosome

2

Retrovirus

Therapeutic DNA
(a)
Vector

(b)
Target cell

(c)
Repaired cell

3

Vectors with
SCID* -repaired genes
(a)

Bone marrow cells
(b)

(c)

Germ-free isolation room

Repaired cells
reproduce
functioning
immune system

*SCID is severe combined immune deficiency.

SOURCE: Larry Thompson, "Figures 1–3. Fundamentals of Gene Therapy," in "Human Gene Therapy: Harsh Lessons, High Hopes," *FDA Consumer*, vol. 34, no. 5, September–October 2000, http://permanent.access.gpo.gov/lps1609/www.fda.gov/fdac/features/2000/500_gene.html (accessed November 7, 2016)

responses or to give rise to tumors. Direct introduction of therapeutic DNA into target cells requires large amounts of DNA and can be used only with certain tissues. Genes can also be delivered via an artificial lipid sphere, called a liposome, with a liquid core. This liposome, which contains the DNA, passes the DNA through the target cell's membrane. DNA can also enter target cells when it is chemically bound to a molecule that in turn binds to special cell receptors. Once united with these receptors, the therapeutic DNA is engulfed by the cell membrane and enters the target cell.

Physical nonviral gene delivery methods involve creating holes in the cell membrane using methods such as rapid injection, particle impact, electric pulse, or ultrasound. The holes or defects in the membrane are temporary but enable the transfer of genes from extracellular space to the nucleus. Chemical vectors, such as cationic lipids and cationic polymers, form complexes with negatively charged DNA. These complexes protect the DNA and facilitate its uptake and intracellular delivery. Other very tiny inorganic particles composed of metals or ceramics have been used to target gene delivery. These tiny particles, known as nanoparticles, are readily transported through cell membranes to the nucleus.

In 1999 the death of the American teenager Jesse Gelsinger after he participated in a clinical gene therapy trial for ornithine transcarbamylase deficiency (an inherited disorder in which ammonia accumulates in the blood, causing damage to the nervous system and liver) shocked and saddened the scientific community and diminished enthusiasm for gene technology. Eleven years after his death, Robin Fretwell Wilson of the Washington and Lee University School of Law recounts in "The Death of Jesse Gelsinger: New Evidence of the Influence of Money and Prestige in Human Research" (*American Journal of Law and Medicine*, vol. 36, nos. 2–3, June 2010) that Gelsinger received a higher dose of the viral vector than he was supposed to have received and was given the viral vector despite the fact that his ammonia level was higher than it should have been. The vector took over and Gelsinger's body immediately launched a massive immune response to it. Four days after receiving the vector, he died from complications that resulted from the immune response.

However, gene therapy trials continued, some with successful outcomes. In "Scientists Use Gene Therapy to Cure Immune Deficient Child" (*British Medical Journal*, vol. 325, no. 7354, July 6, 2002), Judy Siegel-Itzkovich notes that in 2002 an international team of scientists reported curing a child with severe combined immunodeficiency using gene therapy. The child spent the first seven months of her life inside a plastic bubble to protect her from all disease-causing agents because she lacked an immune system. After suppressing her defective bone marrow cells, the researchers introduced, using a GE virus, a healthy copy of the gene she was missing into her purified bone marrow stem cells. The child recovered quickly; within a few weeks she was no longer in isolation and went home healthy. Researchers credited the gene alteration treatment with the cure.

Along with successes, there have been unintended adverse consequences of this type of gene therapy. For example, in some patients the GE virus activated cancer-promoting genes and caused leukemia. However, in 2014 researchers reported successful treatment of eight children with severe combined immunodeficiency. Salima Hacein-Bey-Abina et al. describe in "A Modified γ-Retrovirus Vector for X-Linked Severe Combined Immunodeficiency" (*New England Journal of Medicine*, vol. 371, no. 15, October 9, 2014) a new form of gene therapy that uses a "self-inactivating" virus that minimizes the risk of activating cancer genes.

According to the National Library of Medicine's Genetics Home Reference, in "Is Gene Therapy Safe?" (February 28, 2017, https://ghr.nlm.nih.gov/primer/therapy/safety), because gene therapy is risky, it is only being used to treat diseases that have no other cures. As of February 2017, gene therapy in the United States remained limited to clinical trials and research and the FDA had not approved any human gene therapy product for sale.

The FDA's Cellular, Tissue, and Gene Therapies Advisory Committee (January 21, 2016, https://www.fda.gov/AdvisoryCommittees/CommitteesMeetingMaterials/BloodVaccinesandOtherBiologics/CellularTissueandGeneTherapiesAdvisoryCommittee/default.htm) evaluates "data relating to the safety, effectiveness, and appropriate use of human cells, human tissues, gene transfer therapies and xenotransplantation products." Xenotransplantation is any procedure that involves transplantation, implantation, or infusion into a human recipient of live cells, tissues, or organs from a nonhuman animal source; or human body fluids, cells, tissues, or organs that have had any contact with live nonhuman animal cells, tissues, or organs. It has been used experimentally to treat certain diseases such as liver failure and diabetes, when there are insufficient human donor organs and tissues to meet demand.

Together, the FDA and the NIH jointly oversee regulation of human gene therapy in the United States. The FDA focuses on ensuring that manufacturers produce quality, safe gene therapy products and that these products are adequately studied in human subjects. The NIH evaluates the quality of the science involved in human gene therapy research and funds the laboratory scientists who are involved in the development and refinement of gene transfer technology and clinical studies.

RECENT ADVANCES IN GENE THERAPY. Gene therapy has demonstrated effectiveness as treatment for a variety of genetic conditions including eye disorders and congenital blindness (impaired vision present at birth), hemophilia, and muscular dystrophy. In 2014 researchers discovered a way to prevent the immune system from neutralizing viruses before they deliver their genetic cargo, which has been one of the biggest challenges to using viruses to transport therapeutic genes. In "Plasmapheresis Eliminates the Negative Impact of AAV Antibodies on Microdystrophin Gene Expression following Vascular Delivery" (*Molecular Therapy*, vol. 22, no. 2, February 2014), Louis G. Chicoine et al. find that using plasmapheresis (a process that removes blood from the body, separates the plasma and cells, filters out antibodies, and returns the blood to the patient) just before delivering gene therapy protects the virus long enough for it to enter the cell and deliver the gene.

Lolita Petit, Hemant Khanna, and Claudio Punzo of the University of Massachusetts Medical School report in "Advances in Gene Therapy for Diseases of the Eye" (*Human Gene Therapy*, vol. 27, no. 8, August 2016) that there have been several advances in gene therapy to treat degenerative eye diseases that produce vision loss. The researchers note that clinical trials of gene replacement therapy for Leber's congenital amaurosis (an early-onset form of autosomal recessive retinal degeneration caused by mutations in the RPE65 gene) have been successful. Similarly, there have been promising results from clinical trials of gene replacement therapy for choroideremia (a severe X-linked recessive disorder that results in blindness).

In "Gene Therapy in Pancreatic Cancer" (*World Journal of Gastroenterology*, October 7, 2014), Si-Xue Liu, Zhong-Sheng Xia, and Ying-Qiang Zhong of Sun Yat-sen University describe how gene therapy may be used to treat pancreatic cancer. Approaches include gene augmentation, which aims to transfer therapeutic genes into deficient cells to enable them to prevent or reverse the growth of cancer cells, and gene blockade, which seeks to prevent the transcription and translation of certain cancer-associated genes by using short nucleotide sequences that bind to specific DNA or RNA and can over time induce cell death. The researchers observe that although gene therapy is not yet used in routine clinical practice, its safety and demonstrated benefits make it a promising form of treatment.

EDITING GENOMES. Prior to 2012 genomic engineering of cell lines or animal models was accomplished largely through random mutagenesis or low-efficiency gene targeting. In 2012 researchers reported success using programmable sequence-specific DNA nuclease technologies, such as targeted, high-efficiency modification of genomic sequences, to facilitate genome editing. For example, Tim Wang et al. of the Massachusetts Institute of Technology explain in "Genetic Screens in Human Cells Using the CRISPR-Cas9 System" (*Science*, vol. 343, no. 6166, January 3, 2014) that the clustered regularly interspaced short palindromic repeats (CRISPR) pathway, which functions as an adaptive immune system in bacteria, has been employed to edit mammalian genomes. CRISPRs are segments of DNA that contain short repetitions of base sequences followed by short segments of "spacer DNA" that have been acquired from past exposures to viruses. The CRISPR/Cas9 system is a genome engineering and editing tool that behaves like the search function in a word processing program. It can conduct genome-wide searches for genes with a specific function; introduce, add, or induce change in the sequence of specific genes; and influence gene regulation by activating or silencing specific genes.

Another gene editing technique is transcription activator-like effector nucleases (TALENs). TALENs have a nuclease (an enzyme that cleaves nucleotide chains into smaller units) linked to a DNA recognition section composed of a series of amino acid repeats. Each repeat corresponds to a single nucleotide base (adenine, cytosine, guanine, and thymine), and TALENs may be created with different combinations of repeats to recognize specific genomic sequences. According to Jon Chesnut,

in "Analyzing TALEN vs CRISPR" (GenEngNews.com, October 13, 2016), the choice of using CRISPR or TALENs depends on the application. TALENs are preferable to CRISPR for repairing and inserting mutations and may reduce the risk of unwanted mutations. However, TALENs take more time and are costlier than CRISPR platforms.

In "Methods for Optimizing CRISPR-Cas9 Genome Editing Specificity" (*Molecular Cell Review*, vol. 63, no. 3, August 4. 2016), Josh Tycko, Vic E. Myer, and Patrick D. Hsu note that the CRISPR-Cas9 specificity and efficiency are improving. The researchers assert that the genome editing field will benefit from discussion and consensus about the best ways to minimize, identify, and measure off-target cleavage and reporting on- and off-target activity to shared databases to improve data standardization and enable fair comparison of research results.

MORE APPLICATIONS OF GENETIC RESEARCH

Most applications of genetic biotechnology are agricultural, medical, and scientific. However, geneticists are also engaged in product research and development of related technology and in legal determinations. Involvement with legal matters and the criminal justice system often takes the form of DNA profiling, also known as DNA fingerprinting. Because every organism has its own unique DNA, genetic testing can definitively determine whether individuals are related to one another and whether DNA evidence at a crime scene belongs to a suspect. It can also accurately identify a specific strain of a bacterium.

One example of researchers' use of genetic profiling to identify disease-causing bacteria is reported by Chayapa Techathuvanan, Frances Ann Draughon, and Doris Helen D'Souza in "Real-Time Reverse Transcriptase PCR for the Rapid and Sensitive Detection of Salmonella Typhimurium from Pork" (*Journal of Food Protection*, vol. 73, no. 3, March 2010). The researchers used reverse transcriptase polymerase chain reaction (PCR) to improve the detection of *Salmonella Typhimurium* (a bacterium that causes the second-most commonly occurring bacterial foodborne illness in the United States) in samples of pork chops and sausage.

Another example is engineering plants to monitor the environment. In "Nitroaromatic Detection and Infrared Communication from Wild-Type Plants Using Plant Nanobionics" (*Nature Materials*, vol. 16, no. 2, February 16, 2017), Min Hao Wong et al. of the Massachusetts Institute of Technology report that they embedded carbon nanotubes in spinach plants that enable the plants to detect the odors of chemicals known as nitroaromatics, which are commonly used to make explosives. This technology could also be used to identify and warn about pollutants and environmental threats such as drought.

Forensics

The 19th edition of *Merck Manual of Diagnosis and Therapy* (2011) describes forensic genetics as using molecular genetic techniques to identify an individual's genetic makeup. Forensic genetics relies on the measurement of many different genetic markers, each of which normally varies from individual to individual, and may be used to determine whether two people are genetically related. For example, because half of a person's genetic markers come from the father and half from the mother, analyses of these DNA markers enable laboratory technologists to establish that one person is the offspring of another. Analysis of DNA derived from blood samples can determine whether the supposed parents of a particular child are actually the biological parents. DNA markers may also be used to identify a specimen and establish its origin—that is, definitively determine the individual from whom it came. Biopsies, pathologic specimens, and blood and semen samples can all be used to measure DNA markers.

Forensic investigations often involve analyses of evidence left at a crime scene such as trace amounts of blood, a single hair, or skin cells. Using PCR, DNA from a single cell can be amplified to provide a sample quantity that is large enough to determine the source of the DNA. In PCR the double strand of DNA is denatured into single strands, which are placed in a medium with the chemicals needed for DNA replication. Each single strand becomes a double strand, yielding twice the amount of the initial DNA sample. The double strands are once again denatured to form single strands, and the process is repeated until there is a sufficient quantity of DNA for analysis. Analysis is usually performed using gel electrophoresis, in which DNA is loaded onto a gel and an electrical current is passed through the DNA. Investigators are then able to observe larger molecules migrating more slowly than smaller ones. Examining variable-number tandem repeats (VNTRs) is a procedure that identifies the length of tandem repeats in an individual's DNA. VNTR is an especially useful technique in forensic genetics, because when it is performed in careful laboratory conditions the probability of two individuals having the exact same VNTR results is less than one in a million.

DNA evidence is preferred by forensic specialists because fingerprints can often be erased or eliminated and hair color and appearance may be altered, but DNA is unalterable. It can be used to identify individuals with extremely high probability and is more stable than other biological samples such as proteins or blood groups. Forensic genetics can be a powerful tool and has been used successfully to eliminate suspects and clear their names. Although it is not as useful in proving guilt, forensic genetics provides solid and often convincing evidence when many alleles match.

The first admission of DNA evidence in criminal court occurred in *Florida v. Tommy Lee Andrews* (1987), when the state of Florida used it as part of the prosecution's case to convict a suspect of a series of sexual assaults. In 1989 the Federal Bureau of Investigation (FBI) began accepting work from state forensic laboratories, and in 1996 the National Research Council published *The Evaluation of Forensic DNA Evidence*, which cited the FBI statistic that approximately one-third of primary suspects in rape cases are excluded by using DNA evidence.

Many Americans first became aware of the use and importance of DNA evidence during the widely publicized 1994–95 trial of O. J. Simpson (1947–) for the murders of Nicole Brown Simpson (1959–1994) and Ron Goldman (1968–1994). Nicole Simpson's blood was found in O. J. Simpson's vehicle and house, but the defense discredited the DNA results by claiming the police had conducted a sloppy investigation that caused the samples to be contaminated and, alternatively, that the blood was planted in an attempt to frame Simpson.

In the more than two decades since this sensational case informed Americans about the role of DNA evidence in confirming or refuting innocence or guilt, the use of such evidence has become routine and has been used to successfully exonerate wrongfully convicted people. According to the Innocence Project, in "DNA Exonerations in the United States" (https://www.innocence project.org/dna-exonerations-in-the-united-states/), as of 2016, 349 people had been exonerated by DNA testing, including 20 on death row. In "Access to Post-conviction DNA Testing" (2016, https://www.innocenceproject.org/access-post-conviction-dna-testing/), the Innocence Project reports that although all 50 states have postconviction DNA testing access statutes, many of these testing laws are limited. For example, some state legislation requires adequate safeguards for the preservation of DNA evidence, whereas other states fail to require "full, fair and prompt proceedings" once a DNA testing petition has been filed, which can result in an innocent victim spending long periods being wrongfully imprisoned. Furthermore, despite its ability to prove innocence, some courts still refuse to consider newly discovered DNA evidence after trial.

In 2014 DNA evidence was used to identify Jack the Ripper, who was a serial killer in London in 1888. In *Naming Jack the Ripper: New Crime Scene Evidence, a Stunning Forensic Breakthrough, the Killer Revealed* (2014), Russell Edwards names Adam Kosminski, a 23-year-old hairdresser, as the killer of at least five and as many as 11 women. Edwards tied Kosminski to the crimes via DNA found on a shawl that was worn by Catherine Eddowes, who was one of the Ripper's victims. Edwards bought the shawl at an auction in 2007. Forensic geneticists confirmed the authenticity of the shawl (which had never been washed), traced descendants of both Eddowes and Kosminski, and identified Kosminski as the murderer.

In 2016 DNA analysis confirmed that in 1934 King Albert I of Belgium (1875–1934) died in a rock climbing accident. Jef Akst explains in "Genetic Test Solves Royal Mystery" (The-Scientist.com, October 1, 2016) that because there were no witnesses to the king's death, conspiracy theories had flourished. For example, it was rumored that the king had been killed because of his efforts to keep peace in Belgium during World War I (1914–1918). Genomic analyses equipped to analyze an 80-year-old sample in which the DNA was highly degraded were able to determine that it was the king's blood at the site of the accident.

Nanotechnology

A nanometer (the width of 10 hydrogen atoms laid side by side) is among the smallest units of measure. It is one-billionth of a meter, one-millionth the size of a pinhead, or one-thousandth the length of a typical bacterium. According to the National Nanotechnology Initiative, in "What It Is and How It Works" (2017, https://www.nano.gov/nanotech-101/what), "Nanotechnology is the understanding and control of matter at the nanoscale, at dimensions between approximately 1 and 100 nanometers, where unique phenomena enable novel applications.... Nanotechnology involves imaging, measuring, modeling, and manipulating matter at this length scale."

Nanotechnology is sometimes called molecular manufacturing because it draws from many disciplines (including physics, engineering, molecular biology, and chemistry) and entails the design and manufacture of extremely small electronic circuits and mechanical devices built at the molecular level of matter. K. Eric Drexler (1955–), a former chief technical adviser of Nanorex, a company that develops software for the design and simulation of molecular machine systems, coined the term *nanotechnology* during the early 1980s to describe atomically precise molecular manufacturing systems and their products. He explains in *Engines of Creation* (1986) that it is an emerging technology with the potential to fulfill many scientific, engineering, and medical objectives.

Researchers anticipate many applications combining genetic engineering and nanotechnology such as home food-growing machines that could produce virtually unlimited food supplies and chip-sized diagnostic devices that could revolutionize the detection and management of illness by identifying the genetic variants that are associated with diseases and delivering genetically targeted treatments. The field of genetic nanomedicine encompasses myriad approaches to gene detection and delivery. Researchers envision computer-controlled molecular

tools much smaller than a human cell and constructed with the accuracy and precision of drug molecules. Such tools would enable medicine to intervene in a sophisticated and controlled way at the cellular, molecular, and genetic level by removing obstructions in the circulatory system, destroying cancer cells, or assuming the function of organelles such as the mitochondria. Other potential uses of nanotechnology in medicine include the early detection and treatment of disease via exquisitely precise sensors for use in the laboratory, clinic, and human body, plus new formulations and delivery systems for pharmaceutical drugs.

Some of these nanoscale drug delivery structures are called dendrimers. Because dendrimers can carry different materials on their branches, they can perform various functions simultaneously, such as searching for and detecting diseased and apoptotic cells (apoptosis is the programmed death of a cell), diagnosing and locating gene variants that are associated with specific diseases, delivering drugs, and reporting the responses to treatment.

The technology can be used to develop immediately compatible, rejection-resistant implants made of high-performance materials that respond as the body's needs change. Nanotechnology and genetic engineering will lead to new biomedical therapies as well as to prosthetic devices and medical implants. Some of these will help attract and assemble raw materials in bodily fluids to regenerate bone, skin, or other missing or damaged tissues. Nanotubes that act like tiny straws can circulate in a person's bloodstream and deliver medicines slowly over time or to highly specific locations in the body.

Nano-sized devices are especially applicable to the fields of surgery and regenerative medicine, which is concerned with the improvement and replacement of diseased, aging, or damaged cells, tissues, and organs. Tools such as nanotubes and nanotweezers have been developed for nanosurgery and although the idea of "swallowing the surgeon" may seem like the stuff of science fiction, in the not-too-distant future nanosurgical devices may be able to travel in blood vessels and identify and repair damaged tissue. Regenerative nanomedicine, which is a subspecialty of nanomedicine, uses nanoparticles that contain gene transcription factors and other modulating molecules that facilitate the reprogramming of cells.

DEVELOPMENTS IN NANOMEDICINE. Targeted nanotechnology can be used to control gene delivery in viral and nonviral gene-delivery systems. Rama Mallipeddi and Lisa Cencia Rohan explain in "Progress in Antiretroviral Drug Delivery Using Nanotechnology" (*International Journal of Nanomedicine*, vol. 5, August 9, 2010) that besides their size, which enables them to penetrate through cells and deliver drugs intracellularly, the advantages of nanocarriers (nanoscale particulate drug carriers) include protecting drugs from degrading and targeting drugs to specific sites of action. There are also challenges associated with nanocarrier drug delivery. For example, the smaller the size of the nanocarrier, the harder it is to control the rate of drug release from the nanocarrier. Drugs carried by smaller nanocarriers are released faster, and in bursts, which makes it harder to sustain or control the rate of drug delivery over time.

Mallipeddi and Rohan report that nanocarriers may improve the efficiency of the delivery of antiretroviral drugs that are used to prevent and treat the human immunodeficiency virus (HIV; the virus that causes AIDS). Various types of nanocarriers, such as nanoparticles that are made of polymers, inorganic nanocarriers, liposomes, dendrimers, and cyclodextrins (compounds that are composed of sugar molecules in a ring formation), are currently being studied for the delivery of drugs that are intended for HIV prevention or treatment.

In "Ultra-small Lipid-Polymer Hybrid Nanoparticles for Tumor-Penetrating Drug Delivery" (*Nanoscale*, vol. 8, no. 30, July 2016), Diana Dehaini et al. report that newer lipid-polymer hybrid nanoparticles have the beneficial characteristics of liposomes and nanoparticles, such as long circulation, high drug-loading efficiency, high stability and biocompatibility, and controlled release of drugs. The researchers find that these ultra-small hybrid nanoparticles (less than 25 nanometers) can be loaded with the cancer drug docetaxel and be used to treat tumors in mice. Dehaini et al. suggest that this type of drug delivery application is more effective than the available formulation of the drug.

THE PROMISE OF NANOMEDICINE. Hayley Nehoff et al. of the University of Otago explain in "Nanomedicine for Drug Targeting: Strategies beyond the Enhanced Permeability and Retention Effect" (*International Journal of Nanomedicine*, vol. 9, May 22, 2014) that growing research interest in nanomedicine for the treatment of cancer and other diseases is producing encouraging results. There are, however, obstacles to realizing the potential of nanomedicine, such as its lack of cellular specificity and early release of active agents prior to reaching their targets. New nanomedicine designs are endeavoring to surmount these problems. For example, metal-based magnetic nanoparticles enable targeting of a wide variety of tissues via application of a magnetic field to specifically increase nanoparticle accumulation in a particular restricted area. Magnetic targeting usually involves the use of magnets outside the body, which limits the ability of magnetized nanoparticles to be targeted to deep tissues. This obstacle may be overcome by the use of surgically implanted magnets, which can attract the magnetized nanoparticles to deeper tissues.

Christina von Roemeling et al. describe trends in nanomedicine in "Breaking down the Barriers to Precision Cancer Nanomedicine" (*Trends in Biotechnology*, vol. 35, no. 2, February 2017), explaining that the surface chemistry of nanoparticles is moving from synthetic polymers to coatings that include biologically inspired elements, such as cell membranes to improve uptake. They report that research is under way "to develop smart nanomaterials that can adaptively shape-shift depending on the local environment, inspired by protein conformational changes in response to different stimuli." Furthermore, von Roemeling et al. opine that these responsive nanomaterials "could lead to precision nanomedicine tailored to unique biological conditions."

CHAPTER 11
ETHICAL ISSUES AND PUBLIC OPINION

There is no reason to waste time arguing about whether humans should be genetically engineered. As justifiable as some of the ethical concerns may be, there are simply too many benefits to be gained from preventing hereditary diseases. Those seeking to limit genetic engineering to such efforts would be better off devoting their energies to explaining why eugenics is wrong, rather than attempting to stop the march of progress toward healing the sick and eliminating awful disorders.

—Arthur L. Caplan "Engineering the Better Baby" (*Project Syndicate*, December 9, 2015)

Rapid advances in genetics and its applications pose new and complicated ethical, legal, regulatory, and policy issues for individuals and society. The issues society must consider include the responsible use of genetic engineering technology, the consequences of knowledge about personal genetic information for individuals, and the repercussions of genomic information for groups such as ethnic and racial minorities. For example, African Americans were among the first to call for the inclusion of African American genetic sequences in the human genome's template, and their concerns are vitally important because historically they have been victimized by "genetic inquiries" and research.

Although the potential for discrimination based on genetic information was largely eliminated by the provisions of the Genetic Information Nondiscrimination Act of 2008, a host of ethical issues related to health and medical care remain. These include the use of genetic information to guide reproductive decision making and the application of genetic engineering to reproductive technology. Reproductive applications are an especially emotionally charged topic as many antiabortion advocates staunchly oppose any action to alter the development or course of a naturally occurring pregnancy, such as in vitro fertilization (fertilization that takes place outside the body) and preimplantation intervention as well as

the use of stem cells that are obtained from human embryos. Scientists, ethicists, and others also fear that genetic engineering technology will rapidly progress from enabling prenatal diagnosis and intervention for the prevention of serious disease to preconception selection of the traits and attributes of offspring—the creation of so-called designer babies.

Genetic testing challenges health care professionals to rapidly acquire new knowledge and skills to effectively use new technology and to assist their patients in making informed choices. The quality, reliability, and utility of genetic testing must be continuously reevaluated and regulated. Along with quality-control measures, health care professionals and consumers must determine whether it is appropriate to perform tests for conditions for which no treatment exists and how to interpret and act on test results, such as an individual's increased susceptibility to a disease that is associated with several genes and environmental triggers.

Philosophical, psychological, and spiritual considerations invite mental health professionals, ethicists, members of the clergy, and society overall to redefine concepts of human responsibility, free will, and genetic determinism. Behavioral genetics looks at the extent to which genes influence behavior and the human capacity to control behavior. It also attempts to describe the biological basis and heritability of traits ranging from intelligence and risk-taking behaviors to sexual orientation and drug addiction.

There are also environmental issues associated with the applications of genetic research, such as weighing the benefits and risks to the environment of creating genetically modified animals, crops, and other products for human consumption. Legal and financial issues center on the ownership of genes, deoxyribonucleic acid (DNA), and related data; property rights; patents and copyrights; and public access to research data and other genetic information.

OPINIONS ABOUT GENOMIC DATA, PRIVACY, AND PROTECTING THE PUBLIC INTEREST

In October 2012 the Presidential Commission for the Study of Bioethical Issues released the report *Privacy and Progress in Whole Genome Sequencing* (https://bio ethicsarchive.georgetown.edu/pcsbi/node/764.html), which considers genomics and privacy. In the press release "President's Bioethics Commission Releases Report on Genomics and Privacy" (October 11, 2012, https://bio ethicsarchive.georgetown.edu/pcsbi/node/765.html), the commission states that its goal was "to find the most feasible ways of reconciling the enormous medical potential of whole genome sequencing with the pressing privacy and data access issues raised by the rapid emergence of low-cost whole genome sequencing."

The report summarizes the ethical issues and legal considerations of protecting people from the risks associated with sharing their whole genome sequence data and information and recommends actions the government should take to address these concerns. Chief among these is the recommendation that federal and state governments should institute a consistent level of protection that covers whole genome sequence data independent of how or where they are obtained. For example, if an individual's genome is sequenced at a physician's office, then it is protected by the Health Insurance Portability and Accountability Act of 1996, and the information collected during research is similarly protected. By contrast, DNA obtained from saliva on a fork, spoon, or cup can be sequenced and in many states is not protected by law.

Laws protecting genomic data are intended to prevent discrimination, which could occur if, for example, employers or insurance companies gained access to an individual's genome. In "Panel: Protect Patients Who Use Whole Genome Sequencing" (USAToday.com, October 11, 2012), Janice Lloyd quotes Amy Gutmann (1949–), the chair of the Presidential Commission for the Study of Bioethical Issues, who explains that advancing genomic research depends on providing individuals with these protections: "Without such assurances in place, individuals are less likely to voluntarily supply the data that have the potential to benefit us all with life-saving treatments for genetic diseases. Everyone stands to gain immensely from our society taking the necessary steps to protect privacy in order to facilitate progress in this era of whole genome sequencing."

Yaniv Erlich and Arvind Narayanan confirm the need for protection in "Routes for Breaching and Protecting Genetic Privacy" (*Nature Reviews Genetics*, vol. 15, June 17, 2014). They explain that "prospective participants of scientific studies have ranked privacy of

sensitive information as one of their top concerns and a major determinant of participation in a study." Techniques for breaching genetic data include linking DNA data to demographic data such as date of birth, sex, and zip code to identify study participants and tracing identities by searching for specific Y-chromosome haplotypes (a set of alleles or markers on one of a pair of homologous chromosomes) and surnames in genetic genealogical databases such as online genealogy forums. Although Erlich and Narayanan concede that successfully tracing genetic data requires knowledge of genetics and statistics, they assert that simple measures such as removing obvious identifiers from genetic data sets before publicly sharing them, encrypting data, and stringently controlling access to these data could prevent most data breaches. Erlich and Narayanan conclude, "We as a society can facilitate the development of social and ethical norms, legal frameworks and educational programmes to reduce the chance of misuse of genetic data regardless of the ability to identify data sets."

In April 2016 the U.S. senators Elizabeth Warren (1949–; D-MA) and Michael B. Enzi (1944–; R-WY) introduced the Genetic Research Privacy Protection Act, which would ensure that federally funded researchers cannot reveal genetic data that can identify study participants. The bill was referred to the Senate Committee on Health, Education, Labor, and Pensions.

The Impact of Genetic Testing: Women and Breast Cancer

Many health professionals hope that genetic testing for hereditary cancers will enable early disease detection, offer targeted surveillance, result in effective prevention strategies, and improve health outcomes (how affected individuals fare as a result of treatment). This section describes research that considers the impact of genetic testing on treatment choices made by women with breast cancer and those who are at risk of developing the disease.

Susan M. Domchek et al. report in "Association of Risk-Reducing Surgery in BRCA1 or BRCA2 Mutation Carriers with Cancer Risk and Mortality" (*Journal of the American Medical Association*, vol. 304, no. 9, September 1, 2010) the results of a landmark study of women who carry BRCA mutations. Oophorectomy (surgical removal of the ovaries) has proven effective at preventing ovarian and breast cancer in women with these mutations. Women who carry the harmful BRCA mutations have a significantly increased cancer risk: 15% to 40% will develop ovarian cancer during their lifetime (compared with about 1% of the general population of women), and 60% will develop breast cancer (compared with 12% of the general population of women).

Domchek et al. followed 2,482 women with BRCA1 and BRCA2 mutations for an average duration of three to six years and compared the cancer and mortality outcomes between those who had risk-reducing surgery (172 had a mastectomy [surgical removal of the breasts] and 993 had an oophorectomy) and those who chose not to have surgery. The women who chose not to have the prophylactic (preventive) surgery (about half of the subjects) were offered intensified surveillance for breast cancer—annual mammography and breast magnetic resonance imaging (MRI; a radiological technique that provides detailed images of organs and tissues)—and for ovarian cancer—ultrasound and Ca125 testing every four to 12 months.

The risk of developing ovarian cancer was higher in BRCA1 carriers (7% to 8%) than in BRCA2 carriers (3%) who opted against oophorectomy. No BRCA2 carriers who had an oophorectomy developed peritoneal cancer (cancer of the tissue lining the abdominal and pelvic cavities) during the follow-up. Cancer was diagnosed in the ovaries of nine women who underwent prophylactic oophorectomy.

Prophylactic oophorectomy also protected against breast cancer, reducing the risk by half, from 22% to 11%. Prophylactic salpingo-oophorectomy (removal of the fallopian tubes and ovaries) appears to be more effective in BRCA2 carriers than in BRCA1 carriers; there were no breast or ovarian cancer–related deaths in BRCA2 carriers who had a salpingo-oophorectomy. Interestingly, prophylactic mastectomy reduced the risk of developing breast cancer but did not reduce the mortality from breast cancer. This finding may reflect the fact that breast cancer screening and treatment are effective at reducing mortality.

Intensified screening in the form of more frequent mammograms, sonograms, and MRIs offer BRCA mutation carriers ways to ensure early detection of breast cancer. There are not yet comparably effective screening tests for ovarian cancer; ultrasound and Ca125 tests have not been shown to reduce the mortality from ovarian cancer. Domchek et al. find that surveillance does not confer the same protection against ovarian cancer as does oophorectomy.

Domchek et al. conclude that among women with BRCA1 and BRCA2 mutations, "the use of risk-reducing mastectomy was associated with a lower risk of breast cancer; [and] risk-reducing salpingo-oophorectomy was associated with a lower risk of ovarian cancer, first diagnosis of breast cancer, all-cause mortality, breast cancer–specific mortality, and ovarian cancer–specific mortality."

Even in view of these findings, the decision to have prophylactic surgery is often not straightforward. In "Health Care Provider Recommendations for Reducing Cancer Risks among Women with a BRCA1 or BRCA2 Mutation" (Clinical Genetics, vol. 85, no. 1, January

2014), Kelly Metcalfe et al. observe that although women with a BRCA1 or BRCA2 mutation have lifetime risks of developing breast cancer of up to 87% and 39%, respectively, for ovarian cancer, there is wide variation in the use of cancer risk–reducing options. Metcalfe et al. posit that varying recommendations from health care providers influence patient decisions.

D. Gareth R. Evans et al. report in "The Angelina Jolie Effect: How High Celebrity Profile Can Have a Major Impact on Provision of Cancer-Related Services" (Breast Cancer Research, vol. 16, no. 5, September 19, 2014) that the decision by the actor and human rights activist Angelina Jolie (1975–) to make public her double mastectomy after testing positive for a mutation of the BRCA1 gene more than doubled the number of women in Britain requesting genetic tests for BRCA1 and BRCA2 and inquiring about prophylactic mastectomy. Evans et al. dubbed this increase the "Angelina Jolie effect" and observe that although media attention often produces a short-lived uptick in screening, "The Angelina Jolie effect has been long lasting and global."

Although testing for mutations of BRCA1 and BRCA2 may have increased, the use of medication to reduce risk remains low. In the editorial "Medication to Reduce Breast Cancer Risk: Why Is Uptake Low?" (Annals of Oncology, vol. 27, no. 4, 2016), Phyllis Butow and Kelly-Anne Phillips posit that physicians need training to help them better identify the women who might benefit from risk-reducing medications, such as selective estrogen receptor modulators or aromatase inhibitors, and support them in their decision-making process.

The Impact of Disclosing the Results of Genetic Testing

J. Scott Roberts, Kurt D. Christensen, and Robert C. Green observe in "Using Alzheimer's Disease as a Model for Genetic Risk Disclosure: Implications for Personal Genomics" (Clinical Genetics, vol. 80, no. 5, November 2011) that susceptibility testing for common, adult-onset diseases is likely to grow as genomic testing becomes routine and note that it is important to identify the likely benefits and harms of this testing and the disclosure of results to inform medical practice and health policy. The researchers analyzed the results of several clinical trials that considered psychological and behavioral responses to the disclosure of genetic risk assessment for Alzheimer's disease (AD). Roberts, Christensen, and Green find that delivery of test results does not generally result in adverse psychological effects when it is delivered by trained professionals. They contend that the genetic counseling process can be streamlined without increasing the risk that patients will suffer any loss of understanding or additional emotional distress.

Christensen et al. report similar results to disclosure in "Disclosing Pleiotropic Effects during Genetic Risk

Assessment for Alzheimer's Disease: A Randomized, Controlled Trial" (*Annals of Internal Medicine*, vol. 164, no. 3, February 2, 2016). The researchers looked at how people with close relatives affected by AD responded to disclosure of their genetic risk for the disease. Christensen et al. report that when people at genetic risk for AD were also found to be at genetic risk for heart disease and this additional risk was disclosed to them, they were no more anxious and depressed than people who only received information about their AD risk. The researchers conclude that their findings support the safety of disclosing information about a modifiable condition such as heart disease during genetic assessment for AD and opine that such disclosure may relieve test-related stress even among those who learn they are at increased risk for two, rather than just one, serious disease.

Opinions about Direct-to-Consumer Marketing and Use of Genetic Tests

The widespread availability and direct-to-consumer (DTC) marketing of genetic and genomic tests that do not require a prescription or referral from a physician pose new challenges. Personal genomic tests may be used by people seeking to learn more about their ancestry and their health (specifically their carrier status, disease risk, and response to drugs).

Proponents of such testing maintain that people who learn more about their genetic risks will take steps to improve their health and well-being. For example, in "Genetic Testing Company 23andMe Returns to Market" (Time.com, October 21, 2015), Alice Park notes that 23andMe, a company that offers genetic tests related to carrier status, also provides wellness-related information, such as "how likely you are to like caffeine, whether you're lactose intolerant, if you tend to get red and flushed when you drink alcohol and what your muscle composition is like." Brad Kittredge, the vice president of 23andMe, explained that making this information available to consumers can help them make informed decisions about food and exercise. For example, someone who is lactose intolerant may choose to consume lactose-free dairy products.

Effy Vayena of the University of Zurich opines in "Direct-to-Consumer Genomics on the Scales of Autonomy" (*Journal of Medical Ethics*, May 5, 2014) that DTC genomic testing enhances the autonomy (independence or self-determination) of consumers, enabling them to "make choices from a range of valuable options." Vayena explains that in order for DTC genomics to enhance autonomy, it must provide adequate consumer protection, such as accurate advertisement, informed consent requirements, privacy protections, and the option of obtaining genetic counseling.

Others worry about the stress that such knowledge can create and how the results of genetic testing may be used.

For example, in "Why You Should Think Twice about At-Home Genetic Testing" (USNews.com, July 11, 2016), Pim Suwannarat of the Mid-Atlantic Permanente Medical Group in Washington, DC, observes, "Even when there is a strong rationale to get tested, using a DTC service may trigger more questions than answers." Suwannarat asserts that DTC tests do not provide usable information for consumers' physicians and cautions that although health insurers and employers cannot discriminate on the basis of genetic tests, test results may be used to discriminate against consumers when they purchase life, disability, or long-term care insurance.

CONCERNS ABOUT GENETIC RESEARCH AND ENGINEERING

Genetic research and engineering technologies promise to address some of the most pressing problems of the 21st century, such as cleaning the environment, feeding the world's hungry, and preventing serious diseases. Nevertheless, like all new technologies, they pose risks as well as benefits. In general, Americans support advances in genetic research and technology, and they are optimistic that the outcomes will ultimately be used to reduce sickness and suffering, as opposed to generating legal and ethical controversies.

Some of the fears about genetic engineering presume that the use of such technologies will be alien, impersonal, and technologically difficult. Scientists and advocates of applying genetic engineering to improve human life foresee a future in which genetic testing and new reproductive technologies are routine aspects of medical care. Bioethicists and others concerned about the implications of the routine use of this biotechnology contend that it is possible and advantageous to fund, discuss, and regulate genetics in the same way that society currently considers environmental medicine, nutrition, and public health.

ETHICAL CONSIDERATIONS AND CONCERNS ABOUT NANOTECHNOLOGY AND NANOMEDICINE

Jitendra S. Tate et al. raise in "Addressing Ethical and Safety Issues of Nanotechnology in Health and Medicine in Undergraduate Engineering and Technology Curriculum" (*Global Journal of Engineering Education*, vol. 18, 2016) some of the ethical questions about nanotechnology, such as:

- Can it be used for unsolicited surveillance and monitoring?

- Can nanomachines implanted in patients be reprogrammed to cause harm?

- Should funding for nanomedicine research focus on prevention or cure of disease?

- Who will benefit from developments in nanomedicine? Because development and production of nanomedicine is costly, will it be available only to those who can afford it?

Tate et al. assert, "If the society wants to harness the full potential of nanotechnology in medicine, one needs to assess not just how far one can push the technological advancement envelope, but also how to effectively deliver medicine to everyone who needs it. Nanotechnology has the ability to improve the quality and functionality of the lives of people who suffer from disease and disability. Humanity has a moral obligation to minimise suffering wherever possible and maintain the welfare of patients and the interests of society."

In "Social Implications of Nanotechnology: A Review" (*Asian Journal of Advanced Basic Sciences*, vol. 2, no. 2, July 23, 2014), R. P. Singh observes that intended benefits of nanotechnology may also have unintended consequences. For example, nanomedicine may increase the average human life span, which in turn would be an increase in the proportion of the population that is older, which might require changes in Social Security and an increase in the retirement age. Singh asserts that another potential consequence is the likely increase of inequality in the distribution of wealth that he calls the "nano divide." Participants in the "nano revolution" may become wealthy, whereas people who do not may find it increasingly difficult to afford the technological advances that it brings about.

NOTED AUTHORS ADDRESS THE RISKS AND BENEFITS OF GENETIC ENGINEERING

In *Our Posthuman Future: Consequences of the Biotechnology Revolution* (2002), Francis Fukuyama of Stanford University cautions about the use and misuse of science and biotechnology. He observes that genetic engineering has gained popular acceptance as it is used to prevent or correct selected medical conditions, but he shares the widely held concerns about the unforeseeable results of gene manipulation and the potential for altering the complexion of society through the use of "enhancement technology" to customize the attributes of offspring. Fukuyama fears that those able to afford the genetic interventions to produce offspring that is smarter, stronger, more athletic, talented, and better looking will further widen the chasm between the economic classes. He states that the ways in which society chooses to employ, regulate, and restrict genetic engineering may challenge traditional concepts of human equality, and as a result will change the existing understanding of human personality, identity, and the capacity for moral choice. Furthermore, he contends that genetic engineering technologies may afford societies new techniques for controlling the behavior of their citizens and could potentially overturn existing social hierarchies and affect the rate of intellectual, material, and political progress. Fukuyama predicts that genetic engineering and other applications of biotechnology have the potential to sharply alter the nature of global politics.

Despite his fears about the ability to intentionally modify the human organism as opposed to waiting for evolution, Fukuyama does not consider it necessary or advisable to prohibit actions that alter genetic codes. He warns, "We may be about to enter into a posthuman future, in which technology will give us the capacity gradually to alter [human] essence over time.... We do not have to regard ourselves as slaves to inevitable technological progress when that progress does not serve human ends. True freedom means the freedom of political communities to protect the values they hold most dear, and it is that freedom that we need to exercise with regard to the biotechnology revolution today."

Bioethicist Arthur L. Caplan asserts in *If Gene Therapy Is the Cure, What Is the Disease?* (November 8, 2002, http://public.callutheran.edu/~chenxi/Phil350_061.pdf) that the greatest challenge to securing funding and support for genomic research is public fear of germline engineering (manipulating the human genome to improve the human species), and he acknowledges that this fear is based on the historical reality of horrendous eugenics (hereditary improvement of a race by genetic control) practices in Germany during World War II (1939–1945) and in other countries.

Caplan cites examples of genetic engineering in the United States such as the Repository for Germinal Choice in Escondido, California, also known as the "Nobel Prize sperm bank," that from 1980 to 1999 banked sperm from men known for their scientific, athletic, or entrepreneurial acumen. The banked sperm was available for use by women for the express purpose of creating genetically superior children. Caplan observes that there have been relatively few critics of this practice, whereas the mere suggestion of the possibility of directly modifying the genetic blueprint of gametes (sperm and eggs) has generated fiery debate in professional and lay communities. He contends that the history of eugenically driven social policy is reason enough to question and even protest the actions of the Repository for Germinal Choice but that it does not argue against allowing voluntary, therapeutic efforts using germline manipulations to prevent certain serious or fatal disorders from besetting future generations.

According to Caplan, the decision to forgo germline engineering does not make ethical sense. He laments that "some genetic diseases are so miserable and awful that at least some genetic interventions with the germline seem justifiable.... It is at best cruel to argue that some people must bear the burden of genetic disease in order to allow benefits to accrue to the group or species. At best, genetic diversity is an argument for creating a gamete bank to preserve diversity. It is hard to see why an unborn child has any obligation to preserve the genetic diversity of the species at the price of grave harm or certain death."

Caplan fears that choosing to refrain from efforts to modify the germline will result in lives sacrificed; that is, important benefits will be delayed or lost for people with disorders that might be effectively treated with germline engineering. He recommends responding to concerns about the dangers and potential for abuse of new knowledge that is generated by the genome with frank, objective assessments of the appropriate goals of this application of biotechnology.

In "Ethicist: Fixing Genes Using Cloning Technique Is Worth the Ethical Risk" (NBCNews.com, October 24, 2012), Caplan opines that using cloning to repair faulty genes is ethical. He also believes that experimenting on embryos to prevent or cure disease is morally acceptable. He states, "And while I, too, worry about where genetically engineering eggs might lead, I think doing so to find cures is ethically noble. Those nations that say no to any form of germline engineering, including the U.S., should revisit those policies to permit research that is clearly intended as therapy."

In 2015 Caplan weighed in again on genetic engineering in "Engineering the Better Baby" (*Project Syndicate*, December 9, 2015), an essay about the use of CRISPR to edit genes to reduce the occurrence of inherited disorders. He explains that the benefits of CRISPR will likely be available to those who can afford to pay for the technology and will create yet another health disparity based on income. However, his gravest worry is that the same tool that repairs genetic disorders will be used for eugenics, to create genetically enhanced children. Caplan asserts, "The more time we spend debating whether to adopt a technology that undoubtedly will be adopted, the less we will have to consider more relevant issues. We need to know, for example, how to respond to the promise of taller, smarter, healthier, cuter, stronger, and more loving children before commercial providers begin rolling out their marketing campaigns."

THE ETHICS OF SYNTHETIC BIOLOGY

In "Researchers Say They Created a 'Synthetic Cell'" (NYTimes.com, May 20, 2010), Nicholas Wade reports that in May 2010 the geneticist J. Craig Venter (1946–) and his colleagues announced that they had created a "synthetic cell." The group used the genome of a bacterium that infects goats and then extracted the DNA from it. The researchers created a synthetic genome that was much like the original genome; however, they added a few sequences that would not affect the genome's activity but would enable them to identify the synthetic code. After synthesizing the genome, they transplanted it

into a different type of bacterium to create a self-replicating bacterium. Venter described the bacterium as "the first self-replicating species we've had on the planet whose parent is a computer." He opined that "this is a philosophical advance as much as a technical advance" and suggested that it will reinvigorate the discussion about the nature of life.

That same day *Nature* published in "Sizing Up the 'Synthetic Cell'" (May 20, 2010) the responses of leading scientists and bioethicists to Venter's announcement. Mark Bedau of Reed College called for "precautionary thinking and risk analysis," but also observed that the announcement will "revitalize perennial questions about the significance of life." George M. Church of the Harvard Medical School warned that society must take measures to prevent erroneous applications of this technology as well as its potential use to create bioterrorism threats. By contrast, Caplan compared the accomplishment to "the discoveries of Galileo, Copernicus, Darwin and Einstein" in terms of its ability to fundamentally change deeply held beliefs about the nature of life.

In "Realizing the Potential of Synthetic Biology" (*Nature Reviews Molecular Cell Biology*, vol. 15, no. 4, April 2014), distinguished scientists George M. Church, Michael B. Elowitz, Christina D. Smolke, Christopher A. Voigt, and Ron Weiss discuss the bioethical issues associated with synthetic biology. Smolke, who is the associate chair of education and William M. Keck Foundation faculty scholar in the Department of Bioengineering at Stanford University, opines that ethical and regulatory concerns depend on the potential uses and applications. For example, if the organism is to be used outside of a contained environment (such as an enclosed bioreactor), then potential effects on the environment and existing ecosystems should be addressed, as well as related issues such as evolution, adaptation, containment, and removal. Weiss, a professor in the Departments of Biological Engineering and Electrical Engineering and Computer Science at the Massachusetts Institute of Technology (MIT) and the director of the Synthetic Biology Center at MIT, reminds readers, "Like other responsible scientific fields of endeavour, synthetic biology has an ingrained culture of seeking to reduce risks with existing technological solutions, while seeking to predict and avoid problems with proposed new solutions." Meanwhile, Church exhorts the scientific community to "'keep it real,' focus on the most critical risks seriously and thoughtfully in a way that is grounded in reality, engage with the larger community and make the most of the unique opportunity we now have for fundamental discovery and understanding."

In the podcast "Ethically Sound: New Directions" (October 17, 2016, https://bioethicsarchive.georgetown.edu/pcsbi/node/5897.html), host Hillary Wicai Viers spoke with James Wagner, the vice chair of the Presidential Commission for the Study of Bioethical Issues, and Eleonore Pauwels, to discuss the commission's 2010 report *New Directions: The Ethics of Synthetic Biology and Emerging Technologies* (https://bioethicsarchive .georgetown.edu/pcsbi/synthetic-biology-report.html). Wagner listed the fundamental principles that are described in the report—public beneficence (acting to maximize public benefit and minimize public harm), responsible stewardship (shared obligation among members of the domestic and global communities to act in ways that demonstrate concern for those who are not in a position to represent themselves), intellectual freedom and responsibility, democratic deliberation (collaborative decision making that welcomes respectful debate of opposing views and active participation of citizens), and justice and fairness (the distribution of benefits and burdens across society)—and suggested these principles should continue to inform the advance of this new field.

IMPORTANT NAMES
AND ADDRESSES

Alzheimer's Association
225 N. Michigan Ave., 17th Floor
Chicago, IL 60601-7633
(312) 335-8700
1-800-272-3900
FAX: 1-866-699-1246
E-mail: info@alz.org
URL: http://www.alz.org/

American Cancer Society
250 Williams St. NW
Atlanta, GA 30303
1-800-227-2345
URL: http://www.cancer.org/

**American Diabetes
Association**
2451 Crystal Dr., Ste. 900
Arlington, VA 22202
1-800-342-2383
URL: http://www.diabetes.org/

American Heart Association
7272 Greenville Ave.
Dallas, TX 75231
1-800-242-8721
URL: http://www.americanheart.org/

**American Parkinson Disease
Association**
135 Parkinson Ave.
Staten Island, NY 10305
1-800-223-2732
FAX: (718) 981-4399
E-mail: apda@apdaparkinson.org
URL: http://www.apdaparkinson.org/

**American Society of Gene
and Cell Therapy**
555 E. Wells St., Ste. 1100
Milwaukee, WI 53202
(414) 278-1341
FAX: (414) 276-3349
E-mail: info@asgct.org
URL: http://www.asgct.org/

American Society of Human Genetics
9650 Rockville Pike
Bethesda, MD 20814-3998
(301) 634-7300
FAX: (301) 634-7079
URL: http://www.ashg.org/

Center for Food Safety
660 Pennsylvania Ave. SE, Ste. 302
Washington, DC 20003
(202) 547-9359
FAX: (202) 547-9429
E-mail: office@centerforfoodsafety.org
URL: http://www.centerforfoodsafety.org/

Cold Spring Harbor Laboratory
One Bungtown Rd.
Cold Spring Harbor, NY 11724
(516) 367-8800
URL: http://www.cshl.org/

Cystic Fibrosis Foundation
6931 Arlington Rd., Second Floor
Bethesda, MD 20814
(301) 951-4422
1-800-344-4823
E-mail: info@cff.org
URL: http://www.cff.org/

Genetics Society of America
9650 Rockville Pike
Bethesda, MD 20814-3998
(301) 634-7300
1-866-486-4363
FAX: (301) 634-7079
URL: http://www.genetics-gsa.org/

**Genome Programs of the U.S.
Department of Energy Office
of Science**
URL: http://genomics.energy.gov/

**Huntington's Disease Society
of America**
505 Eighth Ave., Ste. 902
New York, NY 10018

(212) 242-1968
1-800-345-4372
E-mail: hdsainfo@hdsa.org
URL: http://www.hdsa.org/

**International Genetic Epidemiology
Society**
c/o Mayo Clinic
200 First St. SW
Harwick 7
Rochester, MN 55905
E-mail: iges@geneticepi.org
URL: http://www.geneticepi.org/

**International Society for Forensic
Genetics**
URL: http://www.isfg.org/

J. Craig Venter Institute
9714 Medical Center Dr.
Rockville, MD 20850
(301) 795-7000
URL: http://www.jcvi.org/

**Muscular Dystrophy
Association**
222 S. Riverside Plaza, Ste. 1500
Chicago, IL 60606
1-800-572-1717
E-mail: mda@mdausa.org
URL: http://www.mdausa.org/

**National Down Syndrome
Society**
8 E 41st St., Eighth Floor
New York, NY 10017
1-800-221-4602
E-mail: info@ndss.org
URL: http://www.ndss.org/

**National Human Genome Research
Institute**
National Institutes of Health
Bldg. 31, Rm. 4B09
31 Center Dr., MSC 2152
9000 Rockville Pike
Bethesda, MD 20892-2152

(301) 402-0911
FAX: (301) 402-2218
URL: http://www.genome.gov/

National Multiple Sclerosis Society
101A First Ave., Ste. 6
Waltham, MA 02451-1115
1-800-344-4867
FAX: (781) 890-2089
URL: http://www.nationalmssociety.org/

**National Tay-Sachs
and Allied Diseases
Association**
2001 Beacon St., Ste. 204
Boston, MA 02135
(617) 277-4463
1-800-906-8723
E-mail: info@ntsad.org
URL: http://www.ntsad.org/

**Presidential Commission for the Study of
Bioethical Issues**
URL: https://bioethicsarchive.georgetown.edu/

Wellcome Images
215 Euston Rd.
London, NW1 2BE
United Kingdom
(011-44) 20-7611-8348
URL: http://wellcomeimages.org/

RESOURCES

There are many published accounts of the history of genetics, but some of the most exciting versions were written by the pioneering researchers themselves. Although many sources were used to construct the historical overview and highlights contained in this book, James D. Watson's *The Double Helix: A Personal Account of the Discovery of the Structure of DNA* (1968), Francis Crick's memoir *What Mad Pursuit: A Personal View of Scientific Discovery* (1988), and Alfred H. Sturtevant's *A History of Genetics* (1965) provided especially useful insights. Thomas Hunt Morgan's *The Mechanism of Mendelian Heredity* (1915) and *The Theory of the Gene* (1926) offer detailed descriptions of groundbreaking genetic research. Ricki Lewis and Bernard Possidente offer more recent history in *A Short History of Genetics and Genomics* (2003).

Charles Darwin's *On the Origin of Species by Means of Natural Selection, or the Preservation of Favoured Races in the Struggle for Life* (1859) and *The Descent of Man, and Selection in Relation to Sex* (1871) provide historical perspectives on evolution. The discussion of nature versus nurture is drawn from Richard Herrnstein and Charles Murray's *The Bell Curve: Intelligence and Class Structure in American Life* (1994). Ethical issues arising from genetic research and engineering are analyzed in *Our Posthuman Future: Consequences of the Biotechnology Revolution* (2002) by Francis Fukuyama.

The journals *Nature* and *Science* have reported every significant finding and development in genetics, and articles dating from 1953 from both publications are cited in this text, as are articles from *Cell Research, Disease Markers, Nature Biotechnology, Nature Genetics,* and *Nature Medicine.* Research describing genetic testing, disorders, and genetic predisposition to disease is reported in professional medical journals. Studies cited in this book were published in peer-reviewed journals, including *Advanced Drug Delivery Reviews; Advances in Genetics; Alzheimer's & Dementia; American Journal of Pathology; Applied Clinical Genetics; Archives of Disease in Childhood; Archives of Neurology; Asian Journal of Advanced Basic Sciences; Bioassays; Biotechnology Advances; BMC Medicine; Breast Cancer Research; British Journal of Pharmacology; Cell Stem Cell; Clinical Genetics; Diabetes Care; E-biomed: The Journal of Regenerative Medicine; European Journal of Nuclear Medicine and Molecular Imaging; Expert Opinion on Orphan Drugs; Fertility and Sterility; Genome Medicine; Hematology/Oncology Clinics of North America; International Journal of Neuropsychopharmacology; Journal of Allergy and Clinical Immunology; Journal of Alzheimer's Disease; Journal of the American Medical Association; Journal of Clinical Oncology; Journal of General and Family Medicine; Journal of the National Cancer Institute; Journal of Reproductive Development; Life Sciences, Society and Policy; Molecular Biology Reports; Molecular Cytogenetics; Molecular Psychiatry; Nature Communications; Nature Materials; Neuroscience Research; New England Journal of Medicine; Pharmacogenomics Journal; PLoS Biology; PLoS Genetics; PLoS One; Proceedings of the National Academy of Sciences; Science Translational Medicine; Seminars in Respiratory and Critical Care Medicine; Stem Cell Research & Therapy;* and *Twin Research and Human Genetics.*

Ethical and psychological issues and the contributions of genetics to personality and behavior are examined in articles published in the *Archives of Sexual Behavior; Behavioral Genetics; Biological Psychiatry; Biology of Reproduction; Current Directions in Psychological Science; Genes, Brain and Behaviour; Human Brain Mapping; Journal of the American Academy of Child and Adolescent Psychiatry; Journal of Bioethical Inquiry; Journal of Personality and Social Psychology; Natural Resource Modeling;* and *Neuroscience and Psychiatry.*

Besides the U.S. daily newspapers and electronic media, accounts of genetics research and milestones in the Human Genome Project were reported in *Scientific American* and *Time*.

The Human Genome Project Information website, which is operated by the U.S. Department of Energy, describes the ambitious goals and accomplishments of the Human Genome Project since its inception in 1990. The National Institutes of Health provides definitions, epidemiological data, and research findings about a comprehensive range of genetic tests and genetic disorders.

Public opinion data from the following organizations was also helpful: Gallup, Inc., the Genetics and Public Policy Center of the Johns Hopkins University, the International Food Information Council Foundation, and the Pew Research Center. Additionally, many colleges, universities, medical centers, professional associations, and foundations dedicated to research, education, and advocacy about genetic disorders and diseases provided up-to-date information that was included in this edition.

INDEX

Page references in italics refer to photographs. References with the letter t following them indicate the presence of a table. The letter f indicates a figure. If more than one table or figure appears on a particular page, the exact item number for the table or figure being referenced is provided.

A

ACOG (American College of Obstetricians and Gynecologists), 79

Acquired immunodeficiency syndrome (AIDS)/human immunodeficiency virus (HIV), 104, 105, 168

Acquired traits, theory of, 3, 40–41

Adaptation, 40, 43–44

Adenosine deaminase deficiency, 12–13

Adolescent attitudes and behavior, 59

Adopted *vs.* nonadopted children, social attitudes of, 59

Adult stem cells, 147, 150

Adult-onset disorders, genetic testing for, 106, 171

Advanced Cell Technology, Inc., 141, 145, 148

Advertising of direct-to-consumer genetic tests, 107–108

AGC/CTG triplet repeat, 84

Age
Alzheimer's disease, 64
life expectancy by, 37t–38t
mutations and parental age, 46
sex determination, 34

Agriculture and genetic engineering. *See* Genetically modified crops

AKT2 gene, 72, 73, 76(f5.20)

Albany Medical Center Prize, 129

Albert I (king of Belgium), 167

Albert Lasker Basic Medical Research Award, 8

Albinism, 7

Alleles, 28–30, 32, 44

Allis, C. David, 129

Alzheimer's Association, 68–69

Alzheimer's disease, 64–69, 64f–68f, 69t, 105, 171–172

Alzheimer's Disease Sequencing Project, 132

American Academy of Pediatrics, 106

American Association for the Advancement of Science, 43

American Association of University Professors, 43

American College of Medical Genetics, 106

American College of Obstetricians and Gynecologists (ACOG), 79

American Institute of Biological Sciences, 42, 43

American Society for Reproductive Medicine, 148

American Society of Human Genetics, 12, 106, 129

Amino acids, 19–20, 20f, 22–23, 113

Amish population, 44

Amniocentesis, 96–97

Amyloid protein, 64–65

Anatomical structure of humans, 43

Aneuploidy, 46

Animal and Plant Health Inspection Service (APHIS), 156

Animals
cells, 3f
reproductive cloning, 13, 138–142, 143, 158
therapeutic cloning, 145–147

Ankylosing spondylitis, 105

Anthropology, 127

Antidepressants, 104, 162

Antinori, Severino, 142

APC gene, 105

APHIS (Animal and Plant Health Inspection Service), 156

apoE gene, 65, 67f, 105

APP gene, 65f

Applied Biosystems, Inc., 117–118

AquAdvantage salmon, 159

Arabidopsis thaliana, 14, 154

Arber, Werner, 114

Archaeology, 127

Aristotle, 1–2

Array comparative genomic hybridization, 100–101

Artificial chromosomes, 12

Ashkenazi Jews, 79

Asthma, 50–52, 53f

The Astonishing Hypothesis (Crick), 10

Attitudes of adolescents, 59

Audubon Nature Institute, 142

Australia, 141–142

Australian Twin Registry, 55

Autoimmune disorders, 132

Automatic sequencers, 119

Autoradiography, 99

Autosomal dominant genes
affected maternal offspring, 32
genetic disorders, 61

Autosomal recessive genes
carrier identification, 94
genetic disorders, 61
inheritance pattern, 95(f6.1)
sickle-cell disease, 87
Tay-Sachs disease, 90

Autosomes, 32

Avery, Oswald Theodore, 8

AVM Biotechnology, 150

B

Bacteria
biochemical basis of heredity, 8
cystic fibrosis progression, 78, 79
DNA from *Pneumococcus* bacteria, 8f
genome sequence, 13
identifying disease-causing bacteria, 165
making proteins, 18

Balanced polymorphism, 44
BARD1 gene, 71, 74(f5.16)
Barna, Maria, 129
Bateson, William, 7
Baylor College of Medicine, 118
Behavior, 49, 58–59, 64, 169
Bell Curve (Herrnstein and Murray), 56
Berg, Paul, 12
"Bermuda Principles," 119
Biochemistry, 8
Bioengineered food safety, 156–160
Bioethics. *See* Ethical issues
Biogenesis theories, 2
Biological diversity, 155
Biomarkers
    Alzheimer's disease, 68
    genomic medicine, 161
    pharmacogenomic biomarkers, 105
Biomedicine
    Human Genome Project, 119, 124, 129
    nanotechnology, 167
    reproductive cloning, 140
    See also Therapeutic cloning
Biopharming, 160–161
Biotechnology. *See* Genetic engineering and biotechnology
Birth defects, 94–95
Bison, 44
Blackburn, Elizabeth H., 127, 129
Bladder cancer, 105
Blastocysts, 28
Blood disorders, 105, 146
Blood tests
    prenatal genetic testing, 94–95
    Tay-Sachs disease, 90
Bone-producing cells, creating, 149
Boston Biomedical Research Institute, 150
Botstein, David, 129
BoyaLife Group, 141
Brain, structural changes in the, 58
Brain cells, creating, 149
BRCA1 and BRCA2 genes
    depiction of BRCA1 gene, 72(f5.12)
    depiction of BRCA2 gene, 72(f5.13)
    ethics of testing for, 170–171
    identification of, 116
    population screening, 105
    predictive genetic testing, 103
    psychological consequences of genetic testing, 110
Breast cancer
    AKT2 gene, 76(f5.20)
    BARD1 gene, 74(f5.16)
    BRCA1 gene, 72(f5.12)
    BRCA2 gene, 72(f5.13)
    BRIP1 gene, 74(f5.17)
    CASP8 gene, 76(f5.21)
    ESR1 gene, 77(f5.23)
    genetic research, 71–75

identification of determinative gene, 116
    King, Mary-Claire, 12
    Mammaprint test, 104
    NBN gene, 75(f5.19)
    PALB2 gene, 73(f5.14)
    population screening, 105
    predictive testing, 103, 105–106, 170–171
    psychological consequences of genetic testing, 110
    RAD51C gene, 75(f5.18)
    RAD51D gene, 73(f5.15)
    TBX3 gene, 77(f5.22)
    twin and familial studies, 54
Breeding, selective, 40
Brenner, Sydney, 117
BresaGen Limited, 140
Bridges, Calvin Blackman, 7, 8
BRIP1 gene, 71–72, 74(f5.17)
Brown University, 149
Bryan, William Jennings, 41
BT-11 corn, 160
Buffalo, cloned, 141
Bull gaurs, 141
Bulls, cloned, 141
Bush, George W., 149, 151

C

C4 gene, 87
Caenorhabditis elegans, 13, 117–118
CALHM1 gene, 67
California Institute for Regenerative Medicine, 152
Camels, cloned, 141
Canada, 42, 132
Cancer
    cell cycle, 70f
    future of genomic research, 132
    gene therapy, 164, 165
    genetic disorders, 63, 69–71
    genomic medicine, 162
    nanomedicine, 168
    predictive genetic testing, 104
    retinoblastoma, 70, 71f
    therapeutic cloning, 148
    twins studies, 54
    See also Breast cancer
Capacitation, 27
Caplan, Arthur L., 143, 173–174
Carrier identification tests, 91, 94, 95(f6.2), 97, 106
Cartagena Protocol on Biosafety, 159–160
Cartilage repair, 147
CASP8 gene, 72, 73, 76(f5.21)
Cats, cloned, 140
Cattle, cloned, 141
Cavendish Laboratory, 10
Celera Genomics, 14, 119, 120

Cell theory, 2–3, 23–28
Cell-free DNA, 101–102
Cells
    animal cell, 3f
    cancer cells, 70, 70f
Center for Food Safety, 158–159
Center for Food Safety et al. v. Johanns et al., 158
Center for Inherited Disease Research, 118
Centre for Regenerative Medicine, 149
Centromeres, 26, 27f
Cetus Corporation, 115
CFTR gene, 62(f5.1), 77–78, 78f
CFTR modulators, 79–80
Chakrabarty, Diamond v., 12
Charcot-Marie-Tooth disease, 94
Chargaff, Erwin, 10
Charpentier, Emmanuelle, 143
Chase, Martha Cowles, 9
Cheetahs, 142
Chemical genomics, 125, 132
Children, genetic testing of, 106
China, 141, 142
Chorionic villus sampling (CVS), 95, 97
Christianity, 41, 43
Chromosomes
    artificial chromosomes, 12
    cell structure, 24
    centromeres, 27f
    chromosomal disorders, 62
    chromosome theory of inheritance, 7–8
    cytogenetics, 116, 116f
    depiction of, 25f
    first complete sequence of a human chromosome, 14
    Ingram, Vernon, work of, 11
    meiosis, 29f
    microarray, 100, 102f
    mitosis and meiosis, 25–26
    mutations, 46
    noninvasive prenatal testing, 101–102
    Painter, Theophilus Shickel, 9
    sex chromosomes, 35f
    sex determination, 32–34
Church, George M., 174
Cibelli, Jose B., 145
Class-action lawsuits, 109
Classical genetics, 7–8, 8–9
Clinical Laboratory Improvement Amendments, 97
Clinical trials
    diabetes treatment, 82
    gene therapy, 162, 164
    sickle cell disease treatment, 89
Clinton, Bill, 120
Clonaid, 142
Clone Registry, 135
Cloned animals, food from, 158, 159

Q

Quintiles Transnational Inc., 152

R

R groups, 19–20
Rabbits, cloned, 140
Race/ethnicity
  asthma, 52
  cystic fibrosis, 76
  demographic characteristics of births,
    36*t*
  diabetes, 81
  HapMap Project, 130
  Huntington's disease, 83
  intelligence, 56
  life expectancy, 37*t*–38*t*
  phenylketonuria, 87
  sex ratio, 34
  sickle-cell disease, 88
  social Darwinism, 42
  Tay-Sachs disease, 36, 91, 94
RAD51C gene, 71, 75(*f*5.18)
RAD51D gene, 71, 73(*f*5.15)
Radiation, 46–47
Raelians, 142
Random mutagenesis, 165
Rapid-result genetic tests, 105
Recessive genes
  carrier identification, 94
  cystic fibrosis, 76
  genetic disorders, 61
  genetic inheritance, 29
  inheritance pattern, 95(*f*6.1)
  Mendel's laws of heredity, 4
  Punnett squares, 33*f*
  recessive traits, 31(*f*2.19), 32*f*
  sickle-cell disease, 87
  Tay-Sachs disease, 90
  X-linked disease, 35–36
Recloning, 141
Recombinant DNA
  creation of, 113
  gene cloning, 135, 137, 138*f*
  history, 12
  insulin, 161
Regenerative medicine, 167
Regulation
  biotechnology, 156–159
  genetic tests, 108, 108*f*
  therapeutic cloning, 149–151
Relapse, genetic testing for, 104
Reliability of testing, 93–94
Religion and evolution, 41, 42–43
Religious attitudes, 59
Repository for Germinal Choice, 173
Reproduction, 25–28, 30*f*, 32–34, 44, 46
Reproductive cloning, 135, 137–144, 140*f*,
  150
Research participants, 170

Resilience, 58–59
Resistance to transgenics, 155
Restriction enzymes, 114
Restriction fragment length polymorphisms,
  99, 101(*f*6.7), 114, 118, 118*f*
Retinoblastoma, 70, 71*f*
Reverse transcriptase, 113, 165
RGS17 gene, 104
Rhesus monkeys, 139–140
Ribosomal RNA, 22
Ribosomes, 22, 23*f*, 24
Risk assessment, 127
Risk-reducing mastectomy, 106, 110, 171
Risk-taking behavior, 49, 59
RNA, 20–21, 21*f*, 113, 133
Rosalind Franklin Young Investigator
  Award, 129
Roslin Institute, 13, 138
Rothman, James E., 12
Roundup, 156

S

Saccharomyces cerevisiae, 13
Sachs, Bernard, 91
Safety of Genetically Engineered Foods
  (Institute of Medicine), 157–158
Salmon, genetically engineered, 159
San Diego Zoo, 142
Sanger, Frederick, 114
Sanger Centre, 118, 119
Sarafino, Edward P., 50
Schekman, Randy W., 12
Schizophrenia, 87
Schleiden, Matthias Jakob, 2
Schwann, Theodor, 2
Science (journal), 120, 125
SCNT. *See* Somatic cell nuclear transfer
Scopes, John Thomas, 41
Screening. *See* Genetic testing and
  screening
Seeds, terminator, 156
Segregation, law of, 4, 5*f*
Selective breeding, 40
"Self-inactivating" viral vectors, 164
Semiconservative replication, 11
Sensitivity, test, 94
Seoul National University, 141
Sequencing machines, 117–118, 119, 125
Serial somatic cell cloning, 141
Severe combined immunodeficiency, 164
Sex, life expectancy by, 37*t*–38*t*
Sex chromosomes, 7, 35*f*
Sex determination, 32–34
Sex ratio, 34, 36*t*
Sex-linked diseases, misconceptions about,
  36
Sex-linked inheritance, 7, 34–36
Sexual orientation, 55–56
Sheep, cloning of, 138–139, 140, 142

Sherley, James, 150
Sibling studies. *See* Twin and familial
  studies
Sickle-cell disease
  adaptation, 44
  depiction of sickle cell, 88(*f*5.35)
  HBB gene, 62(*f*5.2)
  environmental factors, 49–50
  genetic disorders, 87–89
  Ingram, Vernon, 11
  Pauling, Linus, 9
  recessive traits, 29
Side groups, 19–20
Simpson, Nicole Brown, 166
Simpson, O. J., 166
Single nucleotide polymorphisms
  (SNPs)
  Alzheimer's disease, 67
  breast cancer, 74
  depiction of, 130(*f*7.12)
  HapMap Project, 129–130
  intelligence, 57
  schizophrenia, 87
  in soybean DNA, 115(*f*7.2)
  variation, 130(*f*7.11)
Single-gene disorders, 61
Sinsheimer, Robert Louis, 116
Skin cells, transforming, 148–149
Skin color, 29, 43
Slippery slope arguments, 143
Smith, Hamilton O., 114
Smoking, 52, 59
Smolke, Christina D., 174
SNPs. *See* Single nucleotide polymorphisms
  (SNPs)
Social attitudes, 59
Social Darwinism, 42
Somatic cell nuclear transfer (SCNT)
  reproductive cloning, 138, 141, 142
  therapeutic cloning, 145, 147, 149,
    150
Sooam Biotech, 141
Southern blot analysis, 113, 114*f*
Soviet Union, 117
Spain, 141, 160
Speciation, 47
Species, relatedness between, 37, 39, 47,
  120
Specificity, test, 94
Spectral karyotyping, 98, 101(*f*6.6)
Sperm, 27–28, 33, 46
Spermatogenesis, 26, 30(*f*2.16)
Spindle nuclear transfer, 148
SPOT-Light HER2 CISH test, 104
Stahl, Franklin W., 11
States
  creationism-evolution debate, 42–43
  human cloning laws, 151–152
  reproductive cloning bans, 138

Statistical information
    birth demographics, by race of mother, 36*t*
    cost per genome of sequencing the
      human genome, 127*f*
    genetically modified crops, adoption of,
      154*f*
    Human Genome Project funding,
      117(*t*7.1)
    life expectancy, by race and sex, 37*t*–38*t*
    model organisms sequenced, 117(*t*7.2)
Stem cell therapy, 82, 89
Stem cells
    cloning, 136*f*
    reproductive cloning, 141, 142
    therapeutic cloning, 145–152
Streptococcus pneumoniae, 8
Stress, 54, 58
Structural chromosomal aberrations, 46
Structural Genomics Consortium, 132
Sturtevant, Alfred H., 7, 88
Südhof, Thomas, 12
Surgery
    nanotechnology, 167
    preventive surgery, 106, 110, 171
Survival advantage, 44
Sutton, Walter Stanborough, 7
Sweat test, 78
Switzerland, 116
Synthetic biology, 174–175
Szostak, Jack W., 127, 129

T

TALENs, 165
Tay-Sachs disease, 36, 89–91, 89*f*, 90*f*, 94
TBX3 gene, 73, 77(*f*5.22)
Technology
    automatic sequencers, 119
    CRISPR technology, 174
    editing genomes, 165
    genetically modified crops, 153–155
    Human Genome Project techniques, 118
    sequencing machines, 117–118, 125
Temin, Howard, 113
Temporomandibular joint degeneration, 147
Tennessee Supreme Court, 41
Terminator technology, 156
Terminology
    environment, 49
    genetics, 7, 13
Testing. *See* Genetic testing and screening
Texas A&M University, 140
Theory of acquired traits, 3, 40–41
The Theory of the Gene (Morgan), 7–8
Therapeutic cloning, 143, 145–152, 174
Thomas Hunt Morgan Medal, 129
Thomas More Law Center, 43
Tomatoes, genetically engineered, 14
Tommy Lee Andrews, Florida v., 166
TOP2A FISH pharmDx, 104

Traits, genetic, 4–5, 5*f*, 6*f*, 28–30, 32
Transcription, 21
Transcription activator-like effector
    nucleases (TALENs), 165
Transfer RNA, 22–23
Transgenic crops. *See* Genetically modified
    crops
Transgenic pigs, 142
Transgenic zebrafish, 132
Transposon-induced mutations, 46
Traumatic experiences, 54–55
Treatment
    Alzheimer's disease, 69*t*
    cystic fibrosis, 79–80
    diabetes, 82
    expected outcomes of the Human
      Genome Project, 120
    genetic testing and personalized
      treatment, 104–105
    infertility treatment, 147–148
    muscular dystrophy, 85–86
    phenylketonuria, 86–87
    schizophrenia, 87
    sickle-cell disease, 88–89
Triple-marker screen, 94
Triplets, nucleotide, 11–12, 82, 83
TRUGENE HIV-1 Genotyping Kit, 104
Trump, Donald, 151
Tschermak-Seysenegg, Eric von, 6
Tsui, Lap-Chee, 115–116
Tumors, 70
23andMe, 109–110, 172
Twin and familial studies
    asthma, 50
    behavior, 58
    diabetes, 81
    intelligence, 57
    nature *vs.* nurture research, 53–54
    sexual orientation, 55–56
    social attitudes, 59
Type 1 diabetes, 82
Type 2 diabetes, 80, 80*t*, 81

U

Unanticipated information and results of
    genetic testing, 107
United Kingdom, 148
United Nations Convention on Biological
    Diversity, 159
United States
    Alzheimer's disease, 64
    biotech food exports, 160
    biotechnology regulation, 156–159
    bison population, 44
    crop product, 155
    cystic fibrosis, 76
    gene therapy, 164
    genetic testing regulation, 108–110,
      108*f*

Huntington's disease, 83
muscular dystrophy, 83
public opinion on evolution, 42
sex ratio, 34
Tay-Sachs disease, 91
therapeutic cloning policy, 149–152
University of California, Santa Cruz, 116
University of Idaho, 141
University of Washington, 118
U.S. Circuit Court of Appeals for the
    District of Columbia, 150–151
U.S. Department of Agriculture, 156, 160
U.S. Department of Energy, 13, 116, 122
U.S. Department of Justice, 150
U.S. District Court for the District of
    Columbia, 150
U.S. Environmental Protection Agency
    (EPA), 157
U.S. Food and Drug Administration (FDA)
    bioengineered foods approved by,
      157*t*
    biotechnology regulation, 156–157
    cystic fibrosis treatment, 79
    direct-to-consumer genetic tests,
      108–110
    Elelyso, approval of, 160
    gene therapy, 13, 164
    genetic testing for personalized
      treatment, 104
    genetically modified crops, 158–159
    molecular farming, 161
    sickle cell disease treatment, 89
U.S. Government Accountability Office
    (GAO), 107–108
U.S. House of Representatives, 151
U.S. Senate, 151
Utah State University, 141
Utrophin, 86, 86(*f*5.32)

V

Validity, test, 94
Variable expressivity, 29
Variation, genetic, 26, 43–44, 129–131
Vector-mediated transformation, 154
Vectors, gene therapy, 162–163
Venter, J. Craig, 14, 118–119, 174
Virchow, Rudolf, 2
Viruses
    gene therapy vectors, 162, 164
    nanomedicine, 13
    "Waring blender experiment," 9
Vitamin A deficiency, 14
Voigt, Christopher A., 174

W

Wagner, James, 175
Wallace, Alfred Russel, 2
Warfarin sensitivity, 104

"Waring blender experiment," 9
Warren, Elizabeth, 170
Watson, James D., 10, 13, 113, 116, 117
Weinberg, Wilhelm, 7
Weiss, Ron, 174
Weissmann, August, 2–3
Wellcome Trust, 119, 132
Whitehead Institute for Medical Research, 118
Wilkins, Maurice, 10
William Allan Award, 129
Wilmut, Ian, 13, 138, 146

Wilson, Edward O., 58
Wolf, Don Paul, 139–140
Woolly mammoth, 142
World Health Organization, 106, 156
World Trade Center attacks, 54–55
Worm map, 117–118

X

X chromosomes and sex determination, 32–34
Xenotransplantation, 140
X-linked disorders, 35–36, 84

Y

Y chromosomes and sex determination, 32–34
Yamanaka, Shinya, 147
Yeast genome, 13

Z

Zavos, Panayiotis, 142
Zerhouni, Elias A., 150
Ziagen, 105
Zona pellucida, 27–28
Zwart, Hub, 14

OCT - - 2017

CPSIA information can be obtained
at www.ICGtesting.com
Printed in the USA
FFOW01n2302230917
40285FF

9 781410 325501